PRACTICAL ZOOLOGY

FOR ADVANCED LEVEL
AND INTERMEDIATE STUDENTS

PRACTICAL ZOOLOGY

FOR

ADVANCED LEVEL
AND INTERMEDIATE
STUDENTS

BY

C. J. WALLIS, M.A. (Cantab.)

*Formerly Master-in-Charge of Biology, University College School,
Biology Master, Highgate School and Principal, Eaton and Wallis,
Tutors*

A
LABORATORY MANUAL
Covering the Syllabuses in *Zoology* of the General Certificate of
Education (Advanced Level) and other Examinations of similar
standard.

SIXTH EDITION

LONDON
WILLIAM HEINEMANN MEDICAL BOOKS LTD
1973

FIRST EDITION, 1935
SECOND EDITION, RE-SET, 1947
THIRD EDITION, 1950
FOURTH EDITION, RE-SET, 1957
IN TWO VOLUMES
Volume I: Zoology
Volume II: Botany
FIFTH EDITION, RE-SET, 1965
FIFTH EDITION OF VOLUME I, RE-SET, 1965
(As a separate book)
SIXTH EDITION, RE-SET, 1974

ISBN 0 433 34704 X

Spanish Edition, 1955

By the same author
Practical Botany
Advanced Level Practical Biology
Human Biology

Printed and bound in Great Britain
by R. J. Acford Ltd., Industrial Estate, Chichester, Sussex

FOREWORD TO THE SIXTH EDITION

In the Sixth Edition, the text has been thoroughly revised, enlarged and brought up to date and to meet the requirements of revised Examination Syllabuses many additions have been made. These include the external features and transverse sections of quite a large number of animals, the dissection of the sea anemone and the locust, further experiments in physiology and the inclusion of an entirely new section on Genetics, illustrated by breeding experiments with *Drosophila*. Some of the illustrations have been re-drawn, some taken from my *Practical Biology* and *Human Biology* and many new ones added where it was felt that they would be of assistance to the student. At the same time the original purpose of the book of *giving adequate instructions, illustrated where necessary, but avoiding long and tedious reading* has been maintained.

I am very grateful to those critics, teachers and students who made constructive criticisms of the last edition and I have, as far as possible, embodied their suggestions in the new one.

Once again it gives me great pleasure to record my grateful thanks to my publishers and to Mr. Owen R. Evans and to Mr. Chris. Jarvis in particular for the very considerable help given in the production of the new edition of the book. I am also considerably indebted to Mr. Frank Price for his assistance in preparing my pencil drawings for the block-maker and to my wife for her help with the reading of the proofs and in the tedious process of compiling the index.

WEST QUANTOXHEAD.
1973.   C.J.W.

PREFACE

BIOLOGICAL knowledge, like all scientific knowledge, can be properly acquired only when it is the result of practical investigation. The following scheme of practical work has been drawn up in accordance with the requirements of the syllabuses in *Zoology* of the General Certificate in Education (Advanced Level), and of other examinations of similar standard. Teachers and students can easily discover what should be omitted by reference to the various examinations syllabuses. This book was originally published as part of a manual of practical biology for medical students but the plant and animal biology sections have since been considerably enlarged and separated into two books, *Practical Botany* and *Practical Zoology*, covering the Advanced Level and similar examination syllabuses in the *separate* subjects, and a third book, *Practical Biology*, suited to the subject *Biology* in these examinations has also been published.

Unfortunately, Practical Classes are often unavoidably large and it is impossible for a great deal of individual attention to be given to the students. It is in any case desirable that they should learn by discovering things for themselves, provided they are guided along the right paths, for by this method they not only absorb facts more easily but learn to think and work on scientific lines. This was evident even in the small groups one taught in a tutor's practice in which students worked individually and not as a class, and in which there was adequate time to devote to each student in the group.

An attempt has therefore been made to give *sufficient directions* to enable the student to proceed with his practical work with a minimum of assistance from the demonstrator. At the same time, the *inclusion of elaborate and unnecessary details which make the reading long and tedious has been avoided.* Simple experiments have been included in biochemistry and physiology to emphasise the correlation between structure and function. Drawings and diagrams have been freely inserted when it was considered necessary for the guidance of the student *solely* to assist him in his identifications. It is *essential* that the student should draw *exactly* what he sees in his preparations and that he should *not* copy diagrams from a book.*

For the sake of practical convenience, this manual is divided into seven parts—Microscopical Technique, the Variety of Animals—(Anatomy), Cytology and Histology, Elementary Biochemistry, Physiology, Genetics and Vertebrate Embryology. Introductory

* Many of the illustrations in this manual are intentionally diagrammatic or semi-diagrammatic, though several are drawings from specimens.

notes have been written to each part of the book, giving instructions peculiar to that part, and appendices have been added which contain information which it is hoped will be useful to those in charge of biological laboratories. It is realised that teachers have their own individual methods in practical work, but there is much that is common to all. The book is *essentially a laboratory manual*, and is, of course, intended for use in conjunction with the usual text-books.

When writing the original book I had much pleasure in expressing my gratitude to Sir Frederick Gowland Hopkins, O.M., M.B., F.R.S., Professor of Biochemistry in the University of Cambridge, to Professor A. G. Tansley, M.A., F.R.S., Sherardian Professor of Botany in the University of Oxford and to Dr. L. A. Borradaile, M.A., Sc.D., Lecturer in Zoology in the University of Cambridge, for kindly reading through the manuscripts of the Biochemistry, Plant Biology and Animal Biology sections respectively and for the many helpful suggestions they made.

I was also deeply indebted to Dr. J. H. Woodger, D.Sc., Reader in Biology in the University of London and Lecturer in the Middlesex Hospital Medical School, for reading through the proofs of the entire First Edition and of the vertebrate types of the Second; and to Dr. C. L. Foster, M.Sc., Ph.D., also of the Middlesex Hospital Medical School, for reading through the proofs of the complete Second Edition. In the course of these readings they made several invaluable suggestions, the majority of which I was glad to be able to adopt.

I gratefully appreciate the courtesy of the authors and publishers of two text-books for permission to use or adapt illustrations (acknowledged in each instance) from those books.

Lastly, I gratefully acknowledge my indebtedness to my publishers, and particularly to Mr. Owen R. Evans, for the assistance they have given me in the production of the various editions of the books.

C. J. WALLIS.

ACKNOWLEDGEMENTS

CLARENDON PRESS. *Elements of Zoology*, Dakin. Fig. 50.
HENRY HOLT & CO., New York. *Development of the Chick*, Lillie.
W. WATSON & SONS LTD., Barnet, Herts., for the illustrations of the microscope. Figs. 6 and 7.
Dr. D. KAY and Dr. J. C. F. POOLE of the Sir WILLIAM DUNN School of Pathology, Oxford, for the electronmicrograph of an animal cell, (from the author's *Human Biology*) Fig. 226.

TABLE OF CONTENTS

xi

INTRODUCTION

I. GENERAL DIRECTIONS FOR PRACTICAL WORK

(1) Cleanliness, tidiness and accuracy are of the utmost importance. See that all your dissecting instruments are sharp and that all apparatus is clean before beginning your work. **To sharpen a scalpel** an oilstone is necessary. *Put a drop of oil on the stone and push the scalpel, edge foremost, obliquely across the stone. Then turn the scalpel over and repeat the movement with the other edge. Alternatively, a circular movement may be made, edge foremost, as before. Repeat until the blade is sharp. Then draw it once or twice away from the edge to remove the burr.* Scissors must be sharpened by an expert. *Never use scalpels or scissors for any purpose other than dissection.*

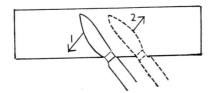

FIG. 1. **Sharpening a scalpel.**

(2) Read the appropriate subject in your text-book *before* you start any practical work.

(3) Read the practical directions carefully before you begin.

(4) Wash, clean, dry and put away all your instruments, apparatus, etc., when you have finished with them.

(5) In the case of *microscopical preparations*, it is advisable to compare your own slides with permanent slides. Always take care to put the permanent slides back *in their proper places* in the trays or boxes; otherwise you (and others) will have difficulty in finding them on a future occasion.

(6) In the case of *dissections*, etc., examination of museum specimens is often very helpful.

(7) Finally, throughout your studies try to *understand the correlation between the structure (anatomy) and functions (physiology) of the various structures, organs and systems.*

1

II. GENERAL DIRECTIONS FOR THE KEEPING OF PRACTICAL NOTEBOOKS

Dissections and Microscopical Preparations

(i) Print the name of the phylum class and animal on top; state whether it is a dorsal or ventral dissection or an entire specimen, and in the latter case which view.

If it is a microscopical preparation, state whether it is a longitudinal or transverse section or entire specimen, and whether it is as seen under the low or high power.

(ii) Write notes of any special directions, *e.g.*, removal or deflection of organs, method of staining.

(iii) Examine the object carefully before you begin to draw it. Then make a **drawing** *or* **diagram** in pencil, roughly to scale. *Draw only what you see* and draw on a *large* scale, showing all the necessary details.

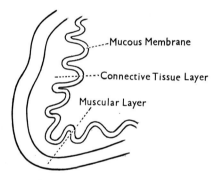

Fig. 2. Low Power Diagram.

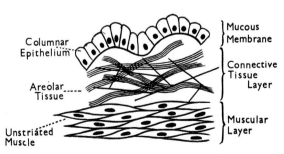

Fig. 3. **High Power Drawing of Tissues Indicated in Low Power Diagram.**
(From Wallis—Practical Biology)

The outline of the animal (or part of it) should be shown where desirable in order to show the position and relationship of parts. The use of shading and colours should be kept down to a minimum, *e.g.*, red for arteries, and blue for veins.

PRINT the names in BLOCK CAPITALS *horizontally* and join them to the corresponding parts by straight lines. Avoid crossing these lines over one another. (The use of letters and a key at the bottom is not desirable.)

When you have a complicated microscopical structure to draw, make a **diagram** (or plan) of the whole structure under the *low power* and detailed **drawings** of small samples of each tissue or of part of the structure under the *high power*, all suitably named.

N.B. Never copy drawings or diagrams from textbooks. By doing so, you learn very little and are therefore wasting valuable time. Draw only what you see in nature. *The illustrations in this book are intended solely to help the student to find and identify the various structures and tissues. Many, though not all, of the figures are dia-grammatic or semi-diagrammatic. This is intentional.*

III. INSTRUMENTS AND APPARATUS REQUIRED

By each Student

(1) A set of *dissecting instruments* in a cloth roll as follows* :—

    1 large all-steel *scalpel* (4·5 cm. blade).
    1 medium all-steel *scalpel* (3·75 cm. blade).
    1 small all-steel *scalpel* (2·5 cm. blade, or less).
    [Alternatively, scalpel handles with detachable blades of various shapes and sizes can be purchased.]

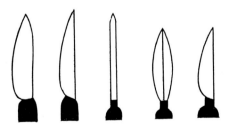

FIG. 4. **Useful Scalpel Shapes.**

*Sizes are approximate.

1 dissecting needle.

1 pair of large *scissors* (12 or 16 cm. overall length).

1 pair of small *scissors* with fine points (9 or 10 cm. overall length).

1 pair of large *forceps*, blunt (12 cm.).

1 pair of small *forceps* with fine points (10 cm.).

3 or 4 mounted *needles*.

1 *seeker*, 1 *camel-hair brush*, 1 *section-lifter*.

Other instruments, *e.g.*, *bone forceps*, may be added as desired.

(2) Large-page practical *note-books* or *files*, with plain pages. The Elementary Biochemistry, Animal Physiology and Genetics can be conveniently kept together in separate parts of the file, preferably with some ruled sheets.

(3) A *hand-lens* (unless supplied by the laboratory).

(4) A *microscope* (unless supplied by the laboratory). See (7) below.

(5) The necessary *drawing materials* and *red* and *blue coloured pencils*.

(6) At least one *white coat* is desirable unless an old jacket is kept for laboratory work.

By the Laboratory

In addition to the usual laboratory apparatus, the following will be needed:—

(1) *Dissecting dishes*. Rectangular enamel trays (about 20 × 15 cm.) with black wax composition in the bottom.

(2) *Dissecting boards* with a rim round the edge (about 60 × 45 cm.) (which may be fitted with hooks or rings at the corners) for larger animals.

(3) *Lenses*. Watchmaker's lenses clamped in small retort stands serve well as dissecting lenses. Larger hand lenses should also be provided.

(4) *Pins*, large and small; and *awls* for large animals.

(5) *Thread* for ligatures.

(6) *Preserving tank* for animal material containing 4 per cent. formaldehyde.

(7) *Microscopes* (unless provided by the students) with 16 mm. and 4 mm. objectives, on a triple nose-piece, and No. 2 (×6) and No. 4 (×10) eyepieces. A few better instruments fitted with sub-stage condensers, Nos. 2 (×6), 4 (×10) and No. 6 (×15) eye-pieces, and a 2 mm. O.I. objective are also desirable. A blue filter to fit below the condenser giving a daylight effect is an advantage when

using artificial light unless lamps with daylight bulbs are used. *All must be kept covered when not in use.*

(8) *Microscope lamps,* preferably fitted with daylight bulbs, unless microscopes with built-in lamps are used.

(9) *A dissecting microscope.*

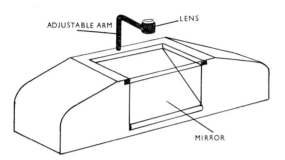

ADJUSTABLE ARM — LENS

MIRROR

Fig. 5. **A Simple Dissecting Microscope.**

(10) *Stains* and *Reagents.* See Appendix I.

(11) Soft *cloths* (*e.g.,* chamois leather) for lenses and objectives.

(12) A *turntable* for ringing slides is useful but not essential.

(13) The *Apparatus* and *accessories* mentioned in the text. Much of this can be made or adapted from other pieces of apparatus at small cost.

PART I

MICROSCOPICAL TECHNIQUE

THE MICROSCOPE

Description

The microscope is a delicately adjusted scientific instrument and *must be handled with care.*

It consists of the following parts (see Fig. 6):—

(i) **The Stand.** This is made up of a heavy **foot** which carries an inclinable **arm**, bearing the **body-tube.** The body-tube can be raised or lowered by the **coarse adjustment which** works by a rack and pinion arrangement and by the **fine adjustment** for more accurate focussing. Both are controlled by milled heads.

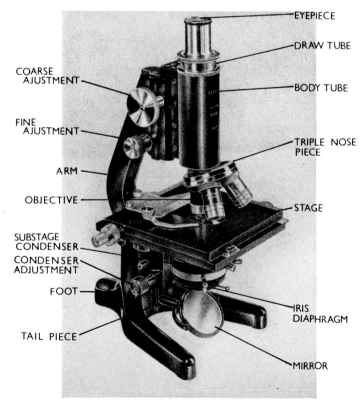

FIG. 6. The Microscope.

6

Most modern microscopes are made with a tube length of 160 mm. This may be increased by raising the **draw-tube,** thus giving greater magnification. The draw-tube is usually graduated. A **nose-piece** (which may be *single but is usually triple*) at the bottom of the tube carries the magnifying lenses or **objectives.** The arm also carries the **stage** on which the slides to be examined are placed and kept in position by springs. An attachable **mechanical stage** provided with vernier scales for moving the slides can be fitted. It is a luxury for ordinary use but a necessity for advanced work. In more expensive instruments, it is built in. The **tailpiece,** into which the **mirror** is fitted, is on the lower part of the arm but many modern microscopes have a built-in **lamp.**

(ii) **The Optical Parts.** These consist of the objectives, the eye-pieces, the mirror and the sub-stage condenser. The **objectives** are small tubes containing a combination of lenses. Those in common use have a focal length of **16 mm. (low power)** and **4 mm. (high power).** A **2 mm. oil immersion** lens is used in bacteriological, cytological, and other work requiring a much higher magnification. The 16 mm. and 4 mm. objectives are used dry, but when using the 2 mm. O.I. objective, a drop of cedar wood oil (of practically the same refractive index as the glass) is put on the coverslip and the objective focussed into it. This increases the illumination. The high power objectives are focussed slightly nearer the object than the 16 mm. objective, and the distance between the objective and the slide is called the **working distance.** In most modern microscopes, once the object has been focussed with the low power, it is almost in focus when the high power objective has been swung into position, about one turn of the fine adjustment being all that is necessary to get it sharply into focus. The **resolving power** of a lens or its power to define detail depends on what is known as its **numerical aperture.** This is constant for any one lens and the higher it is, the greater the resolving power though the working distance is decreased. Good resolution is obtained with a 16 mm. objective of N.A. of about 0·28, with a 4 mm. objective of N.A. of about 0·7 and with a 2 mm. O.I. objective of N.A. of 1·25 to 1·28. Makers always state the N.A. of their objectives. It should be noted that lenses have a curved field and consequently when, under the high power, the object is focussed in the centre of the field, it is only *this* part which is in sharp focus. Flatness of field is only possible with low power objectives. The magnified images produced by the objectives are further magnified by the **eyepieces** which fit into the top of the draw-tube. The magnification of these lenses is marked on them, thus ×4, ×5, ×6, ×8, ×10, ×15 and they may be numbered: No. 0 (= ×4), No. 1 (= ×5), No. 2 (= ×6), No. 3 (= ×8), No. 4 (= ×10), or No. 6 (= ×15). A table of magnifications with different objective and eyepiece combinations is provided by the makers with each instrument. **Binocular eyepieces** are often fitted to the more expensive microscopes. The **mirror** is concave on one side for use when the substage condenser is not in use and plane on the other side for use with the condenser. The **condenser** fits into the underside of the stage, or into a **substage,** and can usually be swung out when not required. It is focussed either by a spiral focussing arrangement or by rack and pinion. It increases the illumination

Fig. 7. **Binocular Microscope with Built-in Lamp and Mechanical Stage.**

when high power objectives are in use but it may be used with the low power also. Accurate centering is essential: this can be attained by means of **centering screws.** The amount of light passing through can be varied by the **iris diaphragm** which is fitted at the base of the condenser.

Magnification

The approximate magnification is found from the formula—

$$m = \frac{l}{f} \times e$$

where m = magnification.

l = length of body tube (usually 160 mm.).

e = magnification of eyepiece.

f = focal length of objective.

As a very *approximate* guide it may be assumed that the—

16 mm. objective magnifies 10 times,

the 4 mm. objective magnifies 40 times,

and the 2 mm. O.I. objective magnifies 80 times.

These magnifications should be multiplied by the magnification of the eyepiece.

A greater magnification can be obtained by raising the draw-tube and the new magnification can then be calculated from the formula

$$M_2 = \frac{m_1 \times l}{160}$$

where M_2 = final magnification.

m_1 = magnification for 160 mm. tube.

l = total length of tube in mm.

160 = length of tube in mm. for which makers calculated m_1 (*i.e.*, length without draw-tube extended).

Measurement

The unit of length under the microscope is 0·001 mm., and is known as 1 micron (μ).* Measurement is made by means of an **eyepiece micrometer** and a **Ramsden's eyepiece,** but this is not required of the elementary student.

Use and Care

(i) Always lift the instrument by the arm.

(ii) Never allow any liquids to get on to the lenses or stage and keep them free from dust. Do not touch the instrument with wet fingers.

* Pronounced *mew.*

(iii) See that the slide and coverslip are *clean* and *dry*.

(iv) Always examine an object as follows:

    (*a*) with the naked eye if visible, otherwise with a hand lens.

    (*b*) with the 16 mm. objective.

    (*c*) with the 4 mm. objective.

In some cases, high power magnification will not be required while in others low power is inadequate.

(v) Never use the high power unless the object is covered with a coverslip.

(vi) Illuminate with the plane mirror when a substage condenser is used and with the concave mirror when it is not used.

(vii) **Focussing.** With the *low* power bring the object clearly into view with the *coarse* adjustment and then focus accurately with the *fine* adjustment.

In using the *high* power, first see that the object is in the centre of the field and accurately focussed under the low power, then swing the high power objective into position without touching the coarse adjustment. About one turn of the fine adjustment will then bring the object into sharp focus. [If the instrument has no nose-piece, with the head at the side of the microscope, slowly lower the tube with the coarse adjustment until the objective is close to the slide: then carefully focus *upwards*, using the fine adjustment as before. You will be less likely to break the slide or damage the objective if this method is used.]

(viii) If a substage condenser is used, it must be accurately focussed. This is usually the case with the Abbé condenser and similar types when the upper lens of the condenser is almost touching the under side of the slide.

(ix) Use the microscope with *both* eyes open (it is only a matter of practice) and get accustomed to using either eye.

(x) Have your note-book on the right of the instrument.

(xi) Study the object carefully before beginning to draw it.

(xii) After using a 2 mm. O.I. objective, carefully clean off the oil from the objective and the slide with xylene on a clean soft cloth or blotting paper. Dry with a soft chamois leather.

(xiii) Clean the outsides of the lenses, if dusty, with a piece of clean soft chamois leather. *Never take an objective to pieces.* Dirt which will not wipe off with a cloth can generally be removed by wiping the lens with a soft cloth dipped in alcohol. To locate dust specks visible when looking through the microscope, rotate the eyepiece. If the specks also rotate, it is obvious that they are there. Move the slide and if the specks move too, they are on the coverslip. Otherwise they will be on the condenser, the mirror or the objective.

(xiv) Always keep the microscope covered when not in use and always keep an eyepiece in the microscope to prevent dust getting into the draw-tube and thus on the inside of the objectives. Dust is the worst enemy of the microscope.

THE PREPARATION OF MICROSCOPICAL SLIDES

Objects to be examined under the microscope are usually mounted on glass **slides** measuring 7·6 × 2·54 cm., and covered by a circular or square **coverslip.** Unless specimens are very small and thin, it is generally necessary to cut **sections.**

Minute organisms, tissues or sections may simply be mounted in water, physiological saline, dilute glycerine, etc., covered and examined **(temporary mounts),** but it is often necessary to use **stains** to show up certain structures; and, again, it is sometimes desirable to make a **permanent mount.** In this case, the object must be subjected to certain processes as follows:—

(1) Killing, Fixing and Hardening.
(2) Staining.
(3) Dehydrating.
(4) Clearing.
(5) Mounting.

The reagents and stains are put into watch glasses and the object is transferred from one to another by means of a section-lifter.

SECTION CUTTING

Sections may be **transverse** or **sagittal** (radial longitudinal), depending on the symmetry of the object from which they are cut. For hand section-cutting a *sharp* **section-cutting razor,** hollow-ground on one side and flat-ground on the other, is necessary for some sections but this will not normally be required in Zoological work and in examinations sections are usually provided already cut. In fact the only section cutting normally required of the student in zoology is of cartilage and this can be done with a sharp scalpel. (For sharpening see p. 1.)

Cut the thinnest possible sections, drawing the scalpel horizontally and obliquely across the material. Try and cut sections of uniform thickness. Several should be cut and transferred to water in a watch glass. Choose the best sections and discard the rest.

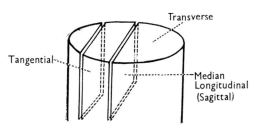

FIG. 8. **Kinds of Sections.**

KILLING, FIXING AND HARDENING*

The organism or tissue is exposed in the living state to a killing or fixing agent. This (i) **kills** the tissues and at the same time **fixes** them so that their histological form does not alter; (ii) prepares the tissue for subsequent treatment with stains and (iii) **hardens** it for section cutting. The tissues must be thoroughly washed in a suitable medium afterwards.

There is a large number of Fixing Agents of which the most important are tabulated below with their uses and washing media.

Hardening and Fixing Agents

The Fixing Agents in more general use are shown in **thick type.**

Fixing Agent	Use	Washing Medium
Alcohol (ethyl), 70% –	Animal tissues.	70% alcohol.
Acetic alcohol – –	Animal tissues.	70% alcohol.
Bouin's fluid – – –	Animal tissues.	50 and 70% alcohol.
Mercuric chloride –	Animal tissues.	Iodine-alcohol. Decolorise with sodium thiosulphate.
Picric acid – – –	Animal tissues.	50 and 70% alcohol.
Acetic acid – – –	Animal nuclei.	50% alcohol.
Corrosive acetic –	Animal tissues.	Iodine-alcohol. Decolorise with sodium thiosulphate.
Formaldehyde – –	Animal tissues.	70% alcohol.
Müller's fluid– – –	Animal tissues.	Water. Decolorise in 1% chloral hydrate.
Osmium tetroxide – ("Osmic acid")	Protozoa.	Water.
Potassium dichromate –	Animal tissues.	Water. Decolorise in 1% chloral hydrate.

* Material is generally supplied to students already fixed by the laboratory.

Tissues may be fixed in bulk or as sections. Immerse the material in several times its own volume of the fixing agent for at least fifteen minutes and preferably, in most cases, for considerably longer. Thoroughly wash out the fixative in the washing medium stated. Use of the wrong medium will cause changes (e.g., precipitates) which will affect the subsequent treatment and results.

STAINING

Staining is the colouring of tissues and structures by the addition of solutions of dyes and its object is to show up tissues and structures which would otherwise be imperfectly seen. Tissues may be stained in bulk but it is better to cut sections first.

Solutions of stains are made up in water or alcohol and the solvent must be known before proceeding with the next process (dehydration). Some stains are acid, some basic, and others are neutral. These terms do not refer to the reactions of the solutions but to the coloured radicles. Consequently, while some stain the nucleus (basic stains), others stain the cytoplasm (acidic stains) or cell contents, but most nuclear stains also stain the cytoplasm to a lesser extent. The converse is also true in some cases. There are **general stains** for ordinary use as opposed to **specific stains** which stain certain tissues only. Owing to this specific nature of some stains, it is possible to use two or even three different stains on the same sections in order to differentiate the tissues more clearly. This is called **Counter-staining** (or **Double or Triple**) **staining.** A list of the commoner stains and the uses to which they are put is given on p. 14.

Methods of Staining*

There are two methods of staining:—

(i) **Progressive staining** in which the tissue is left in the stain until it has reached the required depth of colour.

(ii) **Regressive staining** in which the tissue is intentionally over-stained and then decolorised (**differentiated**) to the required depth of colour.

The beginner should use the method of progressive staining as it is a little difficult for him to judge when the required

* Detailed examples of single and double staining are given at the end of this chapter.

degree of differentiation has been reached. When he has reached some degree of success with this method, he can try the method of regressive staining.

(1) Single Staining

The sections are placed in the stain and left there until they are stained to the required depth (progressive staining). This can be ascertained by placing the watch glass on the stage of the microscope and examining with the low power. Normally, sections take only a few minutes. Excess of stain must afterwards be removed by placing the section in the solvent used for the stain. This will not remove the stain from the tissues. If a section is overstained or if the method of regressive staining is used, the degree of staining can be lessened by placing it in **acid alcohol.** *This is called* **differentiation** *because it differentiates the extent to which the various parts of the tissue are stained. Differentiation should be watched under the low power of the microscope as when staining, and when it is complete the tissue should be washed in 70 per cent. alcohol.*

Stains

Good general stains are in **thick type.**

Stain	Solvent	Use
Borax-carmine – –	**Alcohol.**	**Nucleus—pink.**
Eosin Y – –	**Water or alcohol.**	**Cytoplasm—pink.**
Hæmalum – – –	**Water.**	**Nucleus—blue.**
Hæmatoxylin (Delafield) –	**Alcohol.**	**Nucleus—blue.**
Hæmatoxylin (Ehrlich) –	**Alcohol.**	**Nucleus—blue.**
Leishman's stain – –	Alcohol (methyl).	Blood corpuscles: red-pink: white (nuclei)—blue.
Methyl violet– – –	Water or alcohol.	Nucleus—violet.
Methylene blue – –	Usually alcohol.	Nucleus, blood—blue.
Picro-carmine – –	**Water.**	**Nucleus—red.**
		Cytoplasm—yellow.
Van Geison – – –	Water.	Epithelium—yellow. Connective tissue—red. Muscle—yellow.

(2) Double Staining†

The sections are stained with one stain at a time, the excess being removed as in single staining before placing the section in the second

† Double staining is not permitted in some examinations.

stain. It may be necessary to differentiate and even to partially dehydrate before adding the second stain.

Double Staining

$\left\{\begin{array}{l}\text{Hæmatoxylin (Delafield or Ehrlich) or Hæmalum.}\\ \text{Eosin Y.}\end{array}\right.$

$\left\{\begin{array}{l}\text{Borax-Carmine.}\\ \text{Eosin Y.}\end{array}\right.$

$\left\{\begin{array}{l}\text{Hæmatoxylin or Hæmalum.}\\ \text{Van Geison.}\end{array}\right.$

DEHYDRATION

Owing to the fact that water is immiscible with the oil used for clearing on the one hand and with the solvent of the mounting medium (usually xylene) on the other, it is essential that all traces of water should be removed from the tissue. This removal of water is usually effected by means of ethyl alcohol and is known as **dehydration.** If the sections were placed directly into absolute alcohol, the cells would lose their water so quickly that they would shrink and their shape would be altered. This is avoided by adopting the following method:—

Place the sections in solutions of ethyl alcohol of gradually increasing concentration for one or two minutes. The concentrations generally used are:—

30 *per cent. alcohol.*
50   ,,   ,,
70   ,,   ,,
90   ,,   ,,
Absolute alcohol.

Start with the concentration next above that in which the sections were last placed, i.e., *if the stain was aqueous, begin with 30 per cent. alcohol, but if it was alcoholic, begin with 70 per cent. alcohol.*

To ensure complete dehydration, finally transfer the sections to a second watch glass of absolute alcohol. Avoid breathing on the absolute alcohol in the watch glass and in the bottle; otherwise it will no longer be absolute. It is an advantage, therefore, to cover the watch glass with a second one inverted over it.

(In place of alcohol, a substance known as "Cellosolve" (Ethylene glycol monoethyl ether) can be used for dehydrating sections. This mixes with water, alcohol, oil of cloves and xylene and there is no need to use varying concentrations, as it does not cause shrinkage or alteration in shape; nor is it necessary to clear after using it. But it dehydrates rapidly and because of this it is *not recommended for animal tissues* as it may cause distortion.)

CLEARING

If alcohol has been used as the dehydrating agent, it must now be removed. This process is called **clearing** as it also renders the tissues transparent. For permanent preparations the best results are obtained by using *natural oil of cedar wood* for animal tissues. Xylene can be used but it has a tendency to cause shrinkage.

Leave the sections in the clearing agent until they are transparent. This usually takes two or three minutes. If there is any sign of a white film around the sections, this indicates incomplete dehydration and they should be returned to absolute alcohol and cleared again.

MOUNTING

For final examination under the microscope and for preservation, the sections must now be **mounted** in a suitable medium of about the same refractive index as crown glass (1·5). The following may be used:—

(1) Temporary Mounts

(i) Physiological saline (R.I. = about 1·34).

For Invertebrate tissues and vertebrate blood – – – – – – 0·6 per cent.
Amphibian tissues (except blood) – – 0·75 ,, ,,
Mammalian tissues (except blood) – – 0·9 ,, ,,

or (ii) Ringer's Solution (Invertebrate tissues).

Locke's Solution (Mammalian tissues).

(2) Permanent Mounts

The best Mounting Medium for general use is *Canada Balsam* dissolved in xylene. (R.I. = about 1·524).

Euparal (R.I. = about 1·4) can be used, and in this case it does not matter if dehydration is not complete.

Place a drop of Canada Balsam on the centre of a clean slide. Transfer the section to the balsam with a section lifter. Cover with a clean coverslip by resting the coverslip against the finger and levering it down with a mounted needle (Fig. 9). *This will prevent the entrance of air bubbles. There should be no air bubbles in the balsam. If any appear, they can generally be removed by gently warming the slide over a very small flame. A white film in the balsam indicates incomplete dehydration. Label the slide and leave it to dry. The coverslip may be ringed but this is not by any means essential for ordinary purposes.*

FIG. 9. **Levering Down the Coverslip.**

In labelling slides, state the name of the organism from which the tissue was obtained, the name of the object or tissue and, if it is a section, whether it is L.S. or T.S. It is also an advantage for future reference to add the name of the fixing agent and stain used.

Larvæ, small insects and chitinous structures can be mounted in *Berlese's Medium.** If this is used or if the specimen is thick, the coverslip must be raised from the slide by means of a **cell** of the required height which is affixed to the slide. For some objects, a **cavity slide** is preferable.

RINGING

Place the slide on a turntable (Fig. 10) and fix it in position by the springs provided.

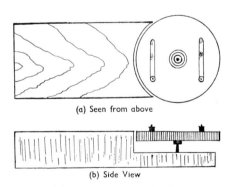

(a) Seen from above

(b) Side View

FIG. 10. **Turntable for Ringing Slides.**

Dip a brush in the ringing cement and start the turntable revolving rapidly by applying the finger to the milled edge. Apply the brush

* This is also a clearing agent for these objects.

to the edge of the coverslip, covering it with a thin layer of the cement. Allow this to dry and repeat the process until a sufficiently thick ring is obtained.

SMEARS

It is obviously impossible to cut sections of such tissues as blood and the contents of the seminal vesicle of the earthworm. The following procedure should therefore be used in these cases:—

*Make a thin film (called a **smear**) either on the slide or on the coverslip. This may be done as follows: Place a drop of the fluid at one end of a slide and hold another slide in contact with the fluid at an angle of about 45°. Draw the second slide over the first to produce a film of even thickness. Allow this to dry in the air or by gentle heat well above the flame of a spirit lamp. Then treat the smear with the stains and reagents in a similar manner to that used for sections.*

IRRIGATION

Minute organisms often require treatment with fluids while still living. This is done on the slide by the process of **irrigation.**

Having made a temporary mount of the object, place a drop of the irrigating fluid against one edge of the coverslip and a small piece of filter paper against the opposite edge. This gradually withdraws the fluid in which the object is mounted and the irrigating fluid enters underneath the coverslip to take its place.

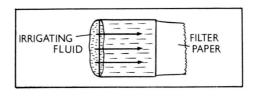

FIG. 11. **Irrigation.**

EXAMPLES OF METHODS OF MAKING PERMANENT STAINED PREPARATIONS†

In the following examples, the times stated must not be taken too rigidly. Some sections will take longer than others to stain. Examination of the tissues under the low power of the microscope will show when

† Specific methods, *e.g.*, for blood, are given in Part III.

*they are suitably stained. Again, sections should be left in the washing
fluid until no further stain comes out and in the clearing agent until
they are transparent.*

These methods are suitable for animal tissues generally.

(i) **Single Staining with Picro-Carmine**
(1) Picro-Carmine (to stain) . . . . 2-5 mins.
(2) Distilled water (to remove excess stain) . 2 mins.
(3) 30% Alcohol (to dehydrate) . . . 2 mins.
(4) 50% Alcohol . . . . . . 2 mins.
(5) 70% Alcohol . . . . . . 2 mins.
(6) 90% Alcohol . . . . . . 2 mins.
(7) Absolute Alcohol . . . . . 2 mins.
(8) Absolute Alcohol . . . . . 1 min.
(9) Natural Oil of Cedar Wood (to clear) . 5 mins.
(10) Mount in Canada Balsam

(ii) **Single Staining with Borax-Carmine**
(1) Borax-Carmine (to stain) . . . . 2-5 mins.
(2) 70% Alcohol (to remove excess stain) . 2 mins.
(3) 90% Alcohol (to dehydrate) . . . 2 mins.
(4) Absolute Alcohol . . . . . 2 mins.
(5) Absolute Alcohol . . . . . 1 min.
(6) Natural Oil of Cedar Wood (to clear) . 5 mins.
(7) Mount in Canada Balsam.

(iii) ***Single Staining with Ehrlich's Hæmatoxylin or Eosin** and
Double Staining with Ehrlich's Hæmatoxylin and Eosin Y
(1) Ehrlich's Hæmatoxylin . . . 10-15 mins.
(2) Acid Alcohol (to differentiate) . . . 3 mins.
(3) 70% Alcohol (to remove excess of stain) . 5 mins.
(4) Tap water (to "blue" Hæmatoxylin) . . 2 mins.
(5) 70% Alcohol . . . . . . 2 mins.
(6) 90% Alcohol . . . . . . 2 mins.
(7) Eosin Y, alcoholic . . . . . 1 min.
(8) 90% Alcohol (to remove excess stain and
dehydrate) . . . . . . 2 mins.
(9) Absolute Alcohol . . . . . 2 mins.
(10) Absolute Alcohol . . . . . 1 min.
(11) Natural Oil of Cedar Wood (to clear) . 5 mins.
(12) Mount in Canada Balsam.

* If **Single Staining with Ehrlich's Hæmatoxylin** is required, simply *omit
processes* (7) *and* (8).
If **Single Staining with alcoholic Eosin,** *omit processes* (1) *to* (6) and after
staining first wash out in 70% alcohol.

PART II

THE VARIETY OF ANIMALS
MORPHOLOGY AND ANATOMY

TAXONOMY—THE CLASSIFICATION OF ANIMALS*

The **Animal Kingdom** is divided into two **Sub-Kingdoms, Protozoa** (non-cellular animals) and **Metazoa** (multicellular animals). The main divisions of these Sub-Kingdoms are the **Phyla** of which the following are represented in this book:—*Protozoa, Porifera, Cœlenterata, Platyhelminthes, Nematoda, Annelida, Arthropoda, Mollusca, Echinodermata,* and *Chordata.*

Each Phylum is divided into **Classes,** the Classes into **Orders** and these into **Families.** Animals in the same Family are arranged in **Genera** and the varieties of animals in the same genus are classified in **Species.** In some cases there is further division into Sub-Phyla, Super-Classes, Sub-Classes, etc.

Every animal is given two names, its generic name and a specific name. Thus the common frog is *Rana temporaria* and the continental edible frog *R. esculenta.* This is known as **binomial nomenclature.**

The Phyla, Classes and Orders to which each of the animals treated in this book belongs are given, together with the *bare outlines* of the characteristics of these groups, at the beginning of the text in each case. Further details of the characteristics of the various groups will be found in the appropriate text-books.

INTRODUCTORY NOTES FOR PRACTICAL WORK

Reference should first be made to the General Directions for Practical Work and to the keeping of notebooks in the Introduction; also to Part I (Microscopical Technique) where necessary.

Personal examination of animals is the only way to learn their physiology and anatomy. The student should avail himself of every opportunity to examine them *alive,* observing how they live by day and night, how they procure their food, breathe, move, grow, reproduce, avoid their enemies, etc. These observations may be made on animals kept in vivaria and aquaria, while visits to zoological gardens and observations made in the country, will be found most beneficial and instructive.

For an accurate knowledge of the anatomy and physiology of the living animal, however, it is necessary to make a more detailed individual study. This usually involves the examination of the dead

* Nomenclature in the classification of animals is frequently undergoing change and some confusion may arise in using different text books.

animal. Small animal organisms are examined under a hand lens or microscope. Larger animals must be dissected.

Microscopic animals should be examined (i) living, if possible, (ii) killed and stained, under the low and high powers of the microscope.*

Larger animals are first examined externally, alive in the first place when possible. A series of dissections is then made on the killed animals to expose the various systems, the **anatomy** of which is studied in detail. A microscopic examination of animal tissues, **histology,*** is made and simple biochemical and physiological experiments performed which assist in the understanding of the **physiology** of the animals. Experiments in animal **genetics** should be performed with the fruit fly *Drosophila* and the **embryology** of certain specified chordate types studied.

The animals are arranged in their systematic positions in this book and drawings should be kept in these places in the practical notebook or file, but it may be necessary to study them in a different order (according to the availability of material).

DISSECTION TECHNIQUE

Dissection is an art which can only be acquired by practice. Do not attempt any practical work until you have thoroughly studied the appropriate subject in your text-book.

    (i) Small animals, such as the earthworm and frog are dissected under water in a dish. This gives support to the organs, keeps them apart, and renders them more easily visible. Always *keep the water clean*: replace it by fresh as soon as it becomes cloudy or dirty.

   (ii) Using pins of suitable size, fix the animal to the bottom of the dissecting dish or, in the case of large animals such as the rabbit and rat, tie the legs to the hooks or rings at the corners of the dissecting board or pin them in position by means of awls. *Keep the parts tightly stretched.*

  (iii) Always fix in the pins or awls *obliquely*, so that they do not get in your way.

  (iv) See that your scalpels and scissors are sharp and that all your instruments are clean.†

   (v) Read the directions carefully *before* beginning your dissection.

  (vi) Choose a scalpel of suitable size and shape and hold it as you would a pen.

* See Part I for detailed instructions in microscopical technique.
† For sharpening of instruments see page 1.

(vii) Never cut anything until you are sure what it is.

(viii) Be gentle. If you are rough with your instruments you may damage the organs or cut blood vessels and your dissection will be a failure.

(ix) Cut through the skin and muscular body wall and pin them down out of your way.

(x) Mop up blood and body fluids with cotton wool.

(xi) You may displace or deflect organs if necessary to expose others to view. Black paper may be placed underneath obscure structures in order to make them more easily visible but it must not be used if it will hide important related structures or organs.

(xii) Always dissect *along* and not across blood vessels and nerves.

(xiii) When the required organs are exposed to view, examine them carefully and *draw exactly what you see in your specimen* and as near as possible to scale, making the organs the right size (in proportion) and the right shape. Make *large* drawings. Be sparing in the use of colours—use red for arteries and blue for veins.*

(xiv) Preserve your specimens, if required for future use, in 10 per cent. formalin or 70 per cent. alcohol.

(xv) The major dissections—those most frequently required in examinations—should be repeated until as near perfection as possible is attained.

(xvi) Thoroughly clean and dry your instruments immediately after use. A light coating of vaseline will prevent their rusting.

(6) The names of all parts printed in **thick type** in the text which follows should be included in the drawings *provided that they can be seen in the specimen*, but in some cases *all* the structures mentioned in the text may not be seen or identified.

(7) Diagrams (or plans) of systems may occasionally be desirable in addition to drawings of dissections (*e.g.*, blood systems). *The general outline of the animal and, where desirable of other organs*, should be indicated by a single or dotted line to show the relationship of the parts.

(8) Write notes on the method of your dissections where desirable and of observations made (*e.g.*, any abnormalities). Do not let your note-book be simply a collection of drawings and diagrams.

---

* As already stated in the Introduction. a number of the illustrations in this book are intentionally diagrammatic or semi-diagrammatic though this does not apply in all instances.

SUB-KINGDOM PROTOZOA
(Microscopic non-cellular animals).

PHYLUM
PROTOZOA
Body not divided into cells but contains specialised parts called organelles.

CLASSES

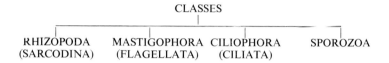

| RHIZOPODA | MASTIGOPHORA | CILIOPHORA | SPOROZOA |
| (SARCODINA) | (FLAGELLATA) | (CILIATA) | |

CLASS RHIZOPODA (SARCODINA)
Lack a firm pellicle and so have no definite body form. Move and feed by pseudopodia. Reproduction asexual or sexual. Most free-living, a few parasitic.

AMŒBA

A. proteus lives in pond and ditch water.

(1) *Examine a drop of water containing* **living Amæbæ.** *Find an Amœba under low power. Draw at intervals of half a minute or so, and examine afterwards under high power.*

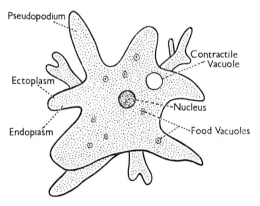

FIG. 12. **Amœba Proetus.**
(*From Wallis—Practical Biology*)

Note the movement and the changes of shape due to the flowing out of promontories of protoplasm called **pseudopodia.** By careful focussing you will be able to distinguish a clear narrow outer layer of protoplasm, the **ectoplasm** (or **plasmagel**), enclosing the granular **endoplasm** (or **plasmasol**). A protoplasmic "membrane", the **plasmalemma,** serves to keep the animal intact. Note that the endoplasm shows a flowing movement. In it may be seen the **nucleus** and spaces, or vacuoles, some of which, **food vacuoles,** may contain particles of food such as algæ. Look for the **contractile vacuole,** which disappears and reappears rhythmically at the same spot: its function is osmoregulation. *If possible, make drawings at timed intervals of an organism* **undergoing binary fission.** During the winter months, **encysted amœbæ** are sometimes found. These are spherical in shape, devoid of pseudopodia and are enclosed in a protective **cyst.** Placing the tube in warm water will often restore activity.

(2) *Irrigate with 1 per cent. acetic acid or $\frac{1}{4}$ per cent. "osmic acid" to kill and fix the organisms; then stain with picro-carmine by irrigation. Alternatively, examine a prepared slide.*

The structures in (1) will be more easily seen, but, of course, the contractile vacuole will be at rest.

CLASS MASTIGOPHORA (FLAGELLATA)

Have a pellicle and therefore a definite shape. One or more flagella for locomotion. Diverse forms, most free living, some parasitic. Reproduction asexual. Some are plant-like and their nutrition is holophytic. Can be considered as "plant-animals".

EUGLENA

Lives in puddles and ditch water rich in organic nitrogenous material. *E. viridis* often regarded as an Alga as nutrition is holophytic. *Peranema,* a closely allied form which is colourless, has holozoic nutrition.

(1) *Mount a drop of water containing* **living Euglena viridis.** *Cover.*

Under *low and high power* note the non-cellular green organisms moving in all directions and *examine the method of movement. By using the iris diaphragm,* movement of the water due to the flagellum may be seen.

(2) *Irrigate with iodine. Also examine a prepared slide.* Under *high power* note the spindle-shaped cell with blunt anterior end bearing a single **flagellum** and pointed posterior end. An elastic **pellicle** encloses a clear **ectoplasm** and a granular **endoplasm.** At the anterior end is a **contractile vacuole** communicating with the exterior by the so-called **gullet** (or **cytopharynx**). The **endoplasm** contains elongated

chloroplasts radiating from a centre or otherwise disposed according to the species, **paramylum granules,** a large **nucleus** containing a **nucleolus** (karyosome) towards the posterior end and a red **pigment-spot** or **stigma** anteriorly.   *Look for stages in* **asexual reproduction.**

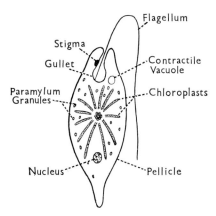

FIG. 13.   **Euglena.**
(From Wallis—Practical Biology)

TRYPANOSOMA

Trypanosoma gambiense is the parasite responsible for the tropical disease, *sleeping sickness,* and is carried by the tsetse fly.   It lives in the blood plasma of man.

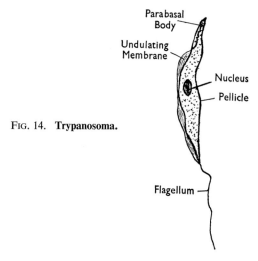

FIG. 14.   **Trypanosoma.**

Examine a slide of **Trypanosoma in human blood,** *preferably under the 2 mm. O.I. objective.* The organism is elongated and spindle-shaped. Its shape is maintained by a **pellicle.** Running along the greater part of its length is an **undulating membrane** and arising from the posterior end is a long **flagellum** which runs along the free edge of this membrane and protrudes beyond the anterior end. A small **parabasal body** lies near the base of the flagellum and a granule known as the **blepharoplast** at its base. The **nucleus** is large and ovoid. Reproduction is by longitudinal binary fission.

CLASS CILIOPHORA (CILIATA)

Have a pellicle. Locomotion is by cilia. Two nuclei present. Reproduction asexual but a sexual process may intervene. Free living.

PARAMŒCIUM

P. caudatum lives in fresh-water ponds in which decomposing plant matter is present.

(1) *Examine a drop of water containing* **living Paramœcia.** *The organisms move rather quickly. Put a drop of weak gelatin sol or a few fibres of cotton wool on a slide and add a drop of the Paramœcium culture. Cover. The organisms cannot move so quickly and their structure can be seen more easily.*

Note the rounded **anterior end** and the pointed **posterior end.** Note also that the animal has a constant shape. This is due to the **pellicle** which surrounds it. Movement is effected by the **cilia,** short hair-like protoplasmic processes which cover the surface.

On one side is a shallow **oral groove,** leading to the **cytopharynx** which runs obliquely backwards to the **cytosome** (or mouth). This is called the **oral side,** as opposed to the other, which is the **aboral side.** In the cytopharynx the cilia are fused, forming the so-called **undulating membrane.**

The **ectoplasm** is a clear narrow outer layer, as in Amœba, but it contains a number of minute capsules, each containing a thread, the **trichocysts.** The **endoplasm** is granular and contains a number of **food vacuoles** and two contractile vacuoles, the **anterior** and **posterior contractile vacuoles,** into which the water is poured from ducts which radiate around them, the **formative vacuoles.** There are two nuclei, a large oval **meganucleus** in which there is a niche on the aboral side containing the smaller spherical **micronucleus.**

Look for stages in **asexual reproduction** *and* **conjugation.**

(2) *Irrigate with iodine or feed with indian ink.*

Note the discharged **trichocysts.**

(3) *Irrigate a fresh drop of the culture with* $\frac{1}{4}$ *per cent. "osmic acid" to kill and fix the organisms. Stain with picro-carmine by irrigation. Alternatively, examine a prepared slide.*

The structures in (1) will be more easily seen.

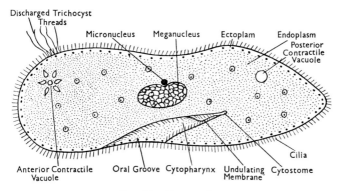

FIG. 15. **Paramœcium.**

CLASS SPOROZOA

Parasitic. Reproduction asexual, sexual or both in a single life-history. Form spores.

MONOCYSTIS

This parasite lives in the seminal vesicles of the earthworm, in a smear from which they may be seen. The two commonest species are **M. agilis** and **M. magna.**

(1) *Examine a prepared slide of* **Monocystis** *in the* **trophozoite stage** *or make a smear of the contents of the seminal vesicle of the earthworm* (see p. 18 and p. 64)*.

Note the cigar-shaped **trophozoite,** surrounded by a **pellicle** and containing a **nucleus.** There is a thin **ectoplasm** not easily recognisable, and a very granular **endoplasm** containing a number of oblong bodies composed of **paramylum** (a carbohydrate allied to starch).

(2) *Examine a prepared slide, showing the* **reproduction of Monocystis,** *and find as many of the following stages as possible under the high power.*

* This should certainly be done when dissecting this animal if not done at this stage.

Look for **gametocytes,** two rounded cells in contact each containing a **nucleus** and enclosed in a chitinous cell wall or **gametocyst (association cyst).** Look for later stages with the **nuclei** round the periphery of each cell and, still later, the pear-shaped **gametes.** After fusion of the gametes, note the ovoid zygotes or **sporoblasts** and the **residual**

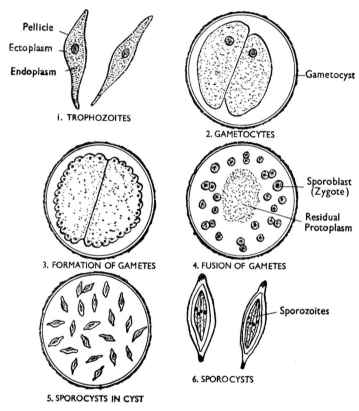

FIG. 16. **Monocystis. Stages in Life History.**

protoplasm. Later the sporoblasts are surrounded by a cyst and are known as **sporocysts** (also called **pseudo-navicellæ** because they are boat-shaped and resemble the plant *Navicella*). Eight minute **sporozoites** (or **falciform bodies),** shaped like sickles, are formed by nuclear division in the sporocyst.

PLASMODIUM

THE MALARIAL PARASITE

The parasite lives in the blood of man but completes its life history in the female *Anopheline* mosquito.

There are four species of plasmodium, **P. vivax** and **P. ovale**, the cause of *benign tertian malaria*, **P. malariæ**, the cause of *quartan malaria*, and **P. falciparum**, the cause of *malignant tertian malaria*.

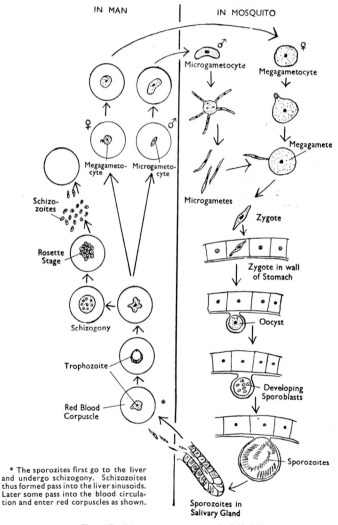

IN MAN | IN MOSQUITO

Microgametocyte
Megagametocyte

♀ ♂
Megagameto- Microgameto-
cyte cyte

Megagamete

Schizo-
zoites

Microgametes

Rosette
Stage

Zygote

Zygote in wall
of Stomach

Schizogony

Oocyst

Trophozoite

Developing
Sporoblasts

Red Blood
Corpuscle

Sporozoites

* The sporozites first go to the liver
and undergo schizogony. Schizozoites
thus formed pass into the liver sinusoids.
Later some pass into the blood circulation and enter red corpuscles as shown.

Sporozoites in
Salivary Gland

FIG. 17. **Plasmodium. Stages in Life History.**

(1) *Examine prepared slides showing one of these species of* **Plasmodium in the blood of man.** *Using the high power, or better a 2 mm. O.I. objective, find as many as possible of the following stages. You will probably succeed in seeing but one or two.*

In the **red corpuscle** note the amœboid, spherical or, at a later stage, vacuolated signet-ring-shaped **trophozoite** with the nucleus to one side, which after nuclear division forms a mulberry-like mass, known as the **rosette stage.** This is an asexual process, **schizogony (or merogony), and the schizozoites (or merozoites)** thus formed are set free into the plasma and enter other corpuscles. Look for the smaller male **microgametocytes** with large nuclei and clear cytoplasm and the larger female **megagametocytes** with small nuclei and granular cytoplasm, globular (*P. vivax* and *P. malariæ*) or sausage-shaped (*P. falciparum*) cells formed in the red corpuscles at a later stage from schizozoites which do not undergo division.

(2) *Examine prepared slides of* **Plasmodium in the mosquito.** *Find as many as possible of the following stages:—*

Long thin **microgametes** and larger spherical **megagametes** are formed from the microgametocytes and megagametocytes respectively. These fuse to form zygotes. **Oocysts** are developed from the zygotes which cause swellings on the "stomach" (crop) of the mosquito. *Examine a slide showing* **oocysts. Sporoblasts** are formed from these by vaculoation and nuclear division. From the sporoblasts arise large numbers of spindle-shaped **sporozoites,** by multiple fission. *Examine a slide of a salivary gland showing* **sporozoites.**

These pass into the blood of a man when he is "bitten" by an infected mosquito, but they quickly leave it and go to the liver where schizogony takes place and the schizozoites enter the sinusoids. This is known as the *pre-erythrocytic stage.* Some of the schizozoites remain in the liver and undergo further schizogony. This is called the *exo-erythrocytic stage* and is responsible for the re-occurrence of the disease after a latent period. Others pass into the blood stream and enter red corpuscles where they develop into trophozoites. This is the *erythrocytic stage.* Schizogony then takes place again as described.

SUB-KINGDOM METAZOA

PHYLUM

PORIFERA

Sponges

Diploblastic aquatic Metazoa, mostly marine. Body wall composed of two layers of cells, an outer pinnacoderm and an inner choanoderm, separated by a gelatinous mesoglœa and enclosing a single gastric cavity with one exhalant opening and several inhalent pores in outer layer. Some radially symmetrical, others irregular.

CLASSES

CALCAREA	HEXACTINELLIDA	DEMOSPONGIA
Skeleton of spicules of CaCO₃, *e.g. Sycon.*	Skeleton of silicious spicules, *e.g. Venus' Flower Basket.*	Skeleton of 4-rayed silicious spicules, *e.g. Bath sponges.*

CLASS CALCAREA

SYCON

(1) *Examine a specimen of* **Sycon.**

This calcareous sponge is marine. The body is composed of a number of branching tubular structures or **cylinders** up to about 7 or

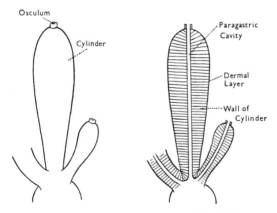

Osculum

Cylinder

Paragastric Cavity

Dermal Layer

Wall of Cylinder

FIG. 18. **Sycon External View and L.S. Cylinder**

8 cm. long and joined at the base by which it is attached to rocks. At the tips of the cylinders is an opening, the **osculum,** out of which water passes. *Examine one of the cylinders with a lens.* The surface

is covered by a large number of elevations and in the grooves between them are tiny pores, ostia, visible only under the microscope. Through these pores water enters the animal and they lead into a central cavity. Support of the body is effected by a skeletal framework made of calcium carbonate in the form of spicules.

(2) *Examine a* **Transverse Section through the wall of a cylinder of Sycon** *under low power and then under high.*

Note the outer **dermal layer,** a single layer of cells bearing spicules. Beneath this lies the **ectoderm (choanoderm)** composed of flattened scales. The wall is traversed by canals of two kinds:—(i) the wider **radial canals** which lead through an aperture, the **apopyle,** by a short **excurrent canal** into the **paragastric cavity** in the centre of the cylinder and (ii) the narrower **incurrent canals** between them which are dilated

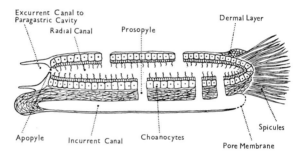

FIG. 19. **Sycon. T. S. Portion of Cylinder Wall.**

at their outer ends where they are covered by a thin membrane in which are several pores. Both kinds of canals end blindly, the radial canals at their outer ends and the incurrent canals at their inner ends, the outer ends being covered by the membrane. The radial canals are lined by columnar flagellated cells known as **choanocytes** and the incurrent canals by cells of the dermal layer. There is communication between the two kinds of canal at intervals by small openings called **prosopyles.**

PHYLUM

CŒLENTERATA

Diploblastic Metazoa, primarily radially symmetrical. Body wall composed of two layers of cells, ectoderm and endoderm, separated by a non-cellular mesoglœa and enclosing a single cavity, the enteron, with a single aperture. Two forms exist, hydroid and medusoid.

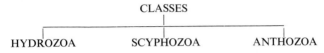

CLASSES

HYDROZOA   SCYPHOZOA   ANTHOZOA

CLASS HYDROZOA

Mouth leads directly into a simple enteron. Mainly marine and colonial, some solitary.

HYDRA

There are two common species—**Hydra viridis**, which is green, and **H. fusca,** which is brown. A third, **H. vulgaris**, is almost colourless. All live in fresh-water ponds attached to weeds or stones and are visible to the naked eye. One form only, the hydroid (or polyp).

(1) *Examine a living specimen of* **H. viridis** *under a hand lens in water in a watch-glass.*

Note the cylindrical **body** bearing a number of **tentacles** on its upper end. The lower end is called the **foot** (or **basal disc**). At the upper end of the body, inside the tentacles, is the conical **hypostome.** A **bud** (young hydra) may be developing from the side of the body.

Stimulate an animal by touching with a mounted needle. Observe the effect.

(2) *Mount in water and examine under the low power, supporting the coverslip on thin pieces of paper to avoid crushing the animal or, better, use a cavity slide.*

The above structures will be more clearly visible. One or more bulges may be visible high up on the body: these are the **testes.** A single similar swelling lower down is the **ovary**; thus the animal is hermaphrodite. You may be able to see the **mouth** on the hypostome.

(3) *Examine* **part of a tentacle** *under the high power.*

Note amongst the **musculo-epithelial cells** (which form the bulk of the cells) the **batteries** of **nematoblasts** (or **cnidoblasts**), pear-shaped cells from each of which projects a short bristle, or **cnidocil.**

(4) *Irrigate with dilute acetic acid or dilute iodine.*

The **nematocyst threads** will be ejected from oval sacs in the nematoblasts called **nematocysts.**

(5) *Examine a prepared slide of a* **T.S. of Hydra** *under the low power.*

Note that the animal is radially symmetrical and that the body wall is composed of an outer layer of cells, the **ectoderm** and an inner layer, the **endoderm,** separated by the structureless **mesoglœa.** The central cavity is called the **enteron.**

(6) *Examine a* **L.S. of Hydra.**

Note the parts as in (5). *Draw a diagram under low power.*

(7) *Examine the* **ectoderm** *under the high power.*

Note the roughly bluntly conical **musculo-epithelial cells** with their narrow ends inwards, from which long **contractile processes (muscle tails)** project both upwards and downwards. The small rounded cells packed between the musculo-epithelial cells are the **interstitial cells,** and the small, less regularly shaped cells with branched processes, lying near the mesoglea, are the **nerve cells.** *Examine a* **nematoblast.** These cells are found amongst the musculo-epithelial cells and occur in batteries on the tentacles as already seen. Note the oval sac, the **nematocyst,** inside which is the coiled **nematocyst thread.** The trigger-like **cnidocil** is attached to the outside of the **nematoblast.** Here and there between the musculo-epithelial cells small **sense cells** will be found.

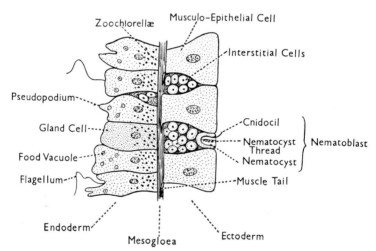

FIG. 20. **Hydra, Cells of Body Wall. High Power.**

(8) *Examine the* **endoderm** *under the high power.*

Note the tall, columnar **endodermal cells,** which are widened towards the enteron. These edges are irregular, and are prolonged into **pseudopodia** in some cells, while others possess **flagella.** The cells contain **vacuoles** and **food particles** and, in *H. viridis*, minute green plants (Algæ) called **zoochlorellæ,** which live symbiotically. (In *H. fusca* the plants are **zooxanthellæ.**) **Interstitial cells** may be found here and there between these cells. Amongst these cells, too, are the **secretory** or **gland cells.**

(9) *Examine a* **T.S.** *or* **L.S. of Hydra through a testis.**

Note the spherical **testis** containing **developing spermatozoa.**

(10) *Examine a* **T.S.** *or* **L.S. of Hydra through an ovary.**

Note the spherical **ovary** containing **stages in development of the ovum** (or an **ovum** with a prominent nucleus and many yolk spherules).

OBELIA

O. **geniculata** is a marine animal and exists in three forms: (i) a branched colony of hydra-like organisms, the **hydranths** or **polyps,** which also bears (ii) **blastostyles** from which develop (iii) free swimming **medusæ,** minute jelly-fish bearing sexual organs. The individual forms are termed **zooids.**

This is sometimes considered as *alternation of generations or metagemesis* but there is no real alternation between the asexual and sexual forms and, furthermore, apart from the gametes, the cells in all forms have diploid nuclei. It is really a case of *trimorphism,* since there are three forms—the hydranth, the blastostyle and the medusa. The existence of many different forms of the same organism is called *polymorphism.*

(1) (*a*) *Examine a prepared slide of the* **hydranth** *colony.*

Note the **nutritive hydranths,** Hydra-like individuals growing alternately on a stalk, the **hydrocaulus** which is attached to seaweeds by a branched **hydrorhiza** at the base (this may not be present). Note the **perisarc,** a non-cellular outer coat covering the entire colony and enclosing a common stalk, the **cœnosarc,** consisting of an outer **ectoderm** and an inner **endoderm** separated by a thin **mesoglea.** In the centre of the cœnosarc is the cavity, the **enteron,** with which the enteron of the hydranths is continuous. Each hydranth fits into a cup-shaped **hydrotheca** (a continuation of the perisarc) and bears at its free end a number of **tentacles** which surround the **hypostome,** on which is situated the **mouth.**

In the axils of the lower hydranths (which are the older ones) are the **blastostyles,** or asexual reproductive zooids. Each is enclosed in a club-shaped continuation of the perisarc, the **gonotheca.** It has no mouth or tentacles, but gives rise to a number of small **medusoid buds,** which develop into free swimming medusæ.

(*b*) *Make a stained permanent preparation of a* **hydroid colony.** *Stain with borax carmine.*

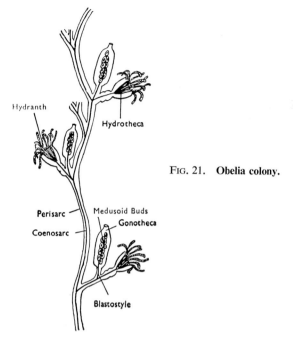

Hydranth

Hydrotheca

FIG. 21.   **Obelia colony.**

Perisarc

Medusoid Buds

Gonotheca

Coenosarc

Blastostyle

(2) *Examine a prepared slide of a* **medusa.**

It is umbrella-shaped and bears round the rim a large number of short **tentacles** with swollen bases. The convex upper side is the **ex-umbrellar surface,** and the concave side is the **sub-umbrellar surface,** from the centre of which is a short stalk, the **manubrium,** bearing on its tip the **mouth** (*specimens are usually mounted to show the sub-umbrellar surface*). A canal leads from this to the spherical **enteron,** from which four **radial canals,** at right angles to one another, lead to the **circular canal** which runs round the edge of the umbrella. In the freshly formed medusa there are sixteen tentacles, four **per-radial tentacles,** one opposite each radial canal, four **inter-radial tentacles,** one between each pair of per-radials, and eight **adradial tentacles,** one between each per-radial and inter-radial but others are formed as it becomes older. Look for the **statocyst,** a small sac containing a small calcareous object, the **statolith,** at the base of each adradial tentacle, and the **ocellus,** a small pigmented spot at the base of the per-radial, inter-radial and adradial tentacles. The **gonads,** or sexual reproductive organs, are rounded sacs, one below each radial canal. The sexes are separate, *i.e.,* the medusæ are *diœcious.*

CLASS SCYPHOZOA

Medusoid form only in most cases. Enteron divided into gastric pouches. *Jelly fish.*

AURELIA

THE JELLY FISH

Unlike the Hydrozoa, the polyp in the *Scyphozoa* is very small and seldom found but the medusa is large, being up to 30 cm. in diameter.

Examine a specimen of the common jelly fish, **Aurelia.**

The animal is radially symmetrical and umbrella-shaped. Around the edge of the **umbrella** is a large number of **tentacles** and from the centre of the sub-umbrella surface is the very short **manubrium** from which four tapering membranes called **oral arms** arise. At the upper end of the manubrium is the **mouth** which leads into the enteron.

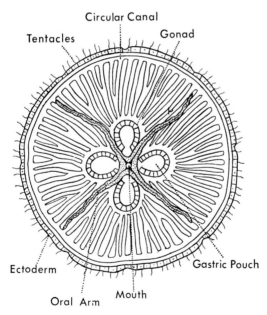

FIG. 22. **Aurelia. Sub-umbrella surface.**

(This is divided into a stomach which leads into four gastric pouches, and a number of branching radial canals which lead from it. These

lead to the **circular canal** around the periphery of the umbrella.)
The four horse-shoe shaped structures on the sub-umbrella surface
(and which project into the enteron) are the **gonads.** They lie on the
floor of the gastric pouches. The sexes are separate. The bulk of the
umbrella is composed of mesoglœa but this is covered externally,
of course, by ectoderm.

CLASS ANTHOZOA

Hydroid form only. *Anemones* and *sea corals.*

ACTINIA

THE SEA ANEMONE

Actinia equina, the beadlet anemone, is found attached to rocks
and seaweeds in rock pools between high and low water levels.
Anemones are usually brightly coloured and very conspicuous.
Examine a specimen of **Actinia equina.**

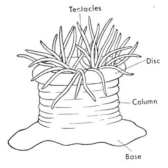

FIG. 23. **Actinia.
External view.**

The body is cylindrical and the diameter, which may be up to
8 cm., is slightly in excess of the height. In a fresh specimen the colour
is red or green but this may have been lost in a preserved specimen.
On top of the central column is the **disc** from the edge of which a
large number of **tentacles** arise. At the bottom of the column is the
base by which the animal is attached to rocks. Clearly this is a
polyp. There is no medusoid stage.

Dissection

(1) *Cut a vertical section of an anemone, passing through the centre,
and examine the exposed half with a lens.*

It will be seen that an aperture in the centre of the disc leads into an infolding of the ectoderm known as the **stomodæum** and that this enters a large **enteron** which is divided into separate compartments by vertically running partitions called **mesenteries** which run radially inwards from the **body wall.** Some reach across to the stomodæum,

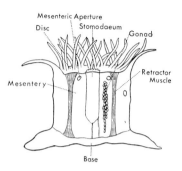

FIG. 24. **Actinia. V.S. through centre of Column.**

others only a short part of the way. The mouth lies at the bottom of the stomodæum. On the edges of the mesenteries are the **gonads** and the sexes are separate. Vertically running **retractor muscles** lie in the mesenteries and are responsible for withdrawal of the tentacles.

(2) *Cut a transverse section of an anemone passing through the upper part of the column and examine with a lens.*

The **mesenteries** will be seen with thickened edges running radially from the body wall and dividing the enteron into **inter-mesenteric**

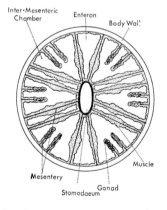

FIG. 25. **Actinia. T.S. through stomodaeum.**

chambers, compartments of varying sizes. The **stomodæum** will be seen in the centre of the section and the **gonads** (if visible) on the tips of the mesenteries.

(3) *Now cut a section across the animal lower down the column near the base.*

It will be seen that the **enteron** occupies the whole of the centre of the column below the stomodæum and that the **inter-mesenteric chambers** are thus continuous with the central part of the enteron.

PHYLUM

PLAYHELMINTHES
Flatworms.

Triploblastic acoelomate Metazoa, bilaterally symmetrical.
Complex reproductive system. Usually hermaphrodite.

CLASSES

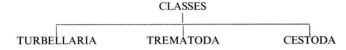

TURBELLARIA   TREMATODA   CESTODA

CLASS TURBELLARIA
Free living. Possess an enteron. Ectoderm ciliated.

PLANARIA and DENDROCŒLUM
These free-living flatworms live in the mud of freshwater ponds.

Examine a living specimen or prepared slides of either **Planaria** *or*
Dendrocœlum.

EXTERNAL STRUCTURE

The animal which may be up to 20 mm. or less in length has a
flattened body and one surface, the **dorsal side,** is always uppermost,
the lower surface being known as the **ventral side.** One end, the

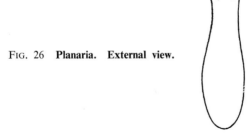

FIG. 26   **Planaria.   External view.**

anterior end, is broader than the **posterior end** which is bluntly
pointed. The colour may be black or grey, though it varies with the
species, in some cases being brown or red. It will be seen that the
body is bilaterally symmetrical longitudinally. Towards the anterior
end lie two black spots; these are sensitive to light and are known as
the **eyes.** About two thirds of the way along the ventral surface is the
mouth. This is situated on the end of a retractable **pharynx** which
can be protruded for feeding.

ALIMENTARY SYSTEM

The **mouth** is situated about two-thirds of the way down on the ventral side and leads directly into a thick-walled protrusible **pharynx** enclosed in a sheath. Continuous with this is the **intestine** which divides first into three main branches, one forwards and two backwards and then these branch repeatedly into **cæca** which end blindly, there being no anus.

EXCRETORY SYSTEM

There are two main **longitudinal excretory ducts,** one on each side, which open on the dorsal side about a third of the way down the animal. There are separate **excretory apertures.**

Examine a slide showing **flame bulbs.**

The cells are large and composed of several **cytoplasmic processes** which are prolongations of the cell. The **nucleus** appears to one side. The centre is occupied by a large cavity containing several **flagella** and a **duct** leads from the cavity.

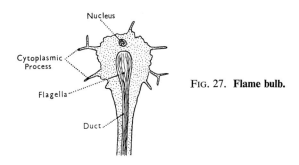

Fig. 27. **Flame bulb.**

REPRODUCTIVE SYSTEM

The animal is hermaphrodite and the **genital aperture,** situated near the posterior end is the opening of a cavity common to both the male and female gonoducts and known as the **genital atrium.**

Male Organs

The **testes** consist of numerous rounded structures near the right and left edges of the animal. A **vas deferens** from each runs backwards and unites with its fellow in the mid-line, traversing the protrusible **penis** which opens into the genital atrium.

Female Organs

There are two small rounded **ovaries** at the anterior end, each with an **oviduct**. These join posteriorly to form a **common oviduct** which opens into the genital atrium. **Vitelline glands** along each side open into the oviducts. The **uterus** and a thick-walled **muscular sac** are also found in the genital atrium.

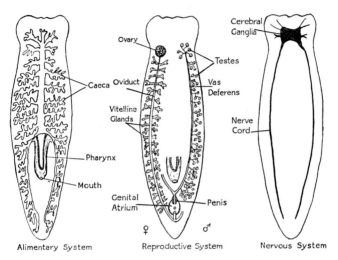

FIG. 28.   **Planaria.**

NERVOUS SYSTEM

At the anterior end of the animal are the **cerebral ganglia** from which two **nerve cords** pass posteriorly. Towards the anterior end are two dark round bodies which are photo-receptors and are consequently known as **eyes.**

TRANSVERSE SECTION

Examine a **T.S.** of a **Planarian.**

The body wall will be seen to consist of an outer **ectoderm** of **ciliated epithelial cells** containing a few clear rod-shaped bodies called **rhabdites.** Beneath this is a layer of **circular**, then **longitudinal** and a **diagonal muscle,** the latter well developed on the ventral side. Inside, lying in the general packing tissue known as **parenchyma,** will be seen cross-sections of the **intestinal caeca** and other structures already indicated. If the section passes through the **pharynx,** this will easily be identified in the centre of the section.

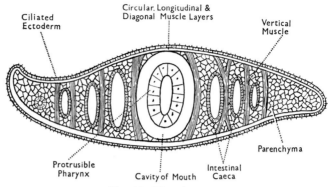

FIG. 29. **Planaria. T.S.**

CLASS TREMATODA

Leaf-shaped parasites. Have one or two intermediate hosts. *Flukes.*

FASCIOLA HEPATICA

THE LIVER FLUKE

This is an endoparasite which lives in the bile duct of the sheep causing a disease known as liver rot. Part of its life history, the larval stages, take place in the water snail *Limnæa.* Triploblastic, metamerically segmented.

EXTERNAL STRUCTURE

Examine a preserved **liver fluke.**

Note the shape and size of the organism. It is protected by a thick cuticle. At the more rounded anterior end is a small projection on the apex of which is the **anterior** or **oral sucker** within which is the **mouth.** On the ventral side, a shorter distance back from the anterior end and in the mid-line is the **ventral sucker.** Between the anterior sucker and the ventral sucker is the **genital pore** and the posterior end of the organism is the **excretory pore.**

Examine a prepared slide of the **liver fluke.** *It will probably be necessary to examine the following systems on the same microscopical preparation.*

ALIMENTARY CANAL

The **mouth** leads into a muscular **pharynx.** Continuous with this is a very short **œsophagus** which leads into the **intestine.** This is

composed of two main branches from which arise a large number of
branched blindly-ending **cæca** which reach all parts of the animal.

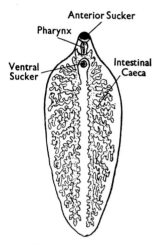

FIG. 30.   **Fasciola.   Alimentary
System.**

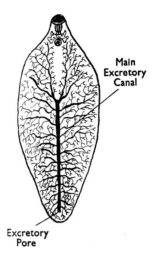

FIG. 31.   **Fasciola.   Excretory
System.**

EXCRETORY SYSTEM

There is a **main excretory canal** into which numerous branched
canals lead from all parts of the body. These small canals originate
in flame bulbs and the main canal leads to the **excretory pore** at the
posterior end of the animal.

Examine a slide showing **flame bulbs** (unless you have already done
so in studying Planaria). See Fig. 27, p. 42.

Note that the cells are large and have several **cytoplasmic processes** prolonged from them. In the centre is a large cavity containing
several **flagella.** A **duct** leads from this cavity. The **nucleus** is found
in the cytoplasm to one side.

NERVOUS SYSTEM

There are two **cerebral ganglia** at the anterior end of the animal.
These are joined by the **nerve ring** round the œsophagus. Two main
lateral nerves will be seen running backwards from the ganglia on
each side of the mid-line. There are no special sense organs.

REPRODUCTIVE SYSTEM

The animal is hermaphrodite.

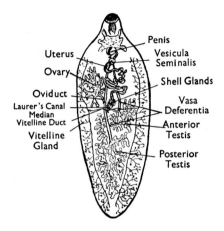

FIG. 32. **Fasciola. Reproductive System.**

MALE ORGANS

There are two **testes** which occupy the greater part of the centre and posterior part of the body, one being posterior to the other. Each consists of a series of much branched tubules and from the central part of each arises a **vas deferens** which runs forwards and then joins its fellow to form a wider tube, the **vesicula seminalis.** This leads by a narrow **ejaculatory duct** into the **penis,** protrusible through the **male genital pore.**

FEMALE ORGANS

There is one **ovary,** anterior to the testes and lying towards one side. This, too, consists of branched tubules and from it a short **oviduct** leads inwards to a point where it is joined by the **median vitelline duct** formed by the union of the two transverse ducts from the **vitelline glands.** These are composed of a series of rounded follicles connected together and occupying the sides of the body. From the point where the oviduct and median vitelline duct join is a wider convoluted tube, the **uterus,** which runs forward to the **female genital pore** lying alongside the male pore. At the base of the uterus is a group of **shell glands** which lead into the oviduct. A further duct, **Laurer's canal,** arises from the point where the oviduct and median vitelline duct join and runs to the surface on the dorsal side of the animal.

TRANSVERSE SECTION

Examine a **transverse section of a liver fluke.**

Note the external **cuticle** bearing small **spines.** Inside this is a layer of **circular muscle** followed by **longitudinal muscle** and then the **vitelline glands. The cæca** will be seen in transverse section.

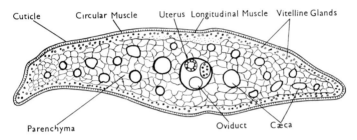

FIG. 33. **Fasciola. T.S.**

Look for **portions of the testes** and **ovary** and their associated structrues, the **vasa deferentia,** the **uterus** and the **oviduct.** Near the outer ends of the sections may be seen the **lateral nerves** and in the centre the **main excretory canal.** Filling up the spaces between these structures and serving as a general packing tissue is **parenchyma.**

LARVAL FORMS

Miracidium

Examine a slide showing **Miracidia.**

These minute larvæ hatch out from the eggs and are roughly conical with a small projection at the anterior end behind which is a pair of **eye-spots.** They bear long cilia on the surface.

Sporocyst

Examine a slide showing **Sporocysts.**

When the miracidium enters the water snail *Limnæa,* which lives in ponds and damp meadows, it develops into a second larva, the sporocyst. This is a simple ovoid structure and may contain **developing rediæ,** a third form of larva which eventually leave the sporocyst.

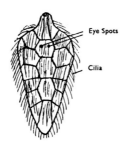

FIG. 34. **Fasciola. Miracidium.**

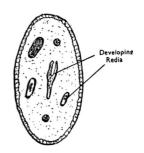

FIG. 35. **Fasciola. Sporocyst.**

Redia

Examine a slide of **Rediæ.**

This larva is an elongated structure with a **mouth** at its anterior end leading into a **pharynx** and wide blindly ending **gut. Secondary rediæ** develop inside during the summer and a fourth kind of larva called a **Cercaria** develops from the redia during the winter.

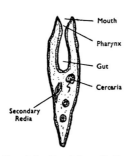

FIG. 36. **Fasciola. Redia.**

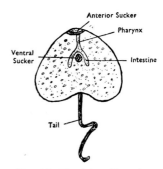

FIG. 37. **Fasciola. Cercaria.**

Cercaria

Examine a slide of **Cercariæ.**

The Cercariæ are roughly heart-shaped. Each has a small **anterior sucker** at the more pointed end. The mouth in this sucker leads to a short **pharynx** behind which is a simple **intestine** shaped like an inverted V. A **ventral sucker** is also present. At the wider posterior end is a long **tail.** The cercariæ leave the snail and encyst in water or on the grass. When eaten by a sheep they develop into adult flukes.

CLASS CESTODA

All parasites. Body composed of a chain of proglottides. Devoid of a gut. One or two intermediate hosts.

TÆNIA

THE TAPE WORM

These endoparasites live in the gut of various vertebrates and need two hosts to complete their life-history.

There are several species of Tænia. **T. serrata** infects the dog and has the rabbit as its secondary host. **T. cœnurus** also infects the dog: its secondary host is the sheep. Another dog tape-worm is **T. echinococcus** which is unusually short: the secondary host is the cow. **T. crassicollis** lives in the cat and the mouse. **T. solium** occurs in man and the pig and **T. saginata** in man and cattle. Dog tape-worms are the commonest and one of these has the dog flea as its alternative host. These are a few examples.

(1) *Examine a* **Tape-worm** *under a hand lens and a slide of the* **scolex** *under the low power.*

The parasite may be several feet in length. Note the minute head or **scolex** bearing on its free end a projection, the **rostellum,** immediately below which is a ring of chitinous **hooks.** On the side of the scolex are four **suckers.** (*T. saginata* has suckers but no hooks.)

Immediately behind the scolex is the narrow **neck** followed by a chain of flattened structures called **proglottides.*** The newly formed

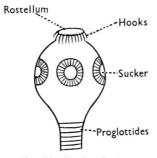

FIG. 38. **Tænia. Scolex.**
(*From Wallis—Practical Biology*)

* These must *not* be called segments as this implies metamerism which does not occur in Tænia.

ones near the neck are small, the older ones behind being much larger.

(2) *Examine a* **mature proglottis** *from about the middle of the animal under the low power, the dissecting microscope or a strong hand lens.*

There is no mouth, alimentary canal, vascular or respiratory system, these being unnecessary, but each mature proglottis has complete sexual organs. It is hermaphrodite, though the male organs develop first and so the anterior proglottides contain these alone.

Note the male organs—the small rounded **testes** scattered throughout the proglottis, particularly towards the anterior end, from which tiny ducts, the **vasa efferentia,** lead and join, forming the **vas deferens.** This leads to the **genital atrium** (or **pit**) which opens to the exterior on one side of the proglottis. The thicker terminal part of the vas deferens is the **cirrus** (or **penis**).

Note also the female organs—two **ovaries,** large, somewhat oval organs in the posterior part of the proglottis, behind which is the yolk-gland or **vitelline gland.** Anterior to this gland in the centre is the small **shell gland.** A **vitelline duct** and a pair of **oviducts** join

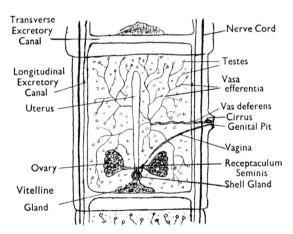

FIG. 39. Tænia. A Mature Proglottis.

behind the shell gland, and from the junction arises the blindly ending **uterus,** which runs up the centre of the proglottis and is at first a simple tube (but later it is much branched) and the **vagina** which runs across to the genital atrium, being wider at the inner end where it is called the **receptaculum seminis.**

The **longitudinal excretory canal** will be seen running along each side and a **transverse excrectory canal** (or **commissural vessel**) uniting the two longitudinal vessels across the posterior end of the proglottis. The **nerve cord** will be found on each side external to the longitudinal excretory duct.

(3) *Examine a* **more mature proglottis** from the posterior end of the animal. Almost the whole of the proglottis is occupied by the much **branched uterus,** the other reproductive organs having degenerated.

(4) *Examine the* **bladder-worm** *under a hand lens and under the low power.*

Note the bladder-worm or **cysticercus,** a bladder-like structure which later bears a narrow protruding neck which has been everted from it. This is the **proscolex** and bears suckers and hooks like the scolex of the tape-worm itself.

(5) *Examine a slide of a* **flame bulb** unless you have already done so in a previous animal (see Fig. 27, p. 42).

Note that the cells are large and have several **cytoplasmic processes** prolonged from them. In the centre is a large cavity containing several **flagella.** A **duct** leads from this cavity. The **nucleus** is found in the cytoplasm to one side.

TRANSVERSE SECTION

Examine a **Transverse Section through a mature proglottis.**

Note the thick external **cuticle** inside which is a layer of **circular muscle** followed by a wider layer of **longitudinal muscle.** The general

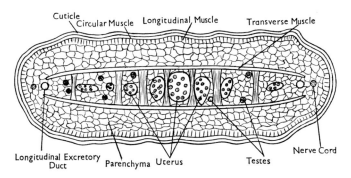

FIG. 40. **Tænia. T.S.**

"packing tissue" is called **parenchyma** and lying across it on each side is a band of **transverse muscle.** In the parenchyma will be found sections of the various parts of the **uterus** and of the **testes** and at each end the **longitudinal excretory canal,** external to which is the **nerve cord.** Other structures, such as the **ovary,** will be seen in a section through the posterior end of a proglottis.

PHYLUM

NEMATODA

Roundworms.

Triploblastic acœlomate Metazoa. No blood or respiratory system. Some free-living, other parasitic.

ASCARIS

This Nematode is parasitic in the intestine of mammals, **A. lumbricoides** occurring in man and the larger **A. megalocephala** in the horse.

EXTERNAL ANATOMY

Examine a specimen of **Ascaris lumbricoides,** *using a hand-lens as necessary.*

This large Nematode is yellowish-white in colour, unsegmented and cylindrical in shape, tapering to a point at each end, though less so anteriorly. It may attain a length of about 15 or more cm. in the male and as much as 22 cm. or more in the female. The posterior end curves more sharply in the male than in the female. The **cuticle** which protects the body is smooth and somewhat transparent and white **dorsal** and **ventral lines** will be seen running along the dorsal and ventral sides respectively and a brownish **lateral line** along each side. Each of the former encloses a nerve cord and in the latter are canals which unite and open by an **excretory pore** a millimetre or two behind the anterior end on the ventral side.

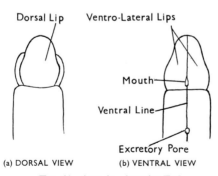

FIG. 41. **Ascaris. Anterior End.**

At the extreme anterior end is the **mouth** bearing three lips, one **dorsal lip** and two **ventro-lateral lips,** all bearing minute teeth.

Examine the mouth in order to determine the dorsal and ventral sides of the animal.

On the ventral side, almost at the posterior end in the *male* is the **cloaca,** the common intestinal and reproductive aperture and protruding from it is a pair of **copulatory setæ** (or **penial setæ.**). In

FIG. 42. **Ascaris. Posterior End.**

the *female* the apertures are separate, the **anus** being in a similar position to that of the cloaca in the male but the **genital pore** is further forward, about a third of the way along from the anterior end. In both sexes the so-called **excretory pore** is found a millimetre or two behind the mouth.

TRANSVERSE SECTION

Examine a **T.S.** of both **male** and **female Ascaris** under the low power.

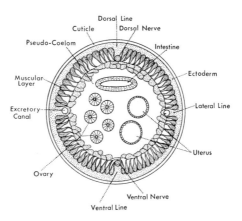

FIG. 43. **Ascaris. T.S. Female.**

The body wall consists of an outer **cuticle** composed of an external membrane inside which is a **protoplasmic layer** or **ectoderm**, containing scattered nuclei. This is followed by a layer of **muscle fibres** which consist of contractile processes which branch externally from a wide cell-like portion containing a nucleus internally placed. The **dorsal, ventral** and **lateral lines** project internally where they show a certain amount of thickening and thus divide the muscular region into four parts. In the dorsal and ventral projections lie the **dorsal** and **ventral nerves** and in the lateral projections are the **excretory canals.** In the centre is a **pseudo-coelom (perivisceral cavity)** in which lies the **pharynx** or **intestine** according to the region of the section and, in the *male*, cross sections of the much coiled **testis.** In the *female* are cross sections of the two coiled **ovaries** and two much larger **uteri.** In sections through the posterior end of the animals, the **vas deferens** and a wider canal, the **vesicula seminalis,** will be seen in the *male* and the **vagina** in the *female*, all in cross section.

PHYLUM

ANNELIDA

Segmented Worms.

Triploblastic, metametrically segmented, coelomate Metazoa. Bilaterally symmetrical. Closed blood system. Excretory organs nephridia. Chaetæ usually present.

CLASSES

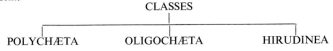

POLYCHÆTA   OLIGOCHÆTA   HIRUDINEA

CLASS POLYCHÆTA

Bristle-worms show distinct cephalisation and metamerism. Chætæ arise from protuberances on bodywall. Diœcious. Marine.

NEREIS

THE RAG WORM

N. diversicolor is one of the Bristle-worms. It lives under stones and burrows in the mud between tide-levels in the sea.

EXTERNAL ANATOMY

Examine a specimen of **Nereis,** *using a hand-lens when necessary.*

The animal is composed of a **head** and about eighty identical **segments,** each bearing bristles called **chætæ.** The dorsal surface is convex, and the ventral surface more or less flat and the colour varies.

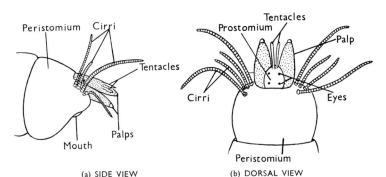

(a) SIDE VIEW   (b) DORSAL VIEW

FIG. 44. **Nereis Head.**

Examine the head. It will be seen to be composed of a **peristomium** surrounding the **mouth** and a bluntly triangular **prostomium** on the dorsal side. At the anterior end of the prostomium two short **tentacles** will be seen on the dorsal side and pair of short **palps,** each consisting of two joints, on the ventral side. Also on the dorsal side four simple **eyes** will be found and, behind the palps, somewhat antero-dorsally, two pairs of fairly long **cirri.**

Examine a **segment** *and a slide of a* **parapodium.**

The segment will be seen to bear a pair of lateral outgrowths known as **parapodia.** Each parapodium is composed of a dorsal bilobed **notopodium** and a ventral bilobed **neuropodium,** each bearing a bunch of **chætæ,** a slender **cirrus** and a short stiff bristle, the **aciculum.**

Parapodium

FIG. 45. **Nereis. Segment.**

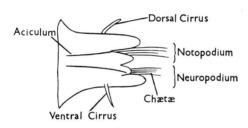

Aciculum

Dorsal Cirrus

Notopodium

Neuropodium

Chætæ

Ventral Cirrus

FIG. 46. **Nereis. Parapodium.**

Finally examine the last segment or **pygidium.**

This is devoid of parapodia and bears the **anus** and a pair of long, slender **anal cirri.**

TRANSVERSE SECTION

Examine a **T.S. of Neresis in the intestinal region.**

The body wall is composed of a thin **cuticle,** composed of a single layer of cells, the **epidermis** and then a layer of **circular muscle.** Inside this are two large bundles of **dorsal longitudinal muscle** and two large bundles of **ventral longitudinal muscle.** The two bands of muscle running aross the section are the **oblique muscles** for moving the **parapodia** which will be seen at the sides. Just inside the parapodia are the **chætæ muscles.** In the centre of the **coelom** is the **intestine** with the **dorsal blood vessel** above and the **ventral blood vessel** below. Ventral to the latter is the **nerve cord.** Parts of **nephridia** may be visible at the sides.

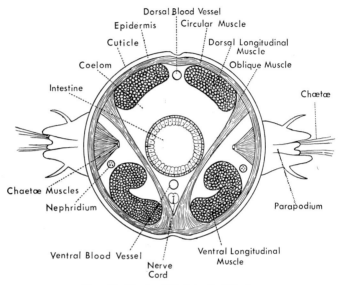

FIG. 47. **Nereis. T.S. Intestinal Region.**

LARVA

Examine a slide of the **free-swimming larva** *or* **trochophore.**

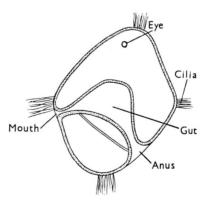

FIG. 48. **Nereis. Larva.**

The larva is more or less rounded and transparent and bears two **bands of cilia** externally. Inside is a simple **gut** and a photoreceptor or **eye.** The larva undergoes a steady metamorphosis into the adult worm.

CLASS OLIGOCHÆTA

Little cephalisation. Distinct metamerism. Chætæ arise from sacs in body wall. Monoecious. No larval stage.

LUMBRICUS

THE EARTHWORM

L. terrestris lives in burrows in the soil.

EXTERNAL ANATOMY

(1) *Examine a* **mature earthworm** *externally.*

Note the shape and colour of the animal and that the body consists of about 150 ring-like segments or **annuli.** The **mouth** is surrounded by the **peristomium** at the anterior end and is protected by a fleshy upper-lip or **prostomium**; the **anus,** or intestinal aperture is at the posterior end on the **pygidium** which is not a true segment. Note the **clitellum,** a thickened band stretching from segments 32 to 37 inclusive.

The **dorsal pores,** in the grooves between all segments after the 9th, except the last, are extremely difficult to see.

Rub the earthworm from back to front between the thumb and first finger.

The roughness on the lower, lighter (ventral) side is due to short bristles or chætæ, which are organs of locomotion.

(2) *Examine the* **first 20 segments on the ventral side,** *using a lens.*

Note the **mouth** and the four pairs of chætæ on each segment except the first and the last, two pairs of **ventral chætæ** and two pairs of **lateral chætæ.** Look in the grooves between segments 9 and 10 and 10 and 11 for the **spermathecal pores,** on segment 14 for the female **oviducal pores** (difficult to see) and on segment 15 for the male **spermaducal pores,** conspicuous on account of their prominent lips. Paired **nephridiopores** occur anterior to the ventral chætæ on all segments except the first three and the last.

(3) *Remove a chæta with a pair of forceps, mount in dilute glycerin and examine under the low power.*

Draw the **chæta.**

INTERNAL ANATOMY

THE ALIMENTARY SYSTEM

Kill a large worm by choloroform but do not allow the liquid to come in contact with the animal as it will make it brittle. (See Appendix II —Killing of Animals.)

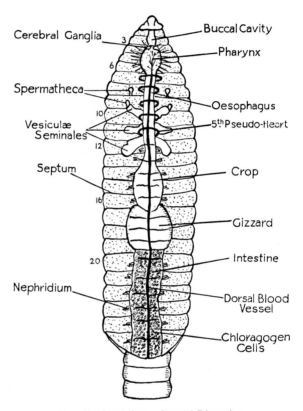

Cerebral Ganglia

Spermatheca

Vesiculæ Seminales

Septum

Nephridium

Buccal Cavity

Pharynx

Oesophagus

5th Pseudo-Heart

Crop

Gizzard

Intestine

Dorsal Blood Vessel

Chloragogen Cells

FIG. 49. **Lumbricus. General Dissection.**

Pin the animal, dorsal side uppermost, in a dissecting dish and cover with water. Put two small pins obliquely through segment 4 or 5 and two through the clitellum, stretching the body as much as possible. Insert the fine scissors in segment 30 or thereabout and cut in the mid-dorsal line up to segment 2. Note that the adjacent segments are separated from each other by a transverse **septum.** *Pin back the* *body wall on each side, cutting through the septa, and examine the viscera with a hand lens.*

These lie in a body cavity or **cœlom.** The mouth leads into the **buccal cavity** which reaches to segment 3. In segments 4, 5 and 6 is the muscular **pharynx** with muscular strands radiating from it. The **œsophagus** is a thin-walled tube extending from segment 6 to segment 12 to 14. In segment 10 is a pair of **œsophageal pouches** and in each of segments 11 and 12 a pair of white **œsophageal glands** (which lead into the pouches), lateral protuberances on the sides of the œsophagus not readily visible until the alimentary canal is removed owing to their ventral position. The **crop** is a dilated sac reaching from the œsophagus to segment 16, and the **gizzard** is a thick muscular part of the alimentary canal extending from segment 16 to segment 18 or 19. The **intestine** stretches from segment 20 to the anus. It bears a longitudinal infolding on the dorsal side (the typhylosole, which will be seen later in the transverse section). The yellow cells lying on the intestine are the **chloragogen cells** which are concerned with nitrogenous excretion.

Note also the paired coiled tubes or **nephridia** (excretory organs) on the ventral side, beginning in segment 4. Some of the **reproductive organs** (spermathecæ and vesiculæ seminales) will also be visible in segments 9 to 12.

THE VASCULAR SYSTEM

Use the worm already dissected.

The **dorsal blood vessel** (in which the blood flows forwards) is the largest and runs along the dorsal side of the alimentary canal. *By means of a camel hair brush, wash away any chloragogen cells hiding it.* On each side of the œsophagus is the **lateral** or **extra œsophageal blood vessel** which branches over the pharynx; it enters the **dorso-subneural** vessel (see below) in segment 10. *By careful examination with a lens these vessels may be seen on the posterior sides of the cut septa where they run.* The **contractile commissural vessels** or **pseudohearts** are paired loops in segments 7 to 11, which connect the dorsal vessel with the **ventral** or **sub-intestinal blood vessel** (in which the blood flows backwards) running on the ventral side of the alimentary canal. *To expose this vessel carefully release a part of the intestine by pulling it aside with forceps and cutting through the septa which hold it in place. Part of the ventral vessel will be seen on the underside of the intestine.*

Most of the following vessels will be seen later in a transverse section. **Ventro-intestinal vessels** in each segment take blood to the wall of the alimentary canal and **ventro-parietal vessels** to the body wall while **dorso-intestinal vessels** may be seen on the surface of the intestine. **Afferent nephridial vessels** run from the ventro-parietals to the nephridia while **efferent nephridial vessels** return the blood to the dorso-subneurals. Vessels from the ventro-parietals also take blood to the reproductive organs in segments 9 to 15. The **subneural vessel** lies ventral to the nerve cord. **Dorso-subneural vessels** connect this to the dorsal blood vessel from segment 12 backwards, these lie in the septa. On the sides of the nerve cord are two **lateral neural vessels.** The other blood vessels are difficult to see.

THE EXCRETORY SYSTEM

(1) *Dissect a worm dry and carefully remove the alimentary canal but nothing else. Holding the intestine in the forceps, cut across it transversely and then, still holding the canal, cut through the septa and connective tissue beneath it. Work forwards, carefully lifting it. Leave the buccal cavity and pharynx in situ (to preserve the cerebral ganglia)*

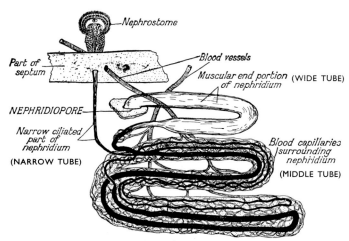

Fig. 50. **Lumbricus. Nephridium.**
(*From Dakin's "Elements of Zoology", Clarendon Press.*)

and before going any further, look for the œsophageal pouches (segment 10) and œsophageal glands (segments 11 and 12) if you were unable to see them earlier.

Now note the **nephridia,** a pair of coiled tubes in each segment except Nos. 1, 2, 3 and the last. Each nephridium opens into the cœlom in one segment, the **pre-septal part,** pierces the septum behind and the **post-septal part** opens to the exterior in the next segment.

(2) *Remove a* **nephridium** *from the animal dissected dry. To remove a nephridium complete with nephrostome is not an easy operation. The largest are in segments* 13 *to* 20. *Working under a lens and using very fine forceps and scissors, remove the nephridium and the septum, cutting transversely under the nephrostome which is anterior to the septum near the nerve cord. Mount in* 0·6 *per cent. physiological saline. Rearrange the specimen with mounted needles if necessary before putting on the coverslip and examine under the low power. Also make a permanent preparation stain with borax-carmine* (see Part I, p. 19).

From the funnel-shaped **nephrostome** (if present) which is the cœlomic opening (and in which you may be able to observe cilia still working), arises the ciliated **narrow tube** which passes through the septum into the wider looped **middle tube** surrounded by **connective tissue** and well supplied with **blood vessels.** The **wide tube** which follows opens to the exterior by the **nephridiopore.**

THE REPRODUCTIVE SYSTEM

This animal is hermaphrodite and the genital organs of both sexes are found in segments 9 to 15.

Male Organs

(1) *Cover the same worm with water and examine with a lens.*

The **anterior vesicula seminalis** is a white lobed body in segment 10, the lobes of which push the septum between segments 9 and 10 forwards. The **mid vesicula seminalis** is similar and lies in segment 11.

The **posterior vesicula seminalis** is in segment 12 and its posterior lobes protrude into segment 13, pushing this septum backwards. The central portions in segments 10 and 11 from which these arise are the **testis sacs.**

Cut open the testis sacs, wash out the contents with a pipette and examine with a lens.

Note the two pairs of digitate **testes** if visible, one pair in segment 10, hanging from the septum between 9 and 10, the other pair in segment 11, hanging from the septum between 10 and 11. They lie towards the mid-line but are seldom visible. Just behind each testis is a **seminal funnel,** the edges of which are convoluted (they are consequently often called **ciliated rosettes**). Each seminal funnel leads to a **vas efferens,** ventral to the posterior vesiculæ seminales and often partly hidden by a nephridium, and each vas efferens runs backwards and outwards on the ventral body wall and unites with its fellow on that side in segment 12 to form the **vas deferens**. This runs straight back to open to the exterior by one of the spermaducal pores on segment 15. The vasa efferentia and the vas deferens are often difficult to see.

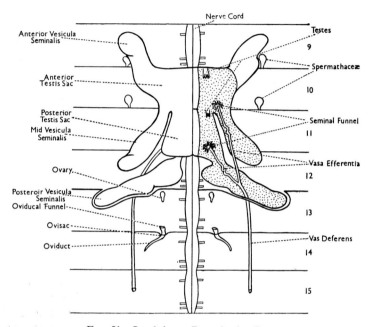

FIG. 51. **Lumbricus. Reproductive System.**

(2) *Dissect a worm dry. Remove a seminal vesicle and put it into a watch glass. Tease with mounted needles. Then make a smear of the milky contents on a coverslip.* Fix with 70 per cent alcohol (5–10 minutes), stain with borax-carmine, picro-carmine or hæmatoxylin. Mount in dilute gylcerine or make a permanent mount.*

Look for **stages in spermatogenesis,** the development of spermatozoa. *Search also for the parasite* **Monocystis**† *which often infects the worm. The trophozoite stage is that most frequently seen.*

Note the ripe **spermatozoa.**

---

* See p. 18. † See p. 27.

Female Organs

(1) *Examine the same worm.*

A pair of **ovaries,** small pear-shaped bodies, are found in segment 13, suspended from the septum between 12 and 13 towards the mid-line on each side of the nerve cord. Behind them in the same segment are the **oviducal funnels,** continuous with which are the short **oviducts,** which open to the exterior by the oviducal pores on segment 14. Each oviduct swells into a sac, the **ovisac** or **receptaculum ovorum** in segment 14.

In each of segments 9 and 10 is a pair of **spermathecæ,** small spherical white sacs which open to the exterior by the spermathecal pores in the grooves between segments 9 and 10, and 10 and 11. They store spermatozoa received from another worm in copulation.

(2) *Remove an ovary with a portion of the septum to which it is attached. Mount in water and examine under the lower power.*

Note the small pear-shaped **ovary** containing developing **ova,** the more mature ones being in the narrow posterior end.

(3) *Mount a drop of the contents of a spermatheca adopting the method used for the seminal vesicle contents (p. 18 and 27).*

THE NERVOUS SYSTEM

Use the worm left from the last dissection, i.e., *with the alimentary canal removed.*

In segment 3 are two white rounded bodies, the **cerebral ganglia.** These lie at the junction of the buccal cavity and the pharynx on the dorsal side. From the side of each arises a loop called the **circum-pharyngeal connective** or **commissure.** The two join and form the **nerve-collar.** In segment 4 where the commissures meet below the œsophagus is a **sub-œsophageal ganglion.** The **ventral nerve cord** is continuous with this and lies in the mid-ventral line, swelling slightly in each segment where three pairs of **segmental nerves** are given off. A pair of **prostomial nerves** arise from the cerebral ganglia and run forwards.

TRANSVERSE SECTION

(1) *Examine a prepared slide of the* **T.S. of the earthworm in the region of the intestine** *under the low power.*

The animal is *triploblastic.*

The **Body-Wall** is composed of an outer protective **cuticle,** then the **peidermis,** the **circular muscular layer,** the **longitudinal muscular** layer, the cœlomic epithelium and the **ventral and lateral chætæ** (if present) in **chætigerous sacs** and provided with **retractor muscles.**

In the **Cœlom** may be seen the **dorsal blood vessel,** the **ventral blood vessel** connected to the alimentary canal blood vessels by the **ventro-intestinal blood vessel,** the ventral **nerve cord** in which three **giant fibres** may be visible, the **sub-neural blood vessel,** and portions of **nephridia.**

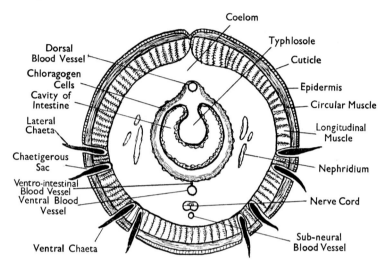

FIG. 52. **Lumbricus. T.S. Intestinal region.**

The **Alimentary Canal,** surrounded by **chloragogen cells,** is made up of the **circular muscular layer,** the **longitudinal muscular layer,** the **intestinal epithelium** and the **gut cavity.** Note that the dorsal side is folded in to form the **typhlosole.**

(2) *You should now examine a* **T.S. in the posterior region of the œsophagus.**

From your knowledge of the anatomy of the animal, you should be able to identify the structures which are visible.

LONGITUDINAL SECTION

Examine a **L.S. of the earthworm in the reproductive region** *under low power.*

The reproductive organs lie in segments 9 to 15. If it is a *sagittal section* the posterior end of the **oesophagus** and the **crop** will be seen in the centre. Beneath it in segments 9 and 10 lies the **anterior seminal vesicle,** in segment 11 the **median seminal vesicle** and in segments 12 and 13 the **posterior seminal vesicle** with overhanging portions of their lobes above the alimentary canal. If the section is not sagittal no part of the œsophagus or crop may be present and larger portions of the seminal vesicles may occupy the segments. Whether any part of the **vas deferens** or the **spermathæcæ** are visible again depends on the line up of the section cutting.

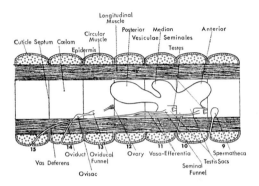

FIG. 53. **Lumbricus. Diagrammatic Representation of entire reproductive system from the side. One spermatheca only visible. Only parts of the system will be seen in the microscopic L.S.**

CLASS HIRUDINEA

Ectoparasites. Limited segmentation. Chætæ usually absent. Diœcious. Suckers at each end of body. *Leeches.*

HIRUDO MEDICINALIS

THE MEDICINAL LEECH

This leech lives in ponds, swamps and slow-flowing streams and attaches itself to fish and frogs from which it sucks blood. The adults will also suck the blood of warm-blooded animals including man.

Examine a specimen of the **Medicinal Leech** *using a hand lens.*

The body is elongated and is about 6–10 cm. in length though it can expand to as much as 15 cm. There are 32 **segments** though there appear to be more because most of them, except at the extremities, consist of five **annuli.** There are 95 in all. *Examine the anterior end on the* **ventral side.** Here will be seen the **anterior sucker** in which lies the **mouth** which has three jaws. *Examine the posterior end.* The

posterior sucker will be found at the other extremity and, just anterior to it, the **anus.** On the second annulus of segment 10 is the
male genital aperture and on the second annulus of segment 11 the
female genital aperture, both situated in the mid-line. Finally on the
last annulus in each of the segments from the 6th to the 22nd is a
pair of **nephridiopores.** Note the absence of chætæ.

Now turn to the **dorsal side** *of the animal.* On the first annulus of
each segment is a row of **sensory papillæ** and on the first segment,
which is the **head,** is a pair of pigmented **eyes.**

The whole body is encased in a thin cuticle which is periodically
shed and replaced. Note that the animal is rather flattened dorso-
ventrally and that the colour varies from black to green and yellow
and is darker above than on the ventral side.

TRANSVERSE SECTION

Examine a **T.S. of Hirudo** under low power.

Beneath a thin **cuticle** lies the **epidermis** and then layers first of
circular and then **longitudinal muscle.** It will immediately strike you
that in contrast to Lumbricus, there is an enormous reduction in the
coelom. The cavity is filled with a tissue composed of a gelatinous

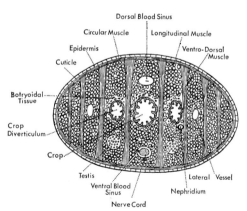

FIG. 54. **Hirudo T. S. Through Crop Region.**

groundwork or matrix in which cells and fibres occur. In the centre
lies the gut and, if the section passes through the **crop,** lateral outgrowths from it called **crop diverticula** will be seen on either side of it.
Large pigmented cells cover the outside of the gut forming what is

known as **botryoidal tissue,** and running across the section from the dorsal to the ventral side are **ventro-dorsal muscles.**

The **nerve cord** is situated in the mid-ventral line inside the **ventral blood sinus.** On the opposite side is the **dorsal blood sinus** and on each side the **lateral blood sinuses.** Portions of **nephridia** may be seen below the crop diverticula and if the section passes through the 11th segment sections of the **ovaries** as well. Between segments 12 and 20 (or 21) lie the **testes** and so these will be visible in cross section beneath the crop diverticula.

PHYLUM

ARTHROPODA

Bilaterally symmetrical, metametically segmented animals possessing an exoskeleton with paired jointed limbs. Definite cephalisation. Body cavity a hæmocœle.

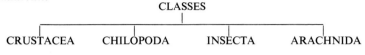

CLASSES

CRUSTACEA   CHILOPODA   INSECTA   ARACHNIDA

CLASS CRUSTACEA

Aquatic. Breathes by gills. Thick exoskeleton. *Crayfish, crabs, shrimps, water fleas, woodlice.*

ASTACUS

THE FRESHWATER CRAYFISH

Crayfish live in rivers in chalky districts, remaining for most of the time in holes in the banks, emerging in order to feed: the animal is omnivorous. The sexes are separate (or diœcious). The common European crayfish is **Astacus fluviatilis.**

OBSERVATIONS ON THE LIVING ANIMALS

(1) *Observe the animal in water* and note its methods of walking, swimming and feeding. Take note also of the way in which it uses its large pincers.

(2) *Place the animal on its back* and observe its movements, noting particularly the mouth parts.

EXTERNAL ANATOMY

Examine a killed specimen of a **crayfish.**

Dorsal View

Note that the body is protected by a hard exoskeleton and that it is divided into a **head, thorax** and **abdomen**: each is segmented and there are nineteen segments in all.* The head and thorax are almost completely fused to form what is known as the **cephalo-thorax,** the division between them being shown by the **cervical groove** running across the body, but here the segmentation is not visible externally.

* There is evidence that there are really twenty segments, a pre-antennal segment, bearing no appendages, preceding that which bears the antennules.

It is clearly seen in the abdomen where each segment is composed of hardened plates called **sclerites,** that on the dorsal side being called the **tergum** and that on the ventral side the **sternum,** while at the sides are the < -shaped **pleura.** There is fusion between the tergum and the pleuron on each side and, to a certain extent, between the sternum and the pleura, but the appendages are inserted here and there is a small sclerite between the appendage and the pleuron on each side called the **epimeron,** seldom evident however.

In the cephalothorax the sclerites are fused to form the **carapace.**

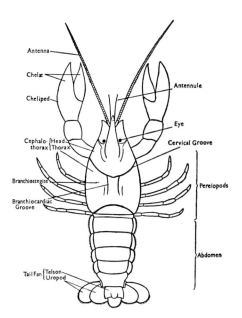

FIG. 55. **Astacus. Dorsal View.**

The Cephalothorax

The **head** consists of five segments (excluding the pre-antennal segment).

Note the pointed **rostrum** projecting anteriorly from the carapace, the short stalked **eyes** lateral to it, the two short **antennules** and the pair of very long **antennæ.**

The **thorax** is composed of eight segments.

Running backwards from the cervical groove are two grooves on the thorax, the **branchiocardiac grooves.** From these the carapace bends down to form the sides of the thoracic cavity and these sides known as the **branchiostegites.** The cavity beneath the two outer portions of the carapace thus formed encloses the gills and is called the branchial chamber while the central portion covers the heart. Note the four pairs of **walking legs** (or **pereiopods**) and the large **chelipeds** with large pincers or **chelæ.**

The Abdomen

This is composed of six movable segments and a structure called the **telson** at the posterior end, on each side of which is a wide **uropod,** the whole structure being called the **tail fan.**

Ventral View

Again note the division into **cephalothorax** and **abdomen.** Note also the **antennules, antennæ, chelipeds,** four pairs of **pereiopods** and the **telson.** Between the bases of the appendages in the centre of the thorax will be seen small **sternal plates.** Find the **anus** on the telson and note the wide **uropods** on each side, forming the **tail-fan.**

THE APPENDAGES

All the body segments except the first one on the head, the preantennal segment, bear paired appendages. There are nineteen pairs in all and they are all built on a common plan, though this is much modified in the various appendages in adaptation to their functions. This common plan must be understood before any attempt to examine the appendages is made. Arising from a basal portion called the **protopodite,** itself composed of a proximal **coxopodite** and a distal **basipodite,** and bearing a small **epipodite,** are two branches (it is therefore said to be *biramous*), an inner **endopodite** and an outer **exopodite,** each composed of several **podomeres.***

Examine the appendages in situ. Then remove them and examine each in detail. Instructions for their removal are given below.

(i) **In situ**

Place the animal, ventral side up, and examine the appendages on each segment.

(ii) **Removal of the Appendages**

Remove the appendages of one side one by one, beginning at the posterior end, by holding the appendage with forceps and cutting

* They should not be called joints.

*through the membrane which joins it to the body at its base. Removal
of the appendages of the thorax may be facilitated if the branchiostegite
is first removed. Extra care is needed with the mouth appendages.*

*As each appendage is removed, immediately fix it in a dissecting dish
or on a card by means of a pin of appropriate size, taking care to keep
the appendages undamaged and in their correct order.*

Region	Seg-ment†	Appendage	Function
Head	1	Antennules.	Sensory (Olfactory).
	2	Antennæ	Sensory (Tactile).
	3	Mandibles.	
	4	Maxillules.	
	5	Maxillæ.	
Thorax	6	1st Maxillipeds.	Mouth parts.
	7	2nd Maxillipeds.	
	8	3rd Maxillipeds.	
	9	Chelipeds.	Pincers—protection.
	10	1st Pereiopods.	Walking legs.—Movement.
	11	2nd Pereipoods.	Bear genital pore in female.
	12	3rd Pereiopods.	
	13	4th Pereiopods.	Bear genital pore in male.
Abdomen	14	1st Pleopods.	Vestigial in female. Modified in male for transference of spermatozoa.
	15	2nd Pleopods.	Swimmerets in female. Modified in male as in 14.
	16	3rd Pleopods.	
	17	4th Pleopods.	Swimmerets.
	18	5th Pleopods.	
	19	Uropods.	Form tail fan with telson.

†The pre-antennal segment is omitted since it bears no appendages.

(1) The Antennule

The **protopodite** consists of three podomeres, the coxopodite being
the largest. (This contains the statocyst.) The two branches constitute a divided endopodite and these have many podomeres. The
coxopodite bears a **statocyst opening** (see later).

(2) The Antenna

The **coxopodite** of the **protopodite** is broad and has a small projection on the ventral side which bears the **excretory pore**; the
basipodite is shorter and bears (i) a short flat triangular **expodite**
with a fringe of hairs called the **scaphocerite** or **scale** and (ii) the
endopodite which is very long and has many podomeres.

(3) The Mandible

The **protopodite** is short, broad and "unjointed" (there is no coxopodite) and is serrated on its inner edge. The **endopodite** is small and is composed of three podomeres; it is known as the **palp.** There is no exopodite.

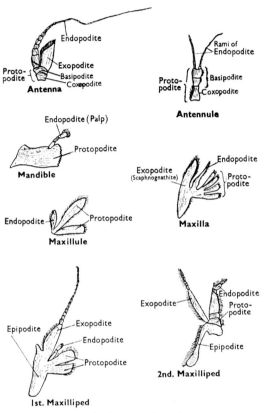

FIG. 56. Astacus. Appendages.

(4) The Maxillule

This is small and consists of **protopodite** of two podomeres with a very small **endopodite** on the outer side. There is no exopodite.

(5) The Maxilla

Though larger than the maxillule, it is still small. The **protopodite** again consists of two podomeres and bears a small, pointed

endopodite and an enlarged **exopodite** called the **scaphognathite** which extends in both directions and which paddles water into the branchial chamber.

(6) The 1st Maxilliped

This is larger and rather flattened. The coxopodite and basipodite of the **protopodite** are somewhat flattened. The **endopodite** is very small and consists of two podomeres. The **exopodite** is large, the distal part being long and slender. Here we find an additional podomere, the **epipodite,** which is a prolongation of the protopodite.

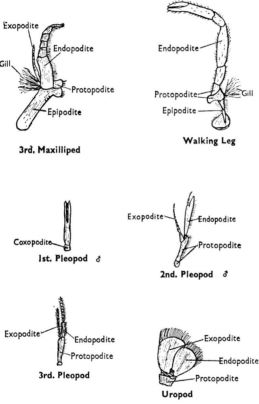

FIG. 57. **Astacus. Appendages.**

(7) The 2nd Maxilliped

This is larger than the 1st Maxilliped and the **protopodite** composed of two podomeres is not flattened. The **epidodite** bears a **gill**. The

endopodite is large and consists of five podomeres; the **exopodite** is smaller and is similar to that on the 1st maxilliped.

(8) The 3rd Maxilliped

This is still larger and is similar to the 2nd maxilliped though the **endopodite** is larger and the **exopodite** small. Both are borne on a **protopodite** with two podomeres.

(9) The Cheliped

(For illustration see Dorsal View p. 71.)

This is the largest appendage of all. The **protopodite** is composed of the usual two podomeres. There is no exopodite but the **endopodite** is very large and consists of five podomeres, the last two being modified into **chelæ** or **pincers.** There is an **epipodite** bearing a **gill.**

(10) The 1st Pereiopod (Walking Leg)

The **protopodite** consists of a short, broad **coxopodite** with a **gill**-bearing **epipodite** and a smaller **basipodite** which bears the **endopodite.** This has five podomeres and the last is pincer-like. There is no exopodite.

(11) The 2nd Pereiopod (Walking Leg)

This is similar to the 1st walking leg but, in the **female,** the **genital pore** opens on the inside of the **coxopodite.**

(12) The 3rd Pereiopod (Walking Leg)

This is again similar to the 1st walking leg except that the endopodite is not pincer-like.

(13) The 4th Pereiopod (Walking Leg)

This differs from the 3rd walking leg only in that it has no epipodite or gill and in the **male,** the coxopodite bears the **genital pore.**

(14) 1st Pleopod

In the **male** this is modified for purposes of reproduction into an incomplete tube, probably composed of a fused protopodite and endopodite, the exopodite being missing. Its function is transference of spermatozoa to the female.

This appendage is vestigial in the female.

(15) 2nd Pleopod

Like the 1st pleopod, this is also modified in the **male** and for a similar purpose. In the protopodite the **coxopodite** is small and the **basipodite** longer, while the **endopodite** has a large proximal "joint" with its distal margin prolonged into an incomplete tube, the distal podomeres being small. A small **exopodite** is present.

In the **female,** the 2nd pleopod is small and may be vestigial. The **coxopodite** is small and the **basipodite** larger as in the male. The **endopodite** is composed of a long proximal podomere and several smaller podomeres and is covered with bristles. The **exopodite** is similar in structure but smaller. This pleopod is called a **swimmeret.** The structure will be better seen in the next three appendages.

(16), (17), (18) 3rd, 4th and 5th Pleopods.

These are **swimmerets** similar in both sexes to the 2nd pleopod in the female, though larger.

(19) Uropod

The protopodite is short and broad and is not composed of the usual two podomeres. The **endopodite** is flat and oval and the **exopodite** similar though larger and composed of two podomeres. The distal edges bear bristles giving the appearance of a fringe.

INTERNAL ANATOMY
All dissections are performed under water on a freshly killed animal.

THE VASCULAR SYSTEM

This is best seen in an injected specimen (see Appendix II).

Hold the animal in the hand dorsal side up. Insert one blade of the large scissors behind the posterior edge of the carapace immediately behind one of the branchiocardiac grooves and, keeping this blade as nearly horizontal as possible so that the internal organs are not damaged, cut forwards as far as the rostrum following the branchiocardiac grooves. Make a similar cut on the other side and then cut immediately behind the rostrum to meet the first cut. Carefully remove the central strip of the carapace.

Now continue the cut along the abdominal terga on both sides, taking the same precaution as before and carefully remove the central strip of the terga. The pigmented epidermis will now be visible: remove this.

The **pericardium,** enclosing a large cavity, the **pericardial cavity** will be seen on the dorsal side of the thorax. Situated inside it is the

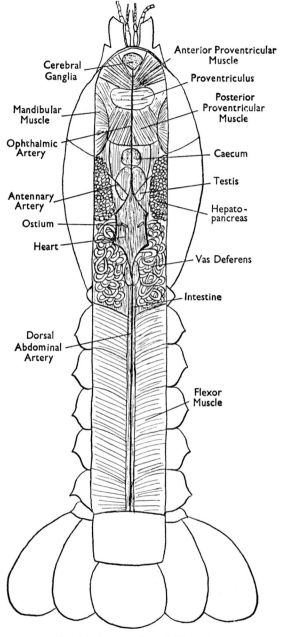

FIG. 58. **Astacus. Dorsal Dissection.**

hexagonal **heart**. There are two **dorsal ostia,** apertures through which the blood enters the heart from the pericardial cavity. (There are also ventral and lateral ostia which will be seen later.) The heart is kept in position in the pericardial cavity by six cords known as the **alæ cordis.**

From the anterior end of the heart arise the median **ophthalmic artery** and the paired **antennary arteries,** posterior and somewhat ventral to which the **hepatic arteries** arise. These run ventralwards. From the posterior end of the heart arises a large median vessel, the **dorsal abdominal artery,** which runs the length of the body, posteriorly. The sternal artery which runs ventrally cannot be seen. These arteries lead to large blood cavities of **hæmocœls** which surround the organs and form the body cavity.

Remove the heart and pericardium by severing all attachments. Examine its under-surface to see the **lateral** and **ventral ostia.**

THE RESPIRATORY SYSTEM

(1) *Expose the* **branchial chamber** *on the side opposite to that from which the appendages were removed by removing the* **branchiostegite** *which encloses it on the lateral side of the thoracic carapace. Place the animal under water otherwise the gills will not be clearly seen.*

Note the **gills** which are arranged in three rows, named according to their position as follows:—

(i) The **podobranchs** are attached to the coxopodites of appendages 7–12 and from the innermost row. These were seen on the epipodites when examining the appendages.

(ii) The **arthrobranchs** are attached to the **arthrodial membranes** between the bases of appendages 7–12 and the body. Appendage No. 7 has one such gill but Nos. 8–12 each bear two. These form the middle row.

(iii) The **pleurobranchs** arise from the pleuron of the thorax. Segments Nos. 10, 11 and 12 bear only rudimentary pleurobranchs but No. 13 bears one fully developed. None is present in segments 6–9.

(2) *Remove and examine one of these* **gills.**

It is composed of a **basal plate** arising from the dorsal side of which is the **stem** which has on its distal end a plate, the **lamina** and at its tip, a structure like a test-tube brush called the **plume.** The plume is composed of thin-walled vascular **branchial filaments.**

THE REPRODUCTIVE SYSTEM
Male

The **testis** is a white sac situated ventral to the pericardium and dorsal to the alimentary canal. It is composed of two anterior lobes and one posterior lobe. The **vasa deferentia** are long coiled tubes, also white, seen on each side of the heart. They lead to the genital pore on the coxopodite of the 4th walking leg on the 13th segment.

Female

The **ovary** is also composed of two anterior lobes and one posterior lobe. The **oviducts** are short, wide, almost straight tubes leading to the genital pore on the coxopodites of the 2nd walking legs on segment 11. They will not be seen until the ovary is removed as they run ventrally from the ovary to the genital pore.

THE ALIMENTARY SYSTEM

Remove the testis (or ovary) and in the male the vasa deferentia. Take care not to remove anything else.

The mouth is on the ventral side and will not be visible. A short œsophagus, also invisible here, leads dorsalwards from the mouth to the **proventriculus,** a large sac in the head and anterior part of the thorax. The œsophagus is immediately ventral to it. A pair of **anterior proventricular muscles** and a pair of **posterior proventricular muscles** are attached to the proventriculus which is divided into an **anterior** (or **cardiac**) **chamber** and a smaller **posterior** (or **pyloric**) **chamber** separated by a transverse constriction. It contains the gastric mill and the filter.

Remove the muscles which partially cover the proventriculus.

The mid-gut or **mesenteron,** continuous with the proventriculus, is short and is protruded dorsally as a small **cæcum.**

Following this is a long narrow tube, the **intestine,** which leads to the anus on the ventral side of the telson.

On each side of the alimentary tract in the thorax is a pair of **digestive glands** (sometimes known as the **hepato-pancreas).**

Remove the alimentary system as follows:—

Cut through the intestine as far back as possible and, holding it with forceps, deflect it forwards, cutting through any connective tissue underneath. It will be necessary to cut through the sternal artery and the vas deferens (if not already removed) or oviduct and, when they are reached, the posterior and anterior proventricular muscles and œsophagus.

Now remove the viscera, detach the digestive gland on one side and lay the whole structure on the other side, dorsal side uppermost.

Note the **œsophagus, proventriculus, mesenteron, cæcum, intestine, vas deferens** (or **oviduct), dorsal abdominal artery** (if visible), **sternal artery** and **hepatic artery** which both run ventralwards, the former

from the posterior corner and the latter from the anterior corner of
the heart.

*Open the chambers of the proventriculus, and thus of the gastric
mill and filter, on the ventral side. Open up the gastric mill and pin it
down to dry. Examine with a lens.*

Note that the anterior chamber of the **gastric mill** is composed of
a series of ossicles. Anteriorly is the **cardiac ossicle** in the centre with
a **pterocardiac ossicle** on each side. Immediately behind is a T-shaped
urocardiac ossicle and behind that a centrally placed **prepyloric
ossicle** followed by the **pyloric ossicle** with **zygocardiac ossicles** on
either side. These move freely on each other and grind up the food.

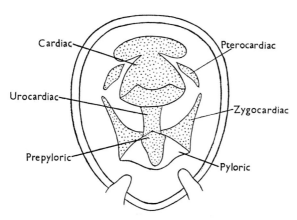

Fig. 59. Astacus. Ossicles of Gastric Mill.

Between the anterior and posterior chambers is a narrow slit-like
opening and the walls of the posterior chamber are provided with
chitinous plates besting hair-like bristles which serve as a strainer.

This is the **filter.**

Two circular calcareous particles may be found at the anterior end
on each side. These are **gastroliths** but they are not always present.

THE EXCRETORY SYSTEM

Remove what remains of the side of the thorax on one side.

The **green gland** will be exposed. *Carefully remove it, cutting
through the duct which connects it to the base of the antennule.
Examine with a lens.*

The **green gland** is rounded, flattened and pale green in colour. It consists of a small **excretory sac** or **end sac,** and an **excretory duct** (already cut) opening to the exterior at the base of the antenna and composed of a **labyrinth,** a **white tube** and a large **bladder,** in that order.

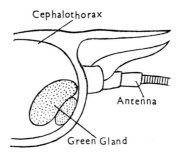

FIG. 60. **Astacus. Green Gland.**

THE ENDOPHRAGMAL SKELETON

Now examine the interior of the thorax and abdomen from which all the organs have been removed.

Note the ingrowth of the exoskeleton (**apodemes**) in the form of uprights from the sterna called **endostermites** bearing approximately horizontal struts, **endopleurites,** at the top reaching to the lateral walls of the thorax. This is known as the **endophragmal skeleton** and serves for the attachment of the leg and abdominal muscles.

THE NERVOUS SYSTEM

If a fresh animal is being used the thorax should be opened by removal of the carapace and branchiostegites. Then the alimentary system must be removed.

The nervous system will be more easily examined if part of the endophragmal skeleton is removed.

Start by cutting through the posterior endosternite of the thorax, then cut through the two endopleurites attached to it. Continue forwards with the rest of the endophragmal skeleton in the thorax and remove it all. Very carefully remove the floor of the rostrum. Then carefully remove the muscles of the abdomen, working backwards, so as to expose the whole nerve cord. Examine the nervous system in situ.

Note the white **cerebral ganglia** situated at the anterior end of the head below the floor of the rostrum. From the posterior end arise two **circumœsophageal commissures** constituting the nerve collar. These loop round the œsophagus and lead to the **subœsophageal ganglia** behind it. **Nerves** arise from the cerebral ganglia to supply the eyes, antennules and antennæ, and from the subœsophageal ganglia to supply the maxillules, maxillæ and 1st and 2nd maxillipeds (*i.e.*, the mouth parts).

From the subœsophageal ganglia the **nerve cord** passes backwards and consists of six **thoracic ganglia** (in segments 8–13) joined by

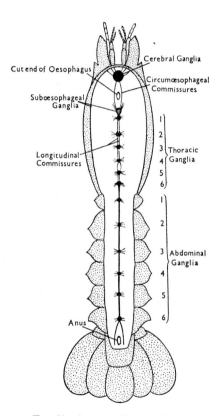

FIG. 61. Astacus. Nervous System.

paired **longitudinal commissures,** and six **abdominal ganglia** also joined by **longitudinal commissures** though the paired nature is not obvious here. The last abdominal ganglia are larger as they supply the telson as well as the last segment. The longitudinal commissures between the 4th and 5th thoracic ganglia are well curved to allow room for the sternal artery. **Nerves** arise from all the ganglia, the last pair passing backwards on each side of the anus to supply the telson.

THE SENSE ORGANS

THE EYE

(i) *Examine the eye in situ.*

Note that it is borne on a short movable stalk.

(ii) *Examine a* **longitudinal section of the eye.**

Note the **ommatidia,** the optical units of which the eye is composed, each terminating in a **corneal facet** or **cuticular lens** surrounded by **pigment cells.** This type of eye is called a **compound eye.** Immediately beneath the lens is the **crystalline cone** with its apex directed backwards. This is formed by the inner walls of a group of four cells, the **vitrellæ.** Behind them are the **retinulæ** the inner walls of which are also refractive and form the **rhabdom.** From these cells nerve fibres pass back to the **optic ganglion** which is situated in the eye stalk and from which the **optic nerve** passes.

THE STATOCYST

(i) *Remove one of the antennules and carefully cut away the walls of the coxopodite on one side.*

Each statocyst consists of a sac lined by sensory hair-like bristles or **setæ** (visible only under the microscope) and containing grains of sand, etc., called **otoliths.** There is a **statocyst opening** on the dorsal side of the coxopodite.

(ii) *Wash out some of the contents of the sac with a small pipette, place on a slide and examine under the low power of the microscope.*

Note the **otoliths.**

THE OLFACTORY SETÆ

Mount and examine the ventral side of the outer branch of the endopodite of the antennule under the low power of the microscope.

Note the **olfactory setæ,** five hair-like structures projecting from this branch of the endopodite.

TRANSVERSE SECTIONS

(i) *Examine* a **Transverse section cut through the thoracic region of Astacus.**

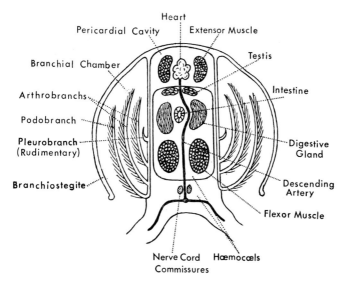

FIG. 62. **Astacus T.S. Thoracic Region (Semi-diagramatic)**

Inside the hard cuticle which forms the **exoskeleton** the following organs will be found in the **hæmocœl.** On the dorsal side lies the **heart** in its **pericardial cavity** with **extensor muscles** of the abdomen on each side. At the sides, covered by the **branchiostegite,** are the **branchial chambers** containing the gills. The external ones are the **podobranchs,** in the centre are the **arthrobranchs** and internally are the **pleurobranchs** which are rudimentary but it will be remembered that all three kinds may not be present, depending on the region of the thorax through which the section passes. It is only in the regions of attachment of the 1st walking legs that all three are present. (Refer to the Respiratory System on page 79.)

In the centre, ventral to the heart the gonads and then the **intestine** will be seen with large **digestive glands** on each side. Below these structures lie the **flexor muscles** of the abdomen which are antagonistic to the extensor muscles above and in the centre is the **nerve cord.** Part of the **endophragmal skeleton** may also be seen in the section.

(ii) *Examine* a **Transverse Section through an abdominal segment of Astacus.**

Note the dorsal **tergum,** the ventral **sternum** and the **pleuron** on each side connecting the two. The greater part of the segment is occupied by muscles, a pair of **extensor muscles** dorsally and large **flexor muscles** below; also a pair of **appendage muscles** situated one on each side somewhat obliquely towards the ventral side. The **intestine** lies in the mid-line just above the flexor muscles and dorsal to it is the **dorsal abdominal artery.** The **ventral abdominal artery** lies in a corresponding position on the ventral side and immediately above it the **nerve cord** will be seen. The **hæmocœl** will also be identified.

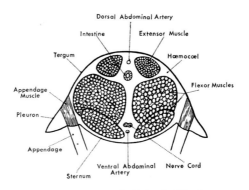

FIG. 63. **Astacus. T.S. Abdominal Region.**

CANCER

THE CRAB

Examine a **crab** *from both the dorsal and ventral aspects.*

In general the structure of the body is similar to that of the crayfish (see pp. 70–86) but it is broader than it is long, the **cephalothorax** being of considerable width whereas the **abdomen** is much reduced and is turned forwards and closely applied to the ventral side of the cephalothorax. In the male only the first two abdominal appendages (**pleopods**) remain and these are used in copulation. In the female five pairs remain for carrying the eggs. When they are present, the abdomen is forced down from contact with the thorax. Unlike the crayfish there is no tail fan. It will be seen that the **antennae** and **antennules** on the head are small and that the **eyes** are borne on stalks and lie in sockets in the **carapace.** The **chelipeds,** which are themselves large, terminate in large pincers or **chelæ.** There are four pairs of **walking legs (periopods)** and these end in claws.

BALANUS

THE ACORN BARNACLE

Barnacles are in a **Sub-Class Cirripedia.** These are sedentary animals in which the head has almost disappeared and the abdomen is much reduced. They are marine and the larval stages are free-swimming. The adult is attached to the bottom of ships and submerged rocks and piers.

(1) *Examine a specimen of the* **Acorn Barnacle** *attached to a substratum.*

The animal's entire body is enclosed in a shell or **carapace** composed of a series of (usually) six calcareous plates which encircle the body with four further plates forming a cover on top. These can open allowing the six pairs of retracted appendages to be protruded. The appendages are divided into two branches apiece (biramous), each many-jointed and covered with bristles which serve as strainers. Food accumulates here when they lash about in the water and is passed to the mouth parts and so into the mouth which is inside the carapace. Attachment to rocks or other substratum is by the head region so that the animal is, as it were, upside down.

(2) *Examine a specimen of the* **Goose Barnacle, Lepas.**

This animal is stalked and is attached to the substratum by the base of the stalk.

ONISCUS

THE WOOD LOUSE

Woodlice belong to the **Sub-Class Malacostraca.**

Examine a specimen of the **Wood Louse** *with a hand lens.*

The body is somewhat flattened dorso-ventrally and is clearly segmented. The **head** and **first thoracic segments** are fused forming a **cephalothorax** and the head bears a pair of **antennæ.** The remaining **thoracic segments** bear paired **appendages** similar to one another and the **abdominal segments** are small and fused. This animal is completely adapted to a terrestial life.

DAPHNIA

THE WATER FLEA

This organism, which serves as food for Hydra, is found abundantly in ponds.

Examine **Daphnia** *in pond water in a small aquarium.*

Note the general appearance of the animals and the curious jerky way in which they row themselves through the water.

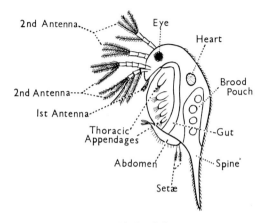

FIG. 64. Daphnia.
(From Wallis—Practical Biology)

Examine a slide of **Daphnia.**

Note that the whole of the body exclusive of the head is enclosed in a **carapace,** though on the ventral side this is incomplete.

On the **head,** which is unusual in shape, are the two **fused compound eyes,** large **antennæ** by means of which the animal rows itself through the water and mouth parts which include **mandibles.** The **thorax** is large and bears **comb-like bristles** which are used for breathing and for collecting food. The **abdomen** is rudimentary. **Eggs** may be visible in what is known as the **brood pouch** towards the posterior end of the body. It is interesting to find that these develop parthenogenetically.

CLASS INSECTA

Body divided into head, thorax and abdomen. Three pairs of legs and, usually, two pairs of wings on thorax. Breathe by tracheæ. Segmented abdomen devoid of appendages. Generally undergo a metamorphosis.

ORDER ORTHOPTERA

Mouth-parts adapted for biting, anterior wings somewhat hardened with chitin and serve as covers for the membranous posterior wings. *Hemimetabolous: cockroaches, locusts, grasshoppers, stick and leaf insects.*

PERIPLANETA

THE COCKROACH

There are two important species, **P. americana** and **Blatta orientalis** which is smaller, darker in colour and the female of which is deficient of wings. The young (known as *nymphs*) are similar to the adults except that they are small and wingless. They grow and undergo a series of moults or *ecdyses*, ultimately developing wings. Metamorphosis is thus incomplete and the insect is said to be hemimetabolous.

The following description refers to *P. americana.*

OBSERVATIONS ON LIVING ANIMALS

Observe the **living insects.**

Note the movements of the antennæ and of the legs in walking.

Place the animal on its back and observe the motions of the mouth parts.

EXTERNAL ANATOMY

Dorsal View

Examine the **cockroach** *externally from the dorsal aspect. Use a lens as necessary.*

The body is divided into head, thorax and abdomen and is protected by a chitinous exoskeleton which consists in each segment of **sclerites,** each composed of a **tergum** (dorsal), a **sternum** (ventral) and **pleura** (lateral).

Note the pear-shaped **head,** usually at right angles to the body and bearing tapering, many-jointed **antennæ** and large **compound eyes.** The **thorax,** separated from the head by a short neck or **cervicum,** is subdivided into the **prothorax** (the **pronotum** or tergum of which hides the neck), the **mesothorax** covered by the **mesonotum** and bearing a pair of **elytra** (the anterior wings somewhat hardened and covering the posterior wings when at rest) and the **metathorax** covered by the **metanotum** and bearing a pair of **membranous wings** strengthened by a framework of **nervures.** *Deflect the wings to the sides and pin them in position.* The segmented **abdomen** is covered by ten **terga** (the 8th and 9th being hidden under the 7th). The two jointed appendages under the lateral edges of the 10th tergum are the **cerci anales** and the small membranes at their bases **podical plates,** *seen by lifting the* 10*th tergum.* Internal to these in the male is a pair of short, many-jointed **styles** on the 9th segment.

Ventral View

Examine the **male** *and* **female** *animals from the ventral aspect. Again use a lens.*

Note the **mouth** on the **head** and the three pairs of jointed **legs,** one pair attached to each segment of the **thorax.** The **abdomen** is covered by **sterna** similar to the terga. Nine are visible in the male and the 9th bears the **styles.** In the **female** only seven will be seen and the 7th is a large boat-shaped process serving as the floor of the **genital pouch.** In both sexes the anus is underneath the 10th abdominal tergum and the **genital pore,** surrounded by complicated structures, the **gonapophyses,** is below the anus. Look for the ten pairs of **spiracles,** the respiratory openings situated in the pleura near the anterior edges of the terga in the first and third thoracic segments and the first eight abdominal segments.

Leg

Remove a **leg** *and examine under a hand lens.*

Note the long proximal **coxa,** then the very small **trochanter,** followed by the longer **femur** and the long thin **tibia,** each of which bear **bristles.** The distal joint is the **tarsus,** which is made up of five segments or **podomeres,** the last of which is often called the **pretarsus** and bears as its tip two **claws** with a pad or **arolium** between them.

Head

This is actually composed of six segments but owing to fusion they are merely visible externally by sutures.

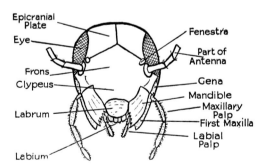

Fig. 65. **Periplaneta. Head. Anterior View.**

Examine the **head from the front,** *under a lens.*

Note the long, many-jointed **antennæ** and the white oval areas near their bases known as **fenestræ.** Above the bases of the antennæ are the curved black **compound eyes. Epicranial plates** cover the top and back of the head, the **frons** and **clypeus** covering the front and the genæ the sides. Attached to the lower edge of the clypeus is the upper lip or **labrum** and three pairs of mouth parts: (i) the long-jointed **maxillary palps**; (ii) the shorter-jointed **labial palps** internal to them and (iii) the toothed **mandibles** posterior and dorsal to the labial palps.

Mouth Parts

This should be left until last if the same insect is to be used for dissection.

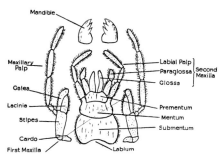

Fig. 66. **Periplaneta. Mouth Parts.**

Cut off the head, boil it gently in 2 per cent. caustic soda for a few minutes to remove the attached muscles. Pour off the caustic soda and wash thoroughly in water then transfer to a watch glass. Separate the parts with a small scalpel or, better, a dissecting needle, starting with the labium, working forwards and finishing with the mandibles. Keeping them in their respective positions, examine under a hand lens. Then make a permanent preparation transferring the parts to a slide with small forceps. Dehydrate in 30, 50, 70, 90 per cent. and absolute alcohol. Clear in natural oil of cedar wood and mount, unstained, in balsam. The mandibles, being thick, will cause rocking of the coverslip. To avoid this, a cell should be made to raise it. Pieces of cardboard of suitable thickness, placed where the edges of the coverslip will rest, will serve this purpose.

Note the **Mandibles** or jaws with their inner margins toothed. The **Maxillæ** (or **First Maxillæ**) arise behind the mandibles and each consists of a **protopodite** made up of two sclerites, a proximal **cardo** and a distal **stipes** at the base of which is a small sclerite, the **palpifer,** the distal end bearing an inner, broad, pointed **lacinia** and an outer, softer, tapering **galea.** External to these is a **maxillary palp** consisting of five podomeres attached to the outside of the stipe.

The **Labium** (or **Second Maxillæ**) shows a structure similar to that of the first maxillæ but is smaller and partly composed of a large proximal podomere, the **submentum,** a smaller median one, the **mentum,** and a distal **prementum,** all being central and really composed of fused paired appendages. An outer **paraglossa** and inner **glossa** (or **lacinia**), constitute the **labial palp** (or **ligula**).

INTERNAL ANATOMY

It is essential that a freshly killed insect be used.

Partially melt a little of the wax in the centre of a dissecting dish by playing a bunsen flame on the surface (the wax must be dry), place the insect, dorsal side upwards, in the soft wax, keeping the edges of the terga above the wax. Allow the wax to cool, when the insect will be fixed. Alternatively the insect may be fixed in the dish with small pins one through the femur of each leg and one through the posterior end. Cover with water. Remove the elytra and wings. Then carefully remove first the abdominal and then the thoracic terga one at a time, working forwards by lifting with forceps and cutting round the edge with small scissors. Examine under a lens.

THE HEART

The **heart** should be visible, enclosed in the pericardium as a long thirteen-chambered tube in the mid-dorsal line of the thorax and abdomen, the chambers corresponding with the segments, and each chamber communicating with the pericardial cavity by a pair of openings, or **ostia,** on its sides. The pericardial cavity is in communication with the hæmocœls around the viscera.

THE ALIMENTARY SYSTEM

Remove the heart and muscles carefully so as to avoid damaging the organs beneath.

Unravel the alimentary canal thus exposed from the white fluffy **fat body** *in which it is enveloped and pin it to one side.*

(1) The **œsophagus** runs into the thorax and joins the mouth to the **crop,** a dilated sac extending into the abdomen. The forked **visceral nerves** will be found on its surface. Note the two **salivary glands,** one on each side of the crop and lying in the thorax. *Take care not to damage them.* The ducts from the glands form a median **salivary duct** which opens into the **mouth.** The **proventriculus** or **gizzard** is a thick-walled muscular sac continuous with the crop, and this is followed by the mid-gut, or **mesenteron,** a narrow tube bearing at its anterior end seven or eight blindly ending tubes, the **hepatic cæca.** The coiled hind-gut is made up of the **small intestine** (sometimes called the ileum), a short tube, followed by the longer and

wider **large intestine** (sometimes called the colon) which leads into the wide **rectum**. This opens to the exterior by the **anus**. At the beginning of the small intestine is a number of fine **Malpighian tubules**. They are not part of the alimentary system and will be considered later.

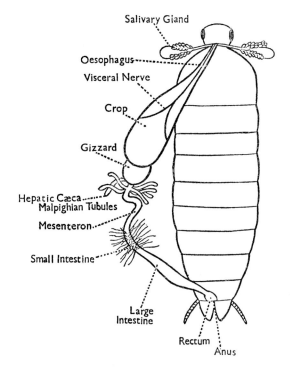

FIG. 67. **Periplaneta. Alimentary System.**
(From Wallis—Practical Biology)

(2) *Cut open the* **proventriculus.**

Note the **cuticular teeth.**

(3) *Carefully remove one of the* **salivary glands** *with its ducts as follows: Remove the alimentary canal by cutting through the œsophagus and rectum. Then cut away the dorsal covering of the neck and head. Deflect the remains of the œsophagus forwards and cut through the salivary duct. Then carefully transfer the freed salivary gland with the duct to a watch glass of 70 per cent. alcohol to fix it. Stain with picro-carmine or Delafield's Hæmatoxylin, dehydrate, clear and mount in balsam. Examine under the lower power.*

Note that each gland consists of two diffuse **glandular portions** and a median sac, the **reservoir**. The duct from each glandular portion and that from each reservoir joins its fellow from the opposite side and the common ducts then unite to form the median **salivary duct**.

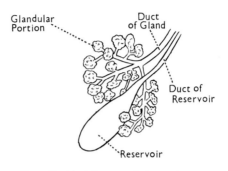

FIG. 68. **Periplaneta. Salivary Gland.**
(From Wallis—Practical Biology)

THE EXCRETORY SYSTEM

A large number of fine tubules at the beginning of the small intestine were seen when examining the alimentary canal. These are the **Malpighian tubules** and are the excretory organs. The excretions are passed into the intestine and are expelled with the fæces.

THE RESPIRATORY SYSTEM

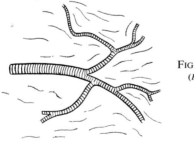

FIG. 69. **Periplaneta. Tracheae.**
(From Wallis—Practical Biology)

This is composed of a number of tubular tracheæ which open to the exterior by the spiracles (already seen) and which terminate internally in minute tracheoles in the tissues.

Remove one of the larger silvery-looking **tracheæ** *which ramify through the tissues of the body. Stain with picro-carmine and mount in dilute glycerine. Examine.*

Under *low power* note the **trachea** with spiral **chitinous lining** and the **cells** which secrete this lining.

Under *high power* note the **cells** as before and their **nuclei.**

THE REPRODUCTIVE SYSTEM

If a fresh insect is used very carefully remove the alimentary canal and the fat body, but be careful not to remove the testes with which the fat body may easily be confused.

Male

The two **testes** and their **vasa deferentia** (which join) are embedded in the fat body with which they may have been removed. Note the two **vesiculæ seminales** constituting the large centrally placed **mushroom-shaped gland** composed of blindly-ending finger-like processes. The vasa deferentia after union lead into this and form the **ejaculatory duct,** a short muscular tube leading to the exterior by the **genital pore,** below the anus. Ventral to the mushroom-shaped gland and the ejaculatory duct is the small whitish elongated **conglobate gland.** *Deflect the mushroom-shaped gland to see this.*

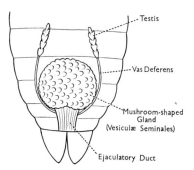

Fig. 70. **Periplaneta.**
Male Reproductive Organs.
(*From Wallis—Practical Biology*)

Testis
Vas Deferens
Mushroom-shaped Gland
(Vesiculæ Seminales)
Ejaculatory Duct

Female

There are two **ovaries** and each consists of eight **ovarian tubules,** showing swellings due to contain ova. The two **oviducts** are short and wide and join to form the **vagina** which opens to the exterior by a vertical pore in the genital pouch on the 8th abdominal sternum.

On the 9th abdominal sternum is a pair of branched tubes, the **colleterial glands** which also open into the **genital pouch.** A pair of **spermatheceæ** of unequal size will also be seen leading into the genital pouch.

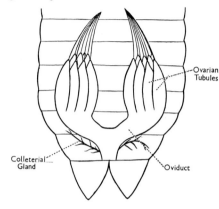

FIG. 71. **Periplaneta.**
Female Reproductive Organs.
(*From Wallis—Practical Biology*)

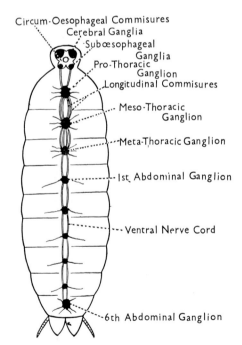

FIG. 72. **Periplaneta. Nervous System.**
(*From Wallis—Practical Biology*)

THE NERVOUS SYSTEM

Carefully remove the alimentary canal and fat body if a fresh insect is used and the dorsal coverings of the neck and head.

Note the two white **cerebral ganglia** and the short, wide **circumœsophageal connectives** which lead from them and join to form the **subœsophageal ganglia.** From this the double **longitudinal commissures** run to the **prothoracic ganglion.** The **meso-** and **metathoracic ganglia** follow and then six **abdominal ganglia,** all being joined by the **ventral nerve cord.** The three thoracic and the last abdominal ganglia are larger than the others. Nerves arise from all the ganglia.

LOCUSTA

THE MIGRATORY LOCUST

The Migratory Locust, **L. migratoria,** inhabits the greater part of the African continent and Central and Southern Asia. The Desert Locust, **Schistocerca** is similar and infests Northern Africa and countries as far east as India. Locusts are voracious feeders and swarms occur which do untold damage to crops over very large areas, costing millions of pounds and imperilling food supplies.

OBSERVATIONS ON LIVING ANIMALS

Locusts are easily reared and can be studied in the Insect Houses of Zoos or in cages kept in the laboratory.

EXTERNAL ANATOMY

Overall Structure

Examine the **locust** *externally from all aspects, using a lens as necessary.*

The body is divided into **head, thorax** and **abdomen** and is enclosed in an exoskeleton of chitin. The body is segmented and each segment consists of a dorsal plate, the **tergum,** and a ventral plate, the **sternum,** between which are the laterally placed **pleura.**

The **head** is longer vertically and bears two many-jointed **antennæ,** two large **compound eyes** and three simple eyes or **ocelli,** two **lateral** and one **median.**

The **thorax** is joined to the head by a **neck membrane.** *Pull the head forwards to see this.* Note that the thorax is composed of three segments, an anterior **prothorax,** a median **mesothorax** and a posterior **metathorax.** *Examine the ventral view* and it will be seen that each segment bears a pair of jointed **legs.** *Examine the dorsal side* and it will be observed that the prothorax has a large tergum called the **pronotum** which bends over rather like a saddle to cover the pleura. It will also be noticed thet the meso- and meta-thorax bear a pair of **wings** apiece. The **anterior** pair are thickened with chitin whereas the **posterior** pair are membranous. Deflect the wings to one side to expose the rest of the body.

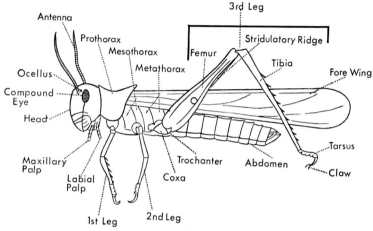

Fig. 73. **Locusta. Left Side. Male.**

The **abdomen** consists of eleven segments each composed of a tergum and a sternum but no pleura are present; nor are there any appendages as in the thorax.

Draw the **Dorsal** *and* **Ventral Views** *of the insect.*

Head

The head shows but little segmentation. *Examine it from the front.* The **median ocellus** lies on the front part of the head which is known as the **frons,** behind which is the **vertex.** The part below the frons is the **clypeus** and the sides of the head are formed by the **genæ** with small **sub-genæ** below. Sutures form the junctions between these plates. The mouth-parts are ventrally situated.

Mouth Parts

Identify as many as possible of the mouth parts from the description which follows.

These consist of an upper lip or **labrum** situated below the **clypeus,** two toothed **mandibles** each with a proximal molar surface for grinding and a distal incisor surface for cutting, a centrally placed **labium** composed of a proximal **submentum,** a smaller medial **mentum** and a distal **prementum** from which a pair of **labial palps** arise. There are two **maxillæ** each consisting of a proximal **cardo** and a distal **stipes.** The latter is divided into two parts, the outer bearing the **galea** to which is attached the **maxillary palp** composed of five joints and the inner part bearing the **lacinia** which ends in two pointed claws. These maxillæ will be found behind the mandibles. To the distal side of the prementum lie two small rounded **glossæ,** but only that on the right will be clearly seen. Beyond these are two large rounded **paraglossæ.**

(i) *Now remove the mouth parts and examine each separately using the following procedure. Cut off the head, boil it gently in 2 per cent. caustic soda for 20 minutes to half an hour to simplify separation of the parts. Then pour off the caustic soda and wash thoroughly with water. Using small forceps and a very small scalpel or, better, a dissecting needle, separate the parts starting with the labium. Lift it carefully* and you will see a tongue-like structure, the **hypopharynx.** *Then, holding the base of each part with forceps, free it very carefully with the scalpel or dissecting needle. As each part is removed place it in water in a watch glass. Examine later using a hand lens.*

(ii) *Make a separate permanent preparation using the following method. Dehydrate in 30 per cent., 50 per cent., 70 per cent., 90 per cent. and absolute alcohol. Clear in clove oil or natural oil of cedar wood and mount the parts in their respective, positions, unstained, in balsam. It will be necessary to make a cell of narrow strips of cardboard of suitable thickness to support the coverslip along its edges as the mandibles, owing to their thickness, will otherwise cause it to rock.*

Thorax

This is composed of three segments as already explained above. Openings of the respiratory tracheæ called **spiracles** will be found one on each side between the pro- and meso-thorax and between the meso- and meta-thorax.

Legs

Note that the posterior pair is longer than the others, being used for jumping. *Remove one of the* **front legs** *and examine under a lens.* The proximal joint is the **coxa** and this is followed in turn by the small **trochanter,** the longer **femur,** the **tibia** and the **tarsus** which is composed of three segments each bearing pads, the first one three and the others one apiece. Two claws and a pad called the **arolium** constitute the terminal **pretarsus.** *Examine one of the* **hind legs** *and* note that there is a ridge on the inner side, the **stridulatory ridge.** This is rubbed against a vein, the stridulatory vein in the forewing of the male and produces the sound made by it.

Wings

Pin down the locust, open out the wings and secure them with pins. Examine the anterior wings or **tegmina** and note their hardness. Compare them with the posterior wings which are membranous. These are supported by **veins** and each has a large **anal lobe.**

Abdomen

Eleven segments comprise the abdomen and on the sides of the first segment, beneath the surface and partly obscured, is a pair of auditory **tympanal organs** covered by **tympanal membranes.** There are eight pairs of **spiracles,** the first being situated anterior to the tympanal membranes on the first segment and the remainder placed on the sides of the next seven segments near their anterior ends and towards the ventral side. In the **female,** unlike the male, the 8th segment is longer than the others and terminates in an **egg guide** situated in the mid-line. A pair of large **dorsal** and **ventral oviposter valves** with smaller inner valves are found on the 8th and 9th segments. *Pull apart the dorsal and ventral valves* to see the **inner valves** between them. The 9th and 10th segments are narrow ones and it is not easy to identify them separately on the ventral side. A pair of short **cerci** bearing **setæ** arise behind the 10th segment on the dorsal side. In the 11th segment the sternum forms what is known as the **paraproct** and the **anus** lies beneath the **epiproct** which is situated above it. In the **male** the posterior part of the 9th segment, which is curved upwards, forms the **subgenital plate** and the **cerci** are longer than those of the female.

Draw the **Abdomen of both a male** and a **female** insect.

INTERNAL ANATOMY

A freshly killed animal should be used for dissection. Remove the wings and pin down the insect, dorsal side up, in a dissecting dish by fixing small pins through the legs and posterior part of the abdomen. Beginning at the posterior end remove first the abdominal terga and then the thoracic terga and working forwards by lifting them with small forceps and cutting through the edges with small scissors. Cover with water and examine with a lens. A fair amount of yellowish

white tissue will be found in the body cavity which is a **hæmocœl.**
This is the **fat body** and *it will be necessary to remove it in order
to see the organs.* The glistening branching tubes are the **tracheæ**
which form part of the respiratory system.

In the abdomen the **hæmocœl** is divided into a dorsally placed
pericardial cavity containing the heart and a **perivisceral cavity** containing most of the viscera by a **dorsal diaphragm.** In both the thorax
and the abdomen lies a **ventral diaphragm** beneath the perivisceral
cavity and this separates off a **perineural cavity** below containing the
nerve cord.

THE HEART

The **heart** will be seen as a tube with dilated chambers lying in
the dorsal mid-line in the abdomen. Each chamber has a pair of
openings called **ostia** on the sides, difficult to identify, through which
blood enters the heart. **Alary muscles** lie to the sides and beneath
the heart. Running anteriorly from the heart in the thorax is the
aorta but it will not be possible to see this at this stage. It may be
seen when dissecting the nervous system.

THE ALIMENTARY SYSTEM

(1) *The alimentary canal lies beneath the heart which must be
removed.* The mouth leads into the pharynx continuous with which
is the **œsophagus** and this enters the dilated **crop** which is followed by
the **gizzard.** These last two structures show no clear demarcation
externally. All these structures constitute the **fore-gut.** The next
region is the **mid-gut** or **mesenteron** into which open six wide elongated
tubular structures called **cæca** at its anterior end. Posterior to this is
the **hind-gut** which consists first of the **ileum,** then a narrower part,
the **colon,** followed by a wider **rectum** which opens at the **anus**
which, as has already been seen, is below the epiprocht. A slight
constriction separates the mid-gut from the hind-gut and immediately anterior to this will be seen a large number of long, fine
thread-like Malpighian tubules. These, however, are not part of the
digestive system but are excretory in function.

(2) *Now cut through the anterior end of the alimentary canal
where it enters the head and either pin the gut to one side or cut
through the rectum and remove it. The object is to expose the salivary
glands.*

The two **salivary glands** lie on each side of the crop and extend
beneath it. Each is composed of a number of small **lobules,** white
in colour, the small ducts from which form a **salivary duct.** These

run forwards and join one another giving rise to a short **median duct** in the hypopharynx which opens into a **salivary cup** on the labrum. *Carefully remove one of the salivary glands with its duct. Fix in 70 per cent. alcohol for about 10 minutes or so, stain with borax-carmine and make a permanent preparation.*

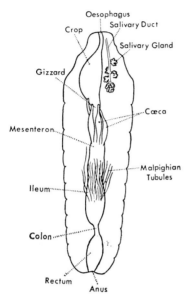

FIG. 74. **Locusta. Alimentary System. Partially deflected to left to show one salivary gland.**

THE EXCRETORY SYSTEM

The large number of **Malpighian tubules** arising at the junction of the mid-gut with the hind-gut have already been seen. These are the excretory organs. They lead into the hind-gut and their excretions are expelled with the fæces.

THE RESPIRATORY SYSTEM

This consists of a number of branched tubular **tracheæ** (see Fig. 69, p. 94) which open to the exterior by the spiracles which were seen when examining the external structure. The tracheæ branch intern-ally into minute tracheoles in the tissues where they give rise to air sacs. The whole system in complicated and difficult to follow. *Remove one of the larger tracheæ which are found in the abdomen. Stain with* *borax-carmine or picro-carmine and mount in dilute glycerine or make a permanent preparation.*

Under low power note the **trachæ** and **air-sac** (if present). Careful focussing may reveal the **spiral chitinous lining** in the wall which *which* keeps the tube open.

THE REPRODUCTIVE SYSTEM

In both sexes cut through the æsophagus and rectum and remove the alimentary canal if still in place in order to expose clearly the sexual organs which lie on its dorsal side. Dispose of any remaining fat body may obscure the view.

Male

The two **testes,** so close to one another as to appear as a single structure, are seen as a yellowish mass of blindly-ending tubules in the mid-line. The tubules are known as **follicles** and from each arises a short narrow vas efferens. All the vasa efferentia from each testis open into a **vas deferens** and the two vasa deferentia lie on the ventral

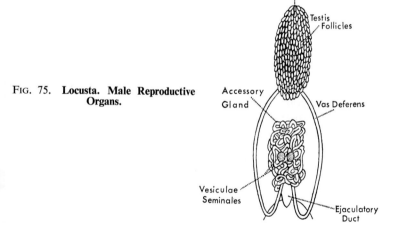

FIG. 75. **Locusta. Male Reproductive Organs.**

side of the testes. They then pass backwards to enter the **ejaculatory duct,** which is situated in the mid-line. A larger number of coiled tubules constituting the **accessory gland** lie behind the testes and open into the ejactulatory duct and two of these tubules terminate in swellings. These are the **vesiculæ seminales.**

Female

The two **ovaries** lie close to one another and each consists of a large number of tubular ovarioles. There are two oviducts into which the ovarioles on their respective sides open. Ova in various stages of development are found in the **ovarioles,** the most mature being at the oviducal end. The two **oviducts** extend anteriorly to form the **accessory glands** and posteriorly to join one another thus

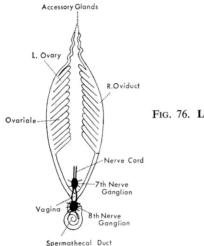

Accessory Glands

L. Ovary

R. Oviduct

Ovariole

FIG. 76. **Locusta. Female Reproductive Organs.**

Nerve Cord

7th Nerve Ganglion

Vagina

8th Nerve Ganglion

Spermathecal Duct

giving rise to the **vagina.** This is partly obscured by the nerve cord. Above the vagina and beneath the nerve cord is the spermatheca with a coiled **spermathecal duct.** The vagina and spermathecal duct open into the genital chamber. *It will be necessary to remove the posterior part of the nerve cord in order to expose these structures but this should be deferred until the nervous system has been examined when the drawing can be completed.*

THE NERVOUS SYSTEM

Before dissecting this system the insect should be left in 70 per cent. alcohol for a few hours. If not already done, remove the alimentary canal and the salivary glands. Also remove the reproductive organs in both sexes but in the male leave the nerves crossing the vasa deferentia which appear as white threads and in the female leave the parts obscured by the nerve cord i.e., the vagina and spermatheca.

Remove any fat body which is causing obstruction. Then, using needles, carefully remove the ventral diaphragm which separates the perivisceral cavity from the perineural cavity in order to expose the nerve cord. Finally remove the dorsal covering of the head by cutting forwards laterally between the compound eyes on each side so that the two incisions join at both ends of the head.

The **aorta** may now be seen in the head and, *if the neck membrane is cut through,* in the neck as well. It is continuous with the heart and passes through the thorax where it is difficult to see. The heart and aorta constitute the **dorsal vessel** which is the only blood vessel in the insect.

The mandibular muscles in the head must be separated in order to expose the cerebral ganglia.

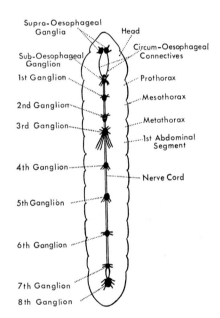

FIG. 77. **Locusta. Nervous System.**

Note the **supra-œsophageal ganglia** from which **circum-œsophageal connectives** form the **sub-œsophageal ganglion.** *Trace the* **nerve cord** *backwards.* Note the **1st** and **2nd ganglia** (**pro-** and **meso-thoracic ganglia**) in the first two thoracic segments and the **3rd ganglion** (**meta-thoracic ganglion**) which is **fused with the first three abdominal ganglia** since the nerves not only supply the thorax but also the first three abdominal segments. The **4th** to **7th ganglia** will be seen in succession associated with the corresponding abdominal segments. The **8th ganglion** in the eighth abdominal segment actually consists of the **fused 8th to 11th ganglia.** Now examine the **nerves** arising from the ganglia noting in particular the three pairs of nerves arising from the 3rd ganglion supplying abdominal segments in addition to those supplying the metathoracic segment.

Now remove the posterior part of the nerve cord in order to expose the remaining parts of the reproductive systems and complete your drawings of them (see p. 104). Then, in the female, unravel the spermathecal duct in order to find the spherical **spermatheca.**

DEVELOPMENT

Eggs are laid in holes about 10 cm in depth in the sand which are dug by the oviposter valves at the posterior end of the female's abdomen. Up to a hundred eggs are laid together with a secretion which hardens to form an elongated **egg pod.** After about a couple of weeks the **nymphs** hatch out and wriggle their way to the surface, shed their coats and form the **1st instars** as they are called. They are referred to as **hoppers** on account of their method of locomotion. There are five subsequent moults or **ecdyses** as the nymphs grow progressively larger and these are the **1st to 5th instars.**

Examine specimens of the **five instars.**

1st Instar: The larva is a miniature wingless form of the adult and in this instar the pronotum is rather square in form.

2nd Instar: The pronotum is more rounded and **wing buds** appear behind it.

3rd Instar: The **wing buds** are more clearly visible and point vertically downwards.

4th Instar: The **wing buds,** though still quite small now point upwards and backwards and the **hind wings** cover the **fore wings**.

5th Instar: The **wing buds** are now very much larger and more wing-like in appearance.
The imago emerges as the result of the ecdysis of this instar.

OMOCESTUS

THE COMMON GREEN GRASSHOPPER

There are several different species of grasshopper and *Omocestus viridulus* is found very widely in grassy habitats in the British Isles. Others differ only in minor details such as colour and the meadow grasshopper has only vestigial posterior wings and is therefore unable to fly.

EXTERNAL ANATOMY

(1) *Examine the external structure of the* **grasshopper.**

The description of the external structure of the locust (pp. 97–100) will also serve for the study of this insect. The colour is bright green or dark brown. The females are larger than the males and their abdomens are rounded posteriorly whereas the males have a more pointed end. The **antennæ** are fairly short and thick. The **hind legs** are long and the **femur** is well developed enabling the insect to jump considerable distances.

(2) *Examine one of the* **hind legs** *under a hand lens* and note the tiny **pegs** on the inside. *Now examine one of the* **fore-wings** *and note the prominent* **nervures.** The characteristic noise made by grasshoppers is effected by rubbing the pegs on the hind legs against the nervures on the fore-wings.

ORDER TRICHOPTERA

Mouth-parts atrophied. Long antennæ. Larvæ aquatic and live in larva cases which they construct from plant material, stones and other pond debris. Holometabolous. *Caddis flies.*

PHRYGANEA

THE CADDIS FLY

P. grandis is one of the largest caddis flies of which there is a very large number of species.

EXTERNAL ANATOMY

Examine a specimen of **Phryganea grandis.**

The mouth-parts are atrophied so that the imago can only suck such liquids as nectar. Note the long **antennæ.** *Examine the* **wings.** They are brown in colour, have two transparent streaks and are covered with hairs. They are folded when the insect is at rest. *Open out the wings and measure them.* You will find that they have a span of about 6 cm.

LARVA

Examine a **caddis fly larva and its case.**

The larva is aquatic and has an elongated body segmented in conformity with the usual insect characteristics. It is wingless. *Examine the* **larva case.** It is spiral in shape and made from plant debris. *If* *larva cases of other species are available examine them* and you will find a great variety in their form and construction.

ORDER COLEOPTERA

Biting mouth-parts. Anterior wings hardened as wing cases. Holometabolous. *Beetles.*

COCCINELLA SEPTEMPUNCTATA

THE SEVEN SPOTTED LADYBIRD

Egg

Examine the **eggs** *and* **larvæ,** *if available, with a hand lens* (they may be found on the leaves of plants infected with *aphis* on which the imagines feed).

The eggs are minute, ovoid in shape and yellow in colour.

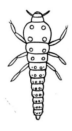

FIG. 78. **Coccinella. Larva.**
(*From Wallis—Practical Biology*)

Larva

These, too, are minute. They are grey in colour with black and yellow spots and the **abdomen** tapers posteriorly. The mouth-parts contain **mandibles** and the three **thoracic segments** a pair of jointed **legs** apiece. There are eight **abdominal segments** which decrease in width as they pass backwards. On the terga are small **tubercles** bearing **setæ.**

Examine the **Imago.**

Note the bright red colour of the ovoid body on the dorsal side of which are seven black spots, three on each wing case (elytrum) and one, the anterior spot, shared between the two. (Other common species have five and two spots respectively and there is one with twenty-two. *C. ocellata* is larger than the others and has orange elytra, with numerous black spots each with a yellow margin.)

The **head** bears sickle-shaped **mandibles, compound eyes** and short **antennæ** and is almost hidden by the pronotum which is black. The **anterior pair of wings** are large (as already seen), strongly thickened with chitin, red with seven black spots (already observed) and serve as wing cases when at rest. These are known as **elytra.** They almost completely cover the body. Beneath the elytra (at rest) is a pair of **membranous posterior wings.**

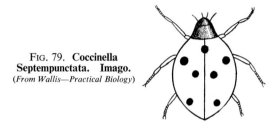

Fig. 79. **Coccinella Septempunctata. Imago.**
(*From Wallis—Practical Biology*)

DYTISCUS MARGINALIS

THE GREAT WATER BEETLE

All stages of this insect's life-history are aquatic though atmospheric oxygen is used in breathing.

Examine the **Larva.**

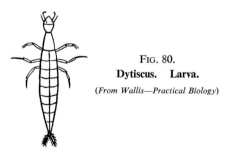

Fig. 80.
Dytiscus. Larva.
(*From Wallis—Practical Biology*)

The larva is about two inches in length and brownish in colour. The **head** is rounded and somewhat flattened and bears a pair of **antennæ** and six **ocelli** on each side. The mouth parts are adapted for biting by means of long sickle-shaped **mandibles** which are grooved on the inside. Each of the three segments of the **thorax** bears a pair of jointed **legs.** The **abdomen** tapers posteriorly and consists of eight segments, the last two having a chitinous fringe. A pair of pointed appendages, also fringed, project from the posterior end. The abdomen is arched in an upward direction and **spiracles** are situated at the posterior end.

Examine the **Imago.**

This large carnivorous insect is oval in shape, the dorsal surface being slightly convex, the ventral side slightly keeled, the entire surface being extremely smooth and the colour brownish-black with a yellow edge.

The **head** bears a pair of **compound eyes,** long **antennæ** and strong toothed **mandibles.** The **thorax** bears a pair of highly polished **anterior wings,** well thickened with chitin which serve as wing cases and are known as **elytra.** In the male these are smooth but those of the female have deep furrows along the greater part of their length. The **posterior wings,** completely hidden by the elytra when the animal is in the water, are membranous. The three pairs of jointed **legs** are used for swimming and the **tibia** and **tarsus** are flattened and bear a

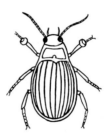

FIG. 81. **Dytiscus. Imago.**

(*From Wallis—Practical Biology*)

fringe of stiff **setæ.** The flat first three podomeres of the tarsus of the front legs of the male are enlarged to form a kind of disc and on the ventral surface of them are several **suckers** (used for gripping the female). Eight pairs of **spiracles** are found dorsally towards the sides. All the thoracic segments are firmly interlocked as is the thorax to the abdomen and the meso- and meta-thorax and the first three abdominal segments are joined together. This gives the body considerable strength.

ORDER HEMIPTERA

Mouth-parts adapted for piercing and sucking. Often parthenogenic stages which are wingless. Hemimetabolous. Many are mechanical vectors of virus diseases. *Bugs*, *Aphids*, *Leaf-hoppers*.

CIMEX LECTULARIUS

THE BUG

The pupal stage is absent. A common species affecting man is the bed bug. This is a night visitant rather than an ectoparasite.

OVA

Examine a slide of the **ova of Cimex.**

The egg is white in colour and ovoid in shape.

LARVA

Examine a slide of the **larva of Cimex.**

The larva is yellowish-white and similar to the adult (see below) but smaller.

IMAGO

Examine a slide of the **imago of Cimex.**

The body is dark brown and is covered with short hairs. It is flattened dorso-ventrally and is about 5 mm. in length. The small

Fig. 82. **Cimex lectularius. Imago.**
(*From Wallis—Practical Biology*)

head fits into a notch in the thorax. There are **compound eyes** but no ocelli and two jointed **antennæ.** The mouth parts are adapted for piercing and sucking. The **thorax,** as already stated, is notched anteriorly and bears on vestigial wings. The metathoracic segment bears stink glands. The **abdomen** is oval and broad, tapering slightly at its posterior end and being slightly narrower in the male than it is in the female.

APHIS

GREEN FLY

The *Aphides* are the well-known *green flies* and *black flies* which are parasitic on plum, apple and rose trees and on the bean plant, feeding on the juices of the plant. They usually have a different plant as host during the Winter months. The pupal stage is absent. Some Aphides are *vectors* carrying virus diseases to crops. The life-history is unusual and complicated and there are variations in different species. In early Summer winged *viviparous females* from the Winter host fly to new plants and produce offspring *parthenogenetically*. The first generation are *wingless females*. When these mature they produce *winged females* which fly away to infect new hosts and give rise to new colonies. Ultimately new *sexual forms* appear. In the Autumn the last generation of these winged females fly to the Winter host plant and give rise to *oviparous females*. These are fertilised by *winged males* and the eggs are laid in the Winter host. In the following Spring these fertilised eggs produce the winged females which fly in early Summer to the Summer host plant.

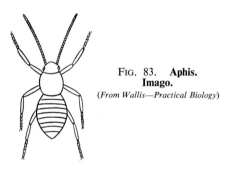

Fig. 83. **Aphis.
Imago.**
(*From Wallis—Practical Biology*)

(1) *Examine a slide of an* **Aphis imago.**

These insects are either black (or brownish-black) or green in colour. The **head** bears long **antennæ** and **compound eyes** and mouth-parts adapted for sucking the plant juices. The three segments of the **thorax** are difficult to distinguish. The **legs** are comparatively long and usually lighter in colour. Both **winged** and **wingless** forms occur during the life-history as stated above. The segments of the **abdomen** are easily identified.

(2) *Examine a slide of the* **mouth-parts of the Aphis imago.**

The labium is modified into a **proboscis** bearing a groove in which lie needle shaped **stylets** when these are not in use. These stylets can be released from the groove and used to penetrate the cells of the plant host.

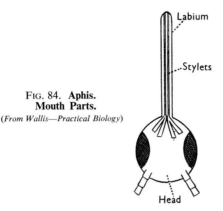

FIG. 84. **Aphis.**
Mouth Parts.
(*From Wallis—Practical Biology*)

ORDER HYMENOPTERA

Mouth-parts for sucking or biting. Short Antennæ. Holometabolous. Many are social insects. Parthenogenesis often occurs. *Bees, wasps, ants, ichneumon flies.*

APIS MELLIFICA

THE HIVE BEE

Examine the **larva** *and* **pupa of the hive bee.**

Larva

This is white in colour, segmented and devoid of limbs.

Pupa

Segmentation can also be seen and the development of structures of the future imago will be clearly visible.

FIG. 85. **Apis mellifica. Larva.**

FIG. 86. **Apis mellifica. Pupa.**

(*From Wallis—Practical Biology*)

Imago

In the Summer there are three kinds of individuals in the hive—one **queen,** a fertile female whose sole duty is to produce ova, a few hundred males, **drones,** which are responsible for fertilising the queen and several thousand **workers,** sterile females in which the gonads do not develop and who build the hive and keep it clean, collect pollen and nectar from flowers, make honey, secrete wax, feed the larvæ and perform other useful functions for the colony.

In consequence of their different functions, there are differences in structure in the three kinds of individual .

Examine and compare the structure of a **queen bee,** *a* **drone** *and a* **worker.**

The queen has a long abdomen and short wings, the drone has a much broader abdomen and large wings while the worker is the smallest of the three and has well developed wings.

The **head** of the worker bears complicated mouth-parts adapted to the obtaining of nectar from flowers, two large **compound eyes** (larger in the drone), three **ocelli** (simple eyes) and a pair of **antennæ.**

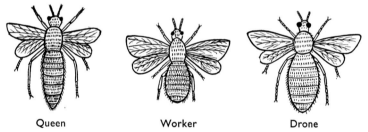

| Queen | Worker | Drone |

Fig. 87. **Apis Mellifica. Imagines.**
(*From Wallis—Practical Biology*)

The **thorax** consists of three segments and each bears a pair of legs composed of the usual podomeres—**coxa, trochanter, femur, tibia** and **tarsus** but these are not identical, being adapted to the performance of different functions. The second and third thoracic segments bear a pair of **wings** which are all membranous, the two on each side being linked together by small hooks to enable them to function as a single wing in flight.

The **abdomen** is segmented and at its posterior end in the queen and the worker is a protrusible **sting** which contains a duct from a poison gland. In the worker it is barbed and can therefore be used only once. In the queen this is known as the **ovipostor** since it is used in the deposition of eggs. Note that the abdomen of the queen is long and narrow and extends beyond the wings when they are folded back.

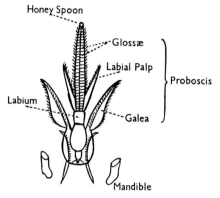

FIG. 88. **Apis Mellifica. Worker. Mouth-parts.**
(*From Wallis—Practical Biology*)

Examine a slide of the mouth-parts of the worker bee. (Refer to the diagram of the mouth-parts of the cockroach on p. 91.)

The **mandibles** are devoid of toothed edges and are somewhat spoon-shaped. They are used for kneading wax. In the first maxilla there is an elongated **blade-like galea** but no lacinia while the maxillary palp is vestigial. In the second maxillae (labium) the two glossæ are much elongated and fused together to form at the tip a small expansion known as the **honey spoon.** External to the glossæ are the elongated **labial palps.** These very long maxillæ form a protective sheath for the sucking tube or **proboscis** formed by the galeae of the first maxillæ and the glossæ and labial palps of the second maxillæ of labium.

Examine a slide of the **first leg of the worker.**

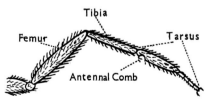

FIG. 89. **Apis Mellifica. Worker. First Leg.**
(*From Wallis—Practical Biology*)

The joint between the tibia and the tarsus is open and the anterior end of the tarsus is provided with a comb-like structure of short setæ, used for removing pollen from the antennæ which can be passed into it. This is known as the **antennal comb.**

Examine a slide of the **second leg of the worker.**

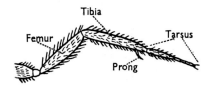

FIG. 90. **Apis Mellifica. Worker. Second Leg.**
(From Wallis—Practical Biology)

The tibia bears a short stiff seta known as the **prong**; it serves to remove pollen from the pollen basket on the third leg.

Examine a slide of the **hind leg of the worker.**

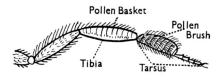

FIG. 91. **Apis Mellifica. Worker. Hind Leg.**
(From Wallis—Practical Biology)

The tibia bears a deep groove lined by bristles; this is the **pollen basket.** The first podomere of the tarsus bears stiff setæ forming the **brush** (or **pollen comb**), used for placing pollen in the pollen basket.

ORDER LEPIDOPTERA

Mouth-parts a proboscis for sucking. Wings covered with scales. Holo-metabolous. *Butterflies and moths.*

PIERIS BRASSICÆ

THE CABBAGE WHITE BUTTERFLY

Examine an **egg** *with a hand lens.*

It is bluntly conical in shape, yellow in colour and bears both vertical and horizontal ridges.

FIG. 92. **Pieris Brassicæ. Egg.**
(*From Wallis—Practical Biology*)

Examine the **larva,** *using a lens where necessary.*

The *caterpillar* consists of a head and thirteen segments and is of a greyish-green colour with a yellow stripe dorsally and a wider one on each side. A number of short bristles protrude from black protuberances on its body.

The spherical **head** has a pair of toothed **mandibles** (which work sideways) posterior to the **labrum** with **maxillæ** underneath. In the centre is the **spinneret** and externally on each side a pair of short **antennæ.** On each side of the head, towards the ventral surface, is a group of six small black **ocelli** (simple eyes).

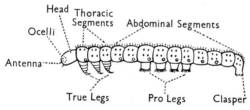

FIG. 93. **Pieris Brassicae. Larva.**
(*From Wallis—Practical Biology*)

The next three segments are the **thoracic segments.** Each bears a pair of five-jointed **true legs,** each of which terminates in a claw.

The remaining ten segments are the **abdominal segments.** No appendages are borne on the first two but each of the next four bears a pair of soft unjointed **pro-legs,** each ending in a pad and a semi-circle of hooks. The segments which follow have no appendages except the last which bears a pair of **claspers** similar to the pro-legs.

Careful examination with a lens will reveal the **spiracles** on the sides of the first thoracic and the first eight abdominal segments.

Examine the **pupa.**

The *chrysalis* is shorter than the caterpillar and is of a greenish colour with black and yellow spots. **Dorsal** and **lateral projections** (which hold the silken girdle in place when pupation takes place), **spiracles,** a segmented **abdomen** and the **outlines of developing** structures of the future imago such as **wings, legs** and **antennæ** will also be seen.

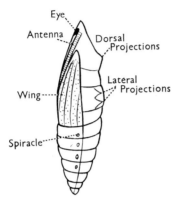

FIG. 94. **Pieris Brassicae. Pupa.**
(*From Wallis—Practical Biology*)

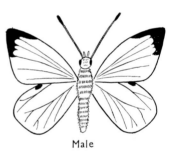

Male

FIG. 95. **Pieris Brassicae. Imagines.**
(*From Wallis—Practical Biology*)

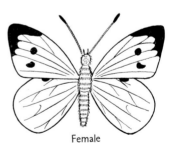

Female

Examine specimens of both **male** *and* **female imagines.**

The body is composed of **head, thorax** and **abdomen,** the thorax and abdomen being covered with hairs which makes it difficult to distinguish the three segments in the former and ten in the latter. The thorax bears three pairs of jointed **legs,** each composed of the usual five podomeres—coxa, trochanter, femur, tibia and tarsus—and two pairs of **wings.**

On the head note the two long, club-shaped and many-jointed **antennæ** and the large **compound eyes** laterally placed.

Examine a slide of the mouth parts.

There is a long coiled **proboscis** and a pair of **labial palps.**

Examine the **wings of both sexes** *from the dorsal and ventral aspects.* On the dorsal side these are creamy white in colour and the anterior wings have black markings in the tips in both sexes. The female also has two black spots on each wing. The posterior wings also have black markings on their anterior edges in both sexes. The ventral surfaces of the wings are pale greenish-yellow with black markings, though less conspicuous than those on the dorsal side.

Examine a slide of **scales from a butterfly's wing.**

These vary considerably in shape, size and colour. When in place on the wing they overlap like tiles on a roof.

ORDER DIPTERA

Mouth-parts for sucking.   Hind-wings modified into balancers or halteres. Short antennæ.   Holometabolous.   Some mechanical others biological vectors. *House-flies, mosquitoes, crane flies, tsetse flies.*

MUSCA DOMESTICA

THE HOUSE FLY

The *Blow Fly* or Blue Bottle (*Calliphora erythrocephala*) is similar to the house fly with minor differences.   These are *mechanical vectors* because, owing to their habit of alighting indiscriminately on manure heaps and other sources of infection and on food, they carry parasites on their bodies, particularly on their legs. Eggs are laid in food such as meat as well as on other organic substrates which serve as food for the larvæ.   *Mosquitoes* and *tsetse flies* such as *Glossina palpalis* also belong to this Order.   The latter live in tropical countries and may carry the protozoal parasite *Trypanosoma* which causes *Sleeping Sickness* in man.   The female *Anopheline* mosquito in the tropics carries the malarial parasite, *Plasmodium.*

Examine a slide of the **Ova of Musca domestica.**

These are only about 1 mm in length, elongated, rounded at the ends and white in colour.

Fig. 96. **Musca Domestica. Egg.**
(From Wallis—Practical Biology)

LARVA

Examine the **larva of M. domestica** *with a hand lens.*

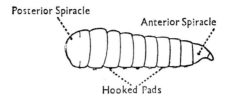

FIG. 97. **Musca Domestica. Larva.**
(From Wallis—Practical Biology)

These are limbless grubs composed of twelve **segments**. They have minute **heads** devoid of eyes and with **hooked mandibles. Spiracles** are found only on the second and last segments and **hooked pads** are present on the ventral side of the 6th to 12th segments.

PUPA

Examine the **pupa of M. domestica** *with a hand lens.*

This is shorter than the larva and is barrel-shaped and brown in colour.

FIG. 98. **Musca Domestica. Pupa.**
(From Wallis—Practical Biology)

IMAGO

(i) *Examine the* **imago of M. domestica** *with a hand lens.*

The **head** bears **compound eyes** and **three ocelli.** The **antennæ** are short and feathery. The labium is modified into a proboscis. The **thorax** bears only one pair of **membranous wings,** the anterior pair, and these are large. The posterior wings are modified into dumb-bell shaped structures which serve as balancers and are known as **halteres.** They are hidden by the anterior wings. The **abdomen** shows the usual segmented structure.

(ii) *Examine a slide of the* **mouth parts of the house-fly** *or* **blow-fly.**

These are adapted for sucking and consist of an extensible **proboscis.** Note the two large expanded lobes (the **labella**) on the under side of which are grooves called **pseudotracheæ** which open down-wards, join and lead into a trunk-like tube which can be protruded. There are no mandibles but **maxillary palps** are present.

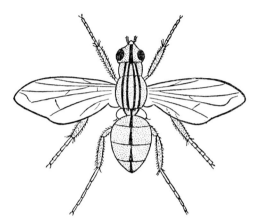

FIG. 99. **Musca Domestica. Imago.**

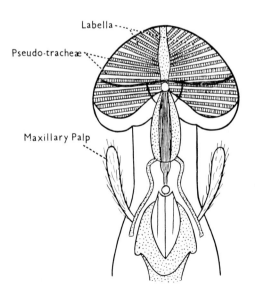

FIG. 100. **Musca Domestica. Mouth Parts.**

CULEX

THE MOSQUITO

This insect is an *ectoparasite*. Though there are other Genera and many other species, the two chief species are *Culex* (the common house mosquito), and *Anopheles*, the female of which is *vector* of the malarial parasite, *Plasmodium*. Virus diseases such as *Yellow Fever* and *Dengue Fever* may be transmitted by other mosquitoes. The following description applies to *Culex*.

CULEX PIPIENS

OVA

Examine a slide of the **egg raft** *or* **isolated ova of Culex.**

The tiny eggs, which are laid in water, are roughly cigar-shaped and are stuck together to form an egg-raft.

FIG. 101. **Culex. Egg-raft.**   FIG. 102. **Anopheles. Eggs.**
(*From Wallis—Practical Biology*)

LARVA

Examine a slide of the **larva of Culex.**

This, too, is aquatic. The large head has two **eyes,** a pair of short **antennæ,** mandibles and maxillæ and two **mouth brushes,** plumed structures for entrapping its food. The three thoracic segments are

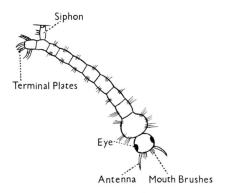

FIG. 103. **Culex. Larva.**

fused together to form a single large **thorax.** The 8th abdominal segment bears a respiratory structure called the **siphon** which opens to the surface of the water. It has five small valves at its tip. The 9th abdominal segment has four plates which serve as a **rudder** but also act as gills and contain the tracheæ. **Setæ** occur on the thorax and the first seven abdominal segments, those on the last segment forming a dense tuft.

PUPA

Examine a slide of the **pupa of Culex.**

This is also aquatic and, unlike most pupæ, it is motile. It is shaped like a large comma. The large rounded anterior portion of the head and thorax has a transparent cuticle through which can be seen the **outlines of the eyes, legs** and **wings of the future imago.** It

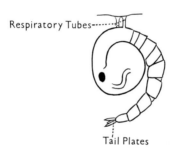

Respiratory Tubes

Tail Plates

Fig. 104. **Culex. Pupa.**
(From Wallis—Practical Biology)

bears two **respiratory tubes** on the dorsal side. The curved, segmented, narrow posterior portion of the pupa is the **abdomen** and it terminates in a pair of **tail plates** forming a paddle and used in swimming.

IMAGO

Examine slides of the **male** *and* **female imagines of Culex.**

The **head** is small but the two **compound eyes** are large. The **antennæ** are slender in the female but feathery in the male. The mouth parts form a **proboscis,** the female sucking blood, the male plant juices. The slender **thorax** bears three pairs of very long, delicate **legs** and one pair of long delicate **wings,** along the edge of which is a fringe of **setæ.** The posterior wings are, as in the house-fly, modified into balancers or **halteres.** The segmented **abdomen** is long and very slender.

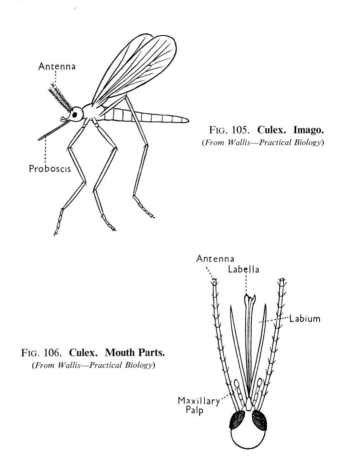

FIG. 105. **Culex. Imago.**
(*From Wallis—Practical Biology*)

FIG. 106. **Culex. Mouth Parts.**
(*From Wallis—Practical Biology*)

Examine a slide of the **mouth-parts of the female mosquito.**

The long **proboscis** is adapted for sucking blood and cannot be retracted. It consists of a long, stiff but flexible **labium,** forked at the tip which is known as the **labella.** Inside is a series of pointed structures for piercing the skin and a tube up which the blood is sucked.

In the male, which feeds solely on plant juices, the proboscis is just a slender tube.

DROSOPHILA MELANOGASTER

A FRUIT FLY

Drosophila melanogaster is a fruit fly which feeds on decaying fruits. There are several species in this Genus but *D. melanogaster* breeds rapidly, produces large numbers of offspring, has a short life-history and shows several contrasting characters. Furthermore the fly is easily cultured and crossing of different types presents little difficulty. It is therefore admirably suited to experimental work in genetics in spite of its small size.

LIFE-HISTORY

The **ova** are minute, white ovoid, and somewhat pointed with a pair of filaments towards one end. They are about 0·5 mm. in length. The **larvae** are transparent, segmented, limbless grubs about 4·5 mm. in length which undergo two ecdyses. The **pupæ** are broader and shorter than the larvæ, about 3 mm. in length and are brownish in colour (white at first). The whole metamorphosis from the hatching of the larvæ from the fertilised ova to the emergence of the imagines from the pupæ takes only about eight days to a fortnight, according to conditions.

EXTERNAL ANATOMY

Examine the **male** *and* **female imagines of the Wild Type** *under a lens on a white tile, under a dissecting microscope or under the low power of the microscope.*

The imago which emerges from the pupa is only about 2 mm. or so in length. The **head** has two red **compound eyes** and three **ocelli** which are dorsally situated. The **antennæ** are feathery and the mouth-parts adapted for sucking. The **thorax** consists of a large anterior segment known as the **dorsum** and a smaller posterior **scutellum.** The three pairs of **legs** terminate in claws. In the *male,* the first of the four podomeres of the tarsus of the first legs. the **matatarsus,** bears what is known as a **sex-comb,** a black hairy structure shaped like a comb. The **anterior wings** are large and membranous and extended beyond the posterior end of the insect. As in all *Diptera* the posterior wings are modified into **halteres.**

The **abdomen** differs in the two sexes. In both it is yellowish in colour with dark transverse bands but it is paler in the *female.* Owing to the general effect, the colour is sometimes referred to as "grey". The abdomen of the *female* is also broader than it is in the male. In fact, the *female* as a whole is slightly larger than the *male.* Furthermore the posterior ends differ in the two sexes. In the *male* where the external genitalia, a penis and two claspers, are situated it is somewhat rounded in side view whereas in the *female* it is more pointed, the external genitalia mostly consisting of a ventrally placed vaginal plate. There are six abdominal segments but, owing to the development of the external genitalia in the *male,* the last segment is so modified that there appear to be only five. These characteristics, with the sex-comb on the first leg of the *male*, enable the sexes to be distinguished and this distinction will be required in the experiments in Genetics (Part VI).

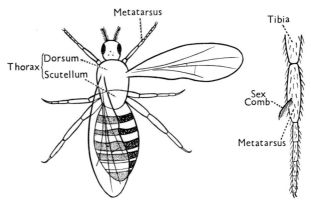

Fig. 107. **Drosophila. Imago. First Leg. Male.**
(*From Wallis—Practical Biology*)

There are several **mutations** in colour of body and eyes, size, shape and form of wings and so on which lend themselves ideally to interbreeding in experimental work in genetics as will be seen in Part VI.

It is interesting to note that **Drosophilia** has only four pairs of chromosomes, three pairs of autosomes and a pair of sex chromosomes, XY in the male and XX in the female. The chromosomes are unusually large in the salivary glands and the actual locations of particular genes have been identified. (Fig. 265, p. 362.)

CLASS CHILOPODA

Body composed of head and trunk. One pair of antennæ. All segments of trunk bear one pair of legs. *Centipedes and millipedes.*

SCOLOPENDRA

THE CENTIPEDE

Examine the external structure of a **Centipede.**

There is a distinct **head** followed by a large number of segments which constitute the trunk. The **head** has a pair of **antennæ** and two groups of **ocelli.** There are mandibles and two pairs of maxillæ, the second pair being fused to form a labium. The **trunk** is composed of some twenty segments, the first of which bears a pair of appendages ending in a claw. A duct from a poison gland leads to this and these maxillipeds act as **poison jaws.** The segments of the trunk are flattened dorso-ventrally and in the soft tissue between the hard terga the spiracles open. Each segment bears a pair of **jointed legs.**

CLASS ARACHNIDA

Body divided into prosoma, mesosoma and metasoma. Four pairs of legs. No wings. Breathe by lung-books or tracheæ. Mostly terrestrial. *Spiders, scorpions, king crabs, mites, water spiders.*

EPEIRA

THE GARDEN SPIDER

Examine a specimen of **Epeira.**

The body consists of a **cephalothorax** which is the **prosoma** and is undivided and an unsegmented **abdomen** composed of the fused mesosoma and metasoma known as the **opisthosoma.** A constriction separates the cephalothorax from the abdomen.

The **head** bears a pair of blade-like structures with toothed inner edges and bearing a pointed fang which has an aperature from the duct of a poison gland. These structures are known as **cheliceræ.** There is also a pair of **pedipalpi** composed of six joints and ending in claws. Between them lies the minute **mouth.** In front of the cephalothorax are eight **simple eyes** but no antennæ. The cephalothorax bears four pairs of **walking legs.** The abdomen in rounded and carries no appendages. At its end is the **anus** and beneath it are six **spinnerets.**

SCARCOPTES

THE ITCH MITE

S. scabii, the Itch Mite, is a parasite which is the cause of scabies in man. Other species cause mange in dogs and other mammals.

Examine a slide of **Scarcoptes scabii.**

The body is somewhat oval in shape and only about 0·3 to 0·45 mm in length in the female and the male is even slightly smaller. The

head, thorax and **abdomen** are unsegmented and fused together so that they are indistinguishable from one another.   The **legs** are short and six-jointed.   The mouth-parts consist of a pair of **pedipalpi** and a pair of **cheliceræ** as in the spider but there is an additional median structure, the **hypostome.**

FIG. 108. **Itch mite.**
(From Wallis—Human Biology)

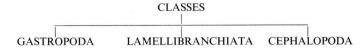

PHYLUM

MOLLUSCA

Unsegmented cœlomates with an exoskeleton, in most cases a shell. Body composed of head, foot and visceral mass enveloped in a respiratory mantle.

CLASSES

GASTROPODA   LAMELLIBRANCHIATA   CEPHALOPODA

Class Gastropoda

Head bears tentacles. Foot flattened. Single shell, often coiled. Univalves. *Snails, limpets, whelks, slugs.*

HELIX

THE SNAIL

Helix pomatia, the Roman snail, is much larger than *H. aspersa*, the garden snail, and is therefore more suitable for dissection.

OBSERVATIONS ON THE LIVING ANIMAL

Note the part of the body which protrudes from the shell and which can be retracted into it. There are two pairs of **tentacles.** *Touch them* and note their rapid withdrawal.

Place the animal on a piece of glass and examine from underneath.

Note the forwardly moving waves of muscular contraction of the foot and the movements of the mouth. Observe the trail of slime left behind.

EXTERNAL STRUCTURE

Before Removal of Shell

Examine a snail which has been freshly killed with the body extended. (See Appendix II.)

The Shell. Note the conical right-handed helix composed of about four and a half turns when completely formed. Each turn almost completely hides the previous one. The oldest part of the shell is the apex and this is known as the **nucleus of the shell.** Observe the coloration of the shell. Find the **shell mouth** and the

umbilicus, an opening on the under surface which leads into the hollow columella, which is the axis around which the shell is coiled.

Cut away one side of the shell and find the **columella.** *Examine the cut edge.*

It will be seen that the **shell** consists of three layers: (i) the **periostracum,** the thin, horny, external layer made of conchiolin; (ii) the **prismatic** (or **middle**) **layer,** thick and densely calcified; and (iii) the smooth, glistening, pearly **nacreous layer** on the inside.

The Head. This is rounded though not distinctly separated from the foot and bears the slit-like mouth, which has distinct lips, on the ventral side and two pairs of tentacles. One pair, the **posterior tentacles,** are long and each bears at its free end a complex **eye,** visible only when the tentacle is fully extended. The **anterior tentacles** are smaller.

The Foot. This is large, oval, flattened and muscular and forms the ventral surface of the body. It bears a **pedal (mucous) gland** on its ventral side just ventral to the mouth.

The Collar is a fleshy structure surrounding the mouth of the shell, somewhat thicker at the sides. It is the free edge of an internal tissue, the mantle.

The **Genital Pore** is situated on the right side, ventral to the optic tentacle. From it the **genital groove** runs backwards. The **pulmonary aperture** (or **pneumostome**) is on the right side of the collar. *Wash away any mucus and pass a seeker into this aperture.* It leads into the pulmonary or mantle cavity. The **excretory pore** is just inside the pneumostome and immediately behind it is the **anus.**

After Removal of Shell

Carefully remove the shell with strong scissors, taking care not to damage the internal organs with the points. Cut upwards at first from the shell mouth and then round the sides of the coils, removing a portion of the shell at a time. Cut through the muscular attachment to the columella and remove the latter.

The Visceral Mass is that part of the snail which remains permanently in the shell. Note that it is coiled, the turns corresponding with those of the shell. It is covered by a thin, transparent tissue through which the internal organs or viscera are visible.

Note the **mantle** (which secretes the shell), a thin layer of tissue forming the roof of the pulmonary (or mantle) cavity, the **kidney,** yellowish-white in colour about half-way round the basal turn of, the visceral mass and the **pericardium,** enclosing the **heart,** lying alongside the anterior side of the kidney. The **digestive gland,** dark reddish brown, reaches from the end of the kidney, beyond the pericardium to the apex of the mass and the **rectum** will be seen passing along the right edge of the basal turn of the mass.

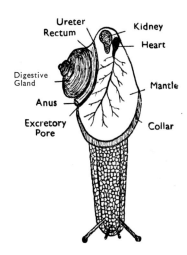

FIG. 109. **Helix. After Removal of Shell. Dorsal View.**

INTERNAL ANATOMY

GENERAL ANATOMY

Holding the animal in the hand, open the mantle cavity as follows: Cut round the collar where it joins the dorsal side of the body and then backwards from each end of the incision to its posterior extremity. Then deflect back the roof of the mantle cavity and pin the animal down in a dissecting dish, putting pins through the head and foot.

Note the network of large blood vessels on the inner surface of the roof and sides of the mantle, which form the large **pulmonary vein** running to the heart. The **pericardium** (already seen) is thin-walled and is situated at the posterior end of the mantle cavity.

Cut open the pericardium and expose the **heart.**

Note the thin-walled **auricle** towards the collar (and which the pulmonary vein enters) and the thick-walled **ventricle** from which the **aorta** leaves. This divides almost at once into **anterior** and **posterior aortæ.**

Find the wide **rectum** on the right side of the mantle cavity and which opens externally at the anus.

Note the **kidney** alongside the heart and the **ureter,** a duct running along the kidney and along the right side of the mantle cavity, dorsal to the rectum and opening to the exterior by the excretory pore.

THE ALIMENTARY SYSTEM

Cut away the collar to the left of the pulmonary aperture and separate the right side from the body wall. Cut along through the mantle to the posterior end of the mantle cavity below the rectum. Particular care must be exercised at this end. This will release the rectum. Deflect it and pin it down. Then cut through the floor of the mantle cavity in the mid-line as far as the head. Holding each half in turn with forceps, dissect away the tissue holding it underneath and pin it down. Now remove the skin covering the visceral mass and unravel the white reproductive organs and the alimentary canal and pin them out on separate sides, taking care not to damage them in the process.

It is now desirable to cover the entire animal with water.

Note the muscular **buccal mass** into which the mouth opens (this will be examined later), the **œsophagus,** a narrow thin-walled tube leading into the **crop** which is a large sac, widest in the middle and narrowest at its posterior end. Behind the crop is the **stomach** and continuous with this the **intestine** but this is largely embedded in the digestive gland. It leads into the **rectum,** already seen.

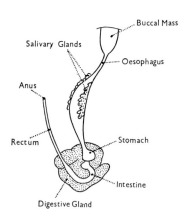

Fig. 110. **Helix. Alimentary Canal.**

Find the large, white, lobed **salivary glands** on the dorsal and lateral sides of the crop, their ducts leading into the anterior end of the œsophagus on each side of the œsophagus.

Examine the large reddish-brown **digestive gland** which surrounds
the stomach and intestine. It consists of two lobes, the **left lobe**
being the larger and being partially subdivided into three lobes. It
has three ducts which join and enter the left side of the stomach.
The **right lobe** which has a coiled appearance has one duct opening
into the right side of the stomach.

THE REPRODUCTIVE SYSTEM

(1) The reproductive system has already been unravelled and
pinned out. The animal is hermaphrodite and therefore possesses
both male and female sexual organs.

Find the yellowish **hermaphrodite gland** or **ovo-testis,** embedded
in the right lobe of the digestive gland. This gland produces both
spermatozoa and ova, hence its name. From it the short, whitish and

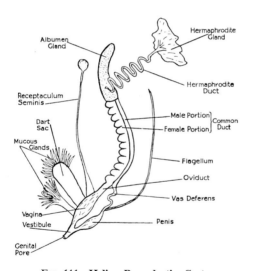

FIG. 111. **Helix. Reproductive System.**

sinuous **hermaphrodite duct** leads into the elongated, yellowish-white
albumen gland, situated between the left lobe of the digestive gland
and the stomach (now deflected). In the gland, near the point of
entry of the hermaphrodite duct arises the **common duct** which runs to
the head. One side of this duct, the **female portion,** is produced into
numerous folds, the other, the **male portion,** is smooth. These two

portions separate at the anterior end into a distinct **oviduct** and **vas deferens.** The former is a short thick-walled tube leading into the **vagina,** also thick-walled and opening into the **vestibule** or **genital atrium** and so to the **genital pore.** The vas deferens is a narrow and slightly convoluted tube running between the female organs and the buccal mass into the **penis,** which is a thick-walled tube also leading to the vestibule and thereby to **genital pore.** In addition to these, are the following accessory male and female organs—the **flagellum,** a long blindly-ending tube arising from the vas deferens near its junction with the penis (it is in this structure that the spermatozoa are formed into the rod-like spermatophore), and the **mucous glands,** consisting of two tufts of blindly-ending digitate structures which open into the vagina. Also opening into the vagina, is a short, wide thick-walled sac known as the **dart sac** and a long narrow tube, lying alongside the common duct, terminating in a small spherical sac; this is the **receptacluum seminis** or **spermatheca.**

(2) *Remove a portion of the* **hermaphrodite gland** *and tease it out on a slide. Mount in water and examine under high power.* Note the **ova** and **spermatozoa** (which develop in the wall and cavity respectively of the digitate follicles of which it is composed).

THE NERVOUS SYSTEM

Examine the **nervous system** *with a lens.*

Find the **nerve collar** which runs round the œsophagus just behind the buccal mass if the head is protruded, but in front of it if the head

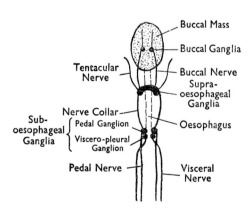

FIG. 112. **Helix. Nervous System. Diagrammatic.**

is retracted. *Now trace the nerve collar dorsalwards and find the* **cerebral** or **supra-œsophageal ganglia** on the dorsal side. From each ganglion arises a **tentacular nerve** running to the tentacles and mouth, a **buccal nerve** running along the side of the œsophagus to the **buccal ganglion** situated near the anterior end of the salivary duct. From each buccal ganglion small nerves arise which innervate the buccal mass and the two ganglia are joined by a transverse commissure.

Now return to the nerve collar and observe that it consists of two large **lateral commissures** on each side with a small **auditory nerve** between them. *If the œsophagus is sufficiently deflected*, a ganglionic mass, the **sub-œsophageal ganglia,** will be seen. Running through the centre is the **anterior aorta** dividing the mass into the anteriorly placed **pedal ganglia** (from which **pedal nerves** run to the foot) and the posteriorly placed **viscero-pleural ganglia** from which **visceral nerves** run to the viscera.

THE SENSE ORGANS

THE EYE

Examine one of the **eyes** *on the end of the optic tentacle.*

It is spherical and pigmented posteriorly. In the anterior surface is the lens. The nerve to the tentacle has already been seen.

THE AUDITORY ORGAN

Completely dissect away with needles the connective tissue in the pedal ganglia and the auditory organs.

There are small sacs embedded in these ganglia and contain otoliths. Small auditory nerves from the cerebral (or supra-œsophageal) ganglia supply these organs.

DISSECTION OF THE BUCCAL MASS

Cut across the œsophagus behind the buccal mass and through the junction of the buccal mass and the mouth. Remove the buccal mass and pin it down in the dish. Then, inserting small scissors into the posterior end, cut through the dorsal wall of the buccal mass in the mid-line. Deflect the two sides outwards and pin them down.

Find the brownish ribbon-like **radula** bearing a large number of teeth, on the floor of the cavity. The posterior end forms a tubular structure and the anterior end is situated on a tissue known as the **radula cushion.** The muscles attached enable the radula to be moved and to be protruded from the mouth. These structures constitute the **odontophore.**

CLASS LAMELLIBRANCHIATA

Head small. Foot wedge-shaped. Shell composed of two parts. Bilaterally symmetrical. =Bivalves. *Mussels, Oysters.*

ANODONTA CYGNEA

THE SWAN MUSSEL

Anodonta cygnea, the Swan Mussel, is found partly buried in the mud at the bottom of rivers, lakes and ponds.

OBSERVATIONS OF THE LIVING ANIMAL

Examine a fresh-water mussel in an aquarium.

Note that it lies, usually obliquely and with valves slightly opened, partly buried in the sand with one end, the posterior end, projecting upwards into the water. It moves slowly by a muscular foot. A stream of water enters the open end on one side, carrying food and oxygen.

Place a few grains of carmine near the partially opened valves.

Note the direction of the water current. It enters by what is known as the inhalent siphon on the ventral side and leaves by the exhalent siphon dorsal to it.

EXTERNAL STRUCTURE

THE SHELL

Note that the **anterior end** is rounded and the **posterior end** more pointed and that the animal is bilaterally symmetrical, the two halves or **valves** being joined along a straight **hinge line** by a **ligament.** This is dorsal. It will be easy, therefore, to determine which are the right and left sides. Each valve shows a series of lines which start from a small elevation near the anterior edge of the hinge line. This is the oldest part of the shell and is called the **umbo**: the lines are **lines of growth.**

Wedge open the shell with the handle of a scalpel.

Note the **mantle lobes** lining the valves and the **adductor muscles** which are for closing the shell.

Place the animal in a dish with either valve uppermost and carefully cut through the mantle lobes and muscles with a scalpel, cutting through the muscles close to the shell. Remove the valve and examine the inside of it.

On the inside of the valve note the following **muscle impressions:**—

The **anterior adductor,** large and oval in shape, near the anterior end of the shell, towards the dorsal side.

The **posterior adductor,** larger than the anterior, near the posterior end of the shell, and near the dorsal edge.

The **protractor of the foot,** small, behind the anterior adductor.

The **anterior retractor of the foot,** also small, beside the anterior adductor but nearer the hinge line.

The **posterior retractor of the foot,** again small, next to the posterior adductor.

Note also the streak between the anterior and posterior adductors, which marks the insertion of the mantle and which is known as the **pallial line.**

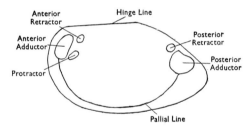

FIG. 113. **Anodonta. Interior of Right Valve. Muscle Impressions.**

Now break the shell near its rounded edge away from the muscle impressions, and examine the broken edge with a lens.

The shell will be seen to consist of three layers:—

(i) a thin, horny outer layer called the **periostracum.** It is made of a substance called conchiolin.

(ii) a **prismatic layer** (or **middle layer**), also made of conchiolin, but partly calcified.

(iii) and inner layer which lines the inner surface except at the edge, also partly calcified. This is the **nacreous layer.**

INTERNAL STRUCTURE

THE MANTLE CAVITY

Dissections should preferably be performed on a freshly killed animal, but preserved material may be used.

Find the **anterior** and **posterior adductor muscles** and the **anterior** and **posterior retractors of the foot.** *They can be identified from the corresponding muscle impressions on the inside of the valve.*

The two **mantle lobes** cover the side of the body and were joined to the valves along the pallial line. At the posterior end below the posterior adductor muscle, the mantle is thickened and pigmented at the edge. Here will be found a ventral slit, the **inhalent or ventral siphon** bounded by small **tentacles.** Immediately dorsal to it and much smaller, is the **exhalent or dorsal siphon,** also known as the **cloacal aperture.** This is not surrounded by tentacles. Between the

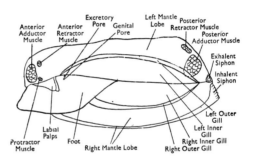

Fig. 114. **Anodonta. After Removal of Left Valve. Left Gills deflected upwards (Semi-Diagrammatic).**

two mantle lobes is the **mantle cavity.** This is divided into a large **branchial chamber** on the ventral side and smaller **supra-branchial passages** on the dorsal side. The two join posteriorly to form the **cloacal chamber.** *Deflect back the right mantle lobe.* In the branchial chamber lie the muscular **foot,** extending downwards, the **visceral mass** which form the upper and larger portion, and the **gills.**

At the sides of the visceral mass and reaching from the posterior adductor muscle to the posterior edge of the mantle cavity are two pairs of flaps known as the **inner gills** and **outer gills** though they do not properly function as such. The two pairs of folds at the sides of the anterior part of the foot between the anterior adductor muscle and the gills are two pairs of flaps called the **labial palps.** Between the anterior adductor muscle and the anterior edge of the foot is the **mouth.**

Remove the mantle lobe by cutting along the bottom of the labial palps, along the anterior end of the gills and along the bottom of the outer gill where it is attached at its base. Examine the gill.

Each **gill** consists of two **lamellæ,** each composed of a large number of vertical gill-filaments and horizontal bars which give it a rather trellis-like appearance.

Examine the **outer gill.**

The **outer lamella** is joined to the inner surface of the mantle and the **inner lamella** to the outer lamella of the inner gill.

Deflect the outer gill back to see this attachment.

Examine the **inner gill.**

The **outer lamella** is joined, as already seen, to the inner lamella of the outer gill.

Deflect the inner gill back and examine its inner lamella.

Note that the anterior part of the dorsal border of the **inner lamella** is joined to the side of the visceral mass, the edge of the middle part, reaching as far as the hinder edge of the foot, is free while the posterior part behind the foot is fused with the gill lamellæ of the opposite side, thus separating the mantle cavity here into inhalent and exhalent divisions.

Insert a seeker into the exhalent siphon and pass it above the bottom of the gills into the wide spaces situated there.

These spaces are the **supra-branchial passages.** One is situated above the outer gill, one above the inner gill at the side of the foot and there are three behind the foot. These three join and form a narrow passage which opens into the cloaca.

Open these passages in turn.

THE PERICARDIAL CAVITY

Remove the body from the valve in which it is lying. Place it dorsal side uppermost and pin back the remaining mantle lobe.

Find the **pericardium,** an elongated sac on the dorsal side above the bases of the gills.

Cut longitudinally through the pericardium and thus expose the **pericardial cavity.** *Make such further dissection as is necessary to expose the contents of this cavity.*

Note the **heart,** composed of a median thick-walled **ventricle** with thin-walled **right** and **left auricles** on either side. The ventricle surrounds the **rectum** which therefore passes through it. Running forwards from the ventricle on the dorsal side of the rectum is the **anterior aorta,** while at the other end of the ventricle will be found the **posterior aorta** below the rectum. The vena cava lies in the midline below the pericardium and receives blood (which is colourless) from the viscera and foot. It will not be seen in this dissection.

The following organs will also be found:—

The **kidneys** or **organs of Bojanus,** a pair of bent elongated structures lying side by side beneath the pericardium.

Keber's organs, one on each side at the anterior end of the pericardial cavity. They store excretory matter.

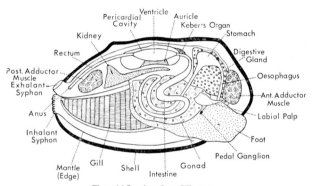

FIG. 115. **Anodon. Viscera.**

Examine one of the kidneys.

It is a thin-walled tube bent back upon itself with the loop posterior. The **ventral portion** (kidney proper) is glandular. *Make a longitudinal incision to see this.* The **dorsal portion** (sometimes called the **ureter**) leads to the **excretory pore** which is small and has distinct lips. Almost immediately below this is the **genital pore** which is much less conspicuous.

To find these apertures, first find the slit between the upper edge of the inner gill and the foot. Cut forwards from the anterior end of this slit and the two pores will be found.

At the anterior end of the pericardial cavity is a pair of curved slits leading from the cavity into the kidneys. These are the **renopericardial openings.** *Locate them and pass a seeker into one of them to see where it leads.*

THE NERVOUS SYSTEM

There are three chief pairs of ganglia, orange in colour, joined by connectives.

Place the animal on its side and deflect the mantle and gills. Find the mouth between the anterior adductor muscle and the anterior edge of the foot. Remove the skin immediately above the mouth.

Note the two **cerebral ganglia,** one on each side of the hinder edge of the mouth, below and in front of the protractor muscle and close to the surface.

Now place the animal ventral side upwards. Cut through the muscular part of the foot in the mid-line and dissect apart.

Find the two **pedal ganglia** lying close together in the anterior third of the foot where the muscular part of it joins the visceral part of the body.

Find also the **cerebro-pedal connectives** which join the cerebral and pedal ganglia.

Finally, dissect open the medium supra-branchial passage by inserting small scissors into the cloaca.

The **visceral ganglia** will be found on the ventral surface of the posterior adductor muscle.

Cut along the mid-line of the dorsal side of the supra-branchial chamber behind the foot.

You should then find the **cerebro-visceral connectives** which join the cerebral and visceral ganglia.

THE REPRODUCTIVE SYSTEM

The animal is diœcious but the reproductive organs are indistinguishable externally in the two sexes and are very simple. The **testes** (or **ovaries,** as the case may be) are large and are situated between the foot and the kidney. The two **gonoducts** converge into the **genital pore,** already seen below the excretory pore.

Find the gonad and tease a piece of it out on a microscopical slide. Mount in water and examine under the microscope. If it is a **testis,** the **spermatozoa** will be difficult to find, but in the case of the **ovary,** the **ova** will be easily identified.

THE ALIMENTARY SYSTEM

Remove some of the digestive gland and gonad until the stomach and intestine are exposed.

Note the mouth, devoid of jaws, and between the palps, the **œsophagus,** a short, straight tube posterior to the anterior adductor muscle, leading into a dilated sac, the **stomach,** which has folded walls. The coiled **intestine** passes backwards into the visceral mass and leads into the wider, straight **rectum,** previously seen in the pericardial cavity. through which it runs, surrounded by the ventricle. It opens into the cloaca by the slit-like **anus.** Surrounding the stomach is a many-lobed **digestive gland.** There is a longitudinal ridge on the ventral wall of the rectum projecting into the inside and known as the **typhosole.**

TRANSVERSE SECTIONS

Leave a specimen in ¼ per cent. chromic acid for a week or so to harden it, wedging the two valves slightly open to enable the hardening fluid to penetrate inside. When it is sufficiently hard, remove the body from the shell, place it on a dissecting board and cut a series of trans-

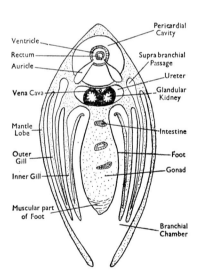

Fig. 116. **Anodonta. T.S. Through Hinder Part of Foot. (Semi-Diagrammatic.)**

verse sections, each about half a centimetre thick. Place the sections in order of their respective positions under water in disecting a dish and examine.

Draw sections as follows, noting in particular the structures mentioned.

(i) **Through the Anterior End of the Pericardium:**
   Ligament, pericardium, ventricle, rectum, kidneys, ureters, Keber's organs, excretory pore, supra-branchial passage, inner and outer gills, branchial chamber, gonad, mantle.

(ii) **Through the Centre of the Pericardium:**
   Ligament, pericardium, ventricle, rectum, auricles, kidneys, ureters, supra-branchial passage, inner and outer gills, branchial chamber, foot, intestine, typhlosole, mantle.

(iii) **Through the Posterior Adductor Muscle:**
   Posterior adductor muscle, rectum, visceral ganglia, supra-branchial passage, branchial chamber, inner and outer lamellæ of the inner and outer gills.

CLASS CEPHALOPODA

Head well developed and bears large eyes. Foot modified into prehensile tentacles. Siphon or funnel for expelling water from the mantle cavity. Shell sometimes present internally. Bilaterally symmetrical. *Cuttlefish*, *squids*, *octopus*.

LOLIGO

THE SQUID

Examine a specimen of **Loligo vulgaris.**

The animal has an internal shell and the body is bilaterally symmetrical. It has a definite **head** with well developed **eyes** and a **mouth** at the extreme end. The anterior part of the foot is in the region of the mouth and is modified into ten **arms** bearing **suckers.** The 4th pair is much longer than the others and these act as **tentacles,** one of them serving as an intromittent organ in the male. The posterior part of the foot is also modified. This serves as the **funnel** or siphon through which water is forced to propel the animal through the water. The rest of the body is the **trunk,** and is rather diamond-shaped, tapering to a point posteriorly. This is due to the **fins** one on each side which serve as swimming organs. The trunk is covered by a thick **mantle** underneath which lies the mantle cavity.

PHYLUM

ECHINODERMATA

Marine triploblastic coelomates. Usually radially symmetrical but larvae bilaterally symmetrical. Calcareous skeleton. Coelom divided into separate sections with varying functions. No blood system. *Starfish, sea-urchins, sea cucumbers, brittle stars, sea lilies.*

CLASSES

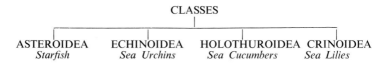

| ASTEROIDEA | ECHINOIDEA | HOLOTHUROIDEA | CRINOIDEA |
| *Starfish* | *Sea Urchins* | *Sea Cucumbers* | *Sea Lilies* |

CLASS STELLEROIDEA (ASTEROIDEA)

Body flattened and star-shaped. Five rayed. Arms bear tube feet on lower surface. Well developed flexible skeleton. *Starfish.*

ASTERIAS RUBENS

THE STARFISH

Examine a specimen of a **starfish.**

This animal (which is not, of course, a fish) is shaped like a star with five arms radiating out from a central **disc.** The upper or **aboral side** is darker than the lower or **oral side** and varies from orange to a purplish colour. In the centre of the disc on the **oral side** is the **mouth** bearing a membranous lip called the **peristome.** Correspondingly on the **aboral side** is the **anus.** Each arm is referred to as a **radius** and between each pair is the **interradius.**

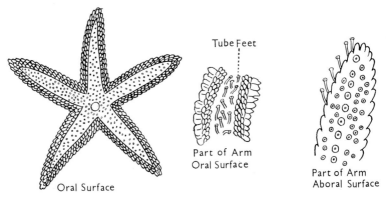

FIG. 117. **Asterias rubens. Starfish.**
(*From Wallis—Practical Biology*)

Examine one of the **arms.**

Note that on the oral side is a deep **ambulacral groove** which can be opened and closed by the animal. The grooves of all the arms meet around the mouth. These grooves contain what are known as **tube-feet** which end in suckers while at the end is a **sensory tentacle.** The body wall contains a large number of rod-shaped **ossicles** from which blunt **spines** arise and between the ossicles are **dermal gills.** Each spine is surrounded by a cushion and on and between these cushions are minute **pincers** (or **pedicellariæ**) which are organs of defence and are of two sizes and forms. On the other, aboral, side are flattened rounded ossicles which bear grooves and in these grooves are **pores** through which water is drawn into the animal.

Examine a **Transverse Section of an arm of Asterias.**

The wall of the arm is thick and bears a number of short **spines** and will be seen to be convex on the upper **aboral side.** This encloses a part of the **cœlom.** The walls of the **ambulacral groove** on the **oral side** contain **ambulacral ossicles** which meet at the top of the groove.

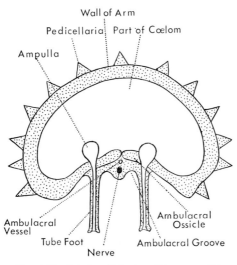

FIG. 118. **Asterias T.S. Arm (diagrammatic)**

On the inside of each row of ambulacral ossicles is a sac or **ampulla** continuous with which is the **tube-foot** on that side. If the tube-foot is extended this rounded sac becomes contracted and lobulated. A small **ambulacral vessel** is seen in cross section in the centre. This is connected to the base of each tube foot but these connections may not be visible in the section. These all form part of the water vascular system. Below the vessel is a **nerve.**

CLASS ECHINOIDEA

Body globular and radially symmetrical but sometimes flattened and bilaterally symmetrical. Covered with moveable spines. Tube feet among spines. Mouth has strong jaws. *Sea urchins.*

ECHINUS

THE SEA URCHIN

Examine a specimen of **Echinus.**

The body is globular though rather compressed in one plane, thus producing two poles. The **oral pole** bears the rounded **mouth** which has powerful jaws provided with **teeth** which may be seen projecting through it. It is surrounded by a **peristomal membrane.** The **anal pole** at the other end bears the much smaller **anus** and this is surrounded by the **periproct.** Apart from these two areas the entire surface is covered with movable pointed **spines** with a number of minute **spinules** between their bases. Amongst the spines here and there are tiny **pedicellariæ** and five double rows of extensible **tube-feet** similar to those in **Asterias.**

PHYLUM CHORDATA

Bilaterally symmetrical animals possessing a dorsal **notochord** in the embryo which persists only in the simpler types, a tubular dorsal nerve cord, a ventral heart and closed blood system, branchial clefts penetrating the wall of the pharynx in the embryo but which persist throughout life in fish, post anal tail usually present.

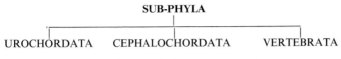

SUB-PHYLUM UROCHORDATA

Acrania (devoid of skull or true brain). Notochord and nerve cord present only in tail of larva. Adults sedentary. Ciliary feeders. Larvæ motile. *Tunicates*. (*Sea Squirts*).

ASCIDIA

THE SEA SQUIRT

These marine animals occur in large numbers attached to rocks.

EXTERNAL ANATOMY

Examine a preserved fixed specimen of **Ascidia.**

The body is broad and roughly cylindrical with a still broader base for attachment to rocks. The outer coat is translucent and is composed of a substance called *tunicin* which seems to be identical with cellulose. At the upper end is a large **inhalant aperture** and lower down on the side is a second opening, the **exhalant** (or **atrial**) **aperture**. The inhalant aperture leads by a wide stomodæum to a large perforated pharynx which communicates with a cavity, the atrium. Water is drawn in through the inhalant aperture and passed out through the exhalant aperture. The outer coat is known as the **tunic**: hence the term *Tunicates* for these animals.

SUB-PHYLUM CEPHALOCHORDATA

Acrania. Small superficially fish-like marine animals. Notochord along entire length of body. Free-swimming. Ciliary feeders. Definite metamerism. *The Lancelets. Amphioxus.*

AMPHIOXUS (BRANCHIOSTOMA)

THE LANCELET

Amphioxus (**Amphioxus lanceolatum**) is a small superficially fish-like chordate which lives in shallow sea-water.

EXTERNAL ANATOMY

Place a preserved specimen of amphioxus on its side in water in a small dissecting dish and examine with a lens.

Note the elongated shape of the body, pointed at both ends. On the dorsal side, extending along its entire length is the **dorsal fin** which, continuing round the **tail,** joins the **ventral fin** which extends for a short distance only. Anterior to this on each side is a downward extension of the body wall called the **metapleural fold.** The two metapleural folds are joined on the ventral side by the **epipleur** and these structures enclose a cavity, the atrium. Just anterior to the ventral fin is its aperture, the **atriopore.** At the anterior end is a projection of the dorsal and lateral body wall called the **oral hood** from the edge of which project a large number of tentacle-like **oral cirri.** The opening of the oral hood is the **mouth.** A short distance from the posterior end of the body on the left side is the **anus.** Note the metamerically segmented >-shaped muscle segments or **myotomes** of the body wall. In segments 25-51 are the metamerically arranged **gonads,** small block-like structures in the atrium: it is not possible to distinguish between testes and ovaries without microscopical examination.

If the specimen has been suitably treated, the **nerve cord** will be seen on the dorsal side and immediately below it, the **notochord.**

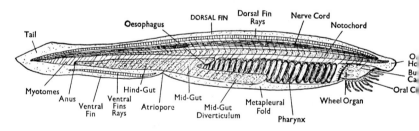

FIG. 119. **Amphioxus. Later View. Transparency.**

INTERNAL ANATOMY

Specimens for dissection must have received special treatment (see Appendix II).

GENERAL STRUCTURE

Place the animal on its left side, still under water, so that you can dissect it from the right side. Fix a very small pin preferably through the posterior end of the body only, behind the anus and towards the ventral side but another may be fixed through the extreme anterior end if necessary. Carefully remove the external cuticle by means of needles and a camel hair brush. Then remove the myotomes along the entire right side in a similar manner, dissecting very gently. The gonads will probably be removed at the same time if great care is not taken.

At the anterior end note the **oral hood** and **oral cirri** and beneath the former, the structure known at the **wheel organ** composed of a complicated series of ridges and bearing cilia. The external opening of the oral hood is the **mouth.** Behind the wheel organ is a partition perforated by the **enterostome.** This is the **velum,** the edges of which bear **velar tentacles** projecting backwards into the large **pharynx.** The walls of the pharynx bears numerous **gill slits** between which are **gill bars** and which are crossed by transverse structures, **synapticula.** These will be seen shortly under the microscope. At the posterior end of the pharynx is the short **œsophagus,** continuous with which is the wide **mid-gut** and this is followed by the **ileo-colonic ring** and the narrow **hind-gut** terminating at the **anus.** A forwardly-projecting, blindly-ending sac, the **mid-gut diverticulum** arises from the junction of the œsophagus and mid-gut and lies alongside the pharynx.

The **atrium** is a cavity surrounding the pharynx and anterior part of the gut on their lateral and ventral sides and into which the gill slits open. It extends on the right side as far back as the anus, *i.e.,* beyond its aperture, the **atriopore.** It will be seen later in transverse sections.

On the dorsal side, note the **nerve cord** and ventral to this the **notochord** running the entire length of the body. The former terminates posteriorly before the notochord.

THE ANTERIOR END

Examine a prepared slide of the anterior end showing the structure of the **pharynx** *and the* **nerve cord.**

Observe the **oral hood, oral cirri, vellum** and **velar tentacles** and the structure of the **pharynx.**

Note the alternating **primary bars,** forked at their ventral ends and the unforked **secondary** (or **tongue**) **bars** between which are the **gill slits.** The **synapticulæ** traverse the gill slits and join the primary and secondary bars together giving a trellis-like appearance.

Note that the **nerve cord** swells slightly at the anterior end to form the **cerebral vesicle** and observe the rod-like **notochord.**

THE NERVE CORD

Very carefully remove the entire nerve cord and notochord together, using needles. Then carefully separate the nerve cord and notochord. Mount the former and examine with a microscope. Also examine a prepared slide.

Again note the **cerebral vesicle** and observe the pigmented spots along the length of the cord; these are **eye-spots** and are sensitive to light.

Entering the anterior end of the cerebral vesicle is a pair of sensory nerves, the **1st nerves,** from the snout. The **2nd nerves,** also from the snout and also sensory, join the nerve cord just dorsal to the vesicle. The succeeding nerves are called **spinal nerves** and they arise alternatively from the dorsal and ventral sides. The **dorsal roots** (sensory) are slightly anterior to the ventral **roots** (motor). These do *not* unite afterwards.

STRUCTURE AS SEEN IN TRANSVERSE SECTIONS

Much of the anatomy is more easily studied and understood by examination of transverse sections.

PHARYNGEAL REGION

Examine a **T.S. through the pharynx** (preferably **in the posterior region).**

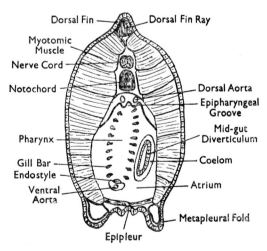

Fɪɢ. 120. **Amphioxus. T.S. Posterior Region of Pharynx.**

Note the **dorsal fin** with **fin ray** inside, **myotomic muscles,** the **nerve cord,** and **notochord.** The **pharynx** with **gill slits** and **gill bars** in the centre reaches from below the notochord to the ventral side and is enclosed by the **atrium.** On the ventral side are the **metapleural folds** with the **epipleur** joining them. The **mid-gut diverticulum** is in the atrium alongside the pharynx. Paired **dorsal aortæ** will be seen just dorsal to the pharynx. The **endostyle** is a cavity at the ventral end of the pharynx on the inner side and the **ventral aorta** is just below it. The **epipharyngeal groove** is at the dorsal end of the pharynx. The **gonads** will be seen in the atrium at the sides. Surrounding the mid-gut diverticulum and in the primary gill bars is the **cœlom.**

INTESTINAL REGION

(a) Anterior to the Atriopore

Examine a **T.S. through the intestinal region anterior to the atriopore.**

Note **dorsal fin, nerve cord, notochord, myotomic muscles** and **atrium** as in the previous section, also the **mid-gut** and the **cœlom**

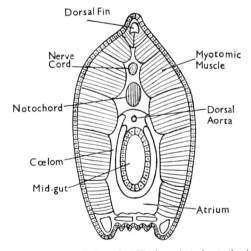

Fig. 121. Amphioxus. T.S. Intestinal Region. Anterior to the Atriopore.

surrounding the mid-gut diverticulum and the mid-gut itself. The single **dorsal aorta** and the **ventral aorta** are on the dorsal and ventral sides of the gut respectively.

(b) Posterior to the Atriopore

Examine a T.S. through the intestinal region, posterior to the atriopore.

Note the **dorsal fin, nerve cord, notochord, myotomic muscles, cœlom, atrium, dorsal aorta** and **ventral aorta** as before; also the **hind-gut** and **ventral fin** containing **fin rays.**

(c) Through the anus

Examine a T.S. through the anus,

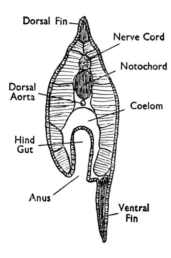

FIG. 122. Amphioxus. T.S. Through Anus.

Note the **hind-gut,** opening at the **anus,** the absence of the atrium, and the **cœlom** on the dorsal and lateral sides of the hind-gut. Other structures as before.

EXCRETORY SYSTEM

Examine a slide of a **nephridium.**

These are situated above the gill slits in the pharyngeal region.

Each nephridium consists of a curved tube with a short upper arm bearing a **pore** (which leads into the atrium) and a long blind lower arm. On the outer side of the structure is a number of branches bearing numerous fine tubules, each terminating in a rounded **solenocyte.**

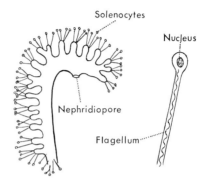

FIG. 123. **Amphioxus. Nephridium.**

ENTIRE NEPHRIDIUM   SOLENOCYTE

SUB-PHYLUM VERTEBRATA

Craniata. (Possess skull and brain). Notochord wholly or partially replaced in adult by vertebral column. Head, brain and sense organs well developed. Brain encased in cranium.

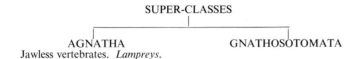

SUPER-CLASSES

AGNATHA   GNATHOSOTOMATA
Jawless vertebrates. *Lampreys.*

SUPER CLASS GNATHOSTOMATA

Cephalisation well developed.   Mouth bounded by jaws.   *Fish, Amphibia, reptiles, birds and mammals.*

CLASS CHONDRICHTHYES*

Sub-class Elasmobranchii. Fish with a skeleton entirely of cartilage. Exposed gill slits. No operculum. No swim bladder. Poikilothermic. Exoskeleton of dermal denticles (placoid scales). Upper and lower jaws each a single cartilage. Mouth ventral. Heterocercal tail. Fertilisation internal. *Dogfish, sharks, skates, rays.*

*This class was formerly known as **Pisces.**

SCYLIORHINUS

THE DOGFISH

The commonest species of dogfish which frequents the British coast is **Scyliorhinus canicula,** the lesser spotted dogfish or Rough Hound. The greater spotted dogfish or Nurse Hound is S. **stellaris** and the spiny dogfish is **Squalus acanthias.** There are certain minor anatomical differences between Scyliorhinus and Squalus, and while the former is **oviparous** (*i. e.,* lays eggs), the latter is **ovo-viviparous** (*i. e.,* the young are hatched before leaving the body of the female).
The following description applies to **Scyliorhinus.**

OBSERVATIONS ON LIVING FISH

If possible, observe living fish in an aquarium.

Note how they swim, and pay particular attention to the way in which the fins and tail are used. Examine the method of breathing, how they eat their food and how they rest.

EXTERNAL ANATOMY

(1) *Place a complete, freshly killed or preserved* **dogfish** *on a dissecting board and examine its external structure**

Note the shape and colour of the body, which is divided into **head, trunk** and **tail.** Note the **lateral line,** marking the position of a sense organ running along each side of the head and body.

The Head

The head bears on the ventral side a large crescentic **mouth,** in front of which are the two **nostrils,** circular apertures connected to the mouth by the **oro-nasal grooves.** The **eyes,** with immovable upper and movable lower lids, are at the sides of the head, and immediately behind each is a small round aperture, the **spiracle,** which is a modified gill-cleft. The five slits on each side behind the eyes are the **gill-slits** (or **clefts**). The last of these marks the posterior end of the head. *Pass a seeker through the spiracle and gill-slits into the pharynx.*

On the surface of the head are rows of minute apertures of the **sensory canals** or **ampullary canals,** the latter being abundant on the snout. These apertures contain a gelatinous substance. *Squeeze the head to see this substance exude.*

---

* Specimens preserved for dissection generally have the greater part of the tail removed and the ventral body wall is sometimes cut open. Fresh specimens are better and are sometimes obtainable.

The Trunk

The trunk bears paired and unpaired fins. The anterior pair of paired fins are the **pectoral fins.** They are roughly triangular in shape and project horizontally from the ventral side. The posterior pair of paired fins are the **pelvic fins,** somewhat similar to the pectoral fins, though smaller. In the **male,** a pair of so-called **claspers,** muscular rods strengthened by cartilage internally, will be seen between the pelvic fins. Their function is that of an intromittent organ in copulation. The pelvic fins mark the posterior end of the trunk. Between their bases is the **cloacal aperture,** just posterior to which on either side are the **cloacal pouches** bearing **abdominal pores.**

The Tail

The unpaired fins are the **anterior** and **posterior dorsal fins** which project vertically from the animal's back, the **ventral fin** projecting vertically from the ventral side, and the **caudal fin** which surrounds the end of the **heterocercal tail** (a tail with its axis directed upwards and having a large part of the caudal fin on its ventral side).
Make drawings of (i) **the lateral view** of the entire fish, (ii) the **ventral view of the head,** (iii) the **ventral view of the male and female pelvic fins,** noting the structures above.

(2) *Examine the surface of the body with a lens and rub the finger along the skin both backwards and forwards.*
Embedded in the skin are small tooth-like structures, **dermal denticles** (or **placoid scales**).
Open the mouth as far as possible and note the three or four rows of teeth on the jaws, similar in structure to the dermal denticles.
(3) *Examine a prepared slide of a* **vertical section of a dermal denticle** *under the low power* (see p. 317).
It is composed of a **basal plate** which is embedded in the skin and which bears a backwardly projecting **spine.** The basal plate is made of **cement** and the spine of **dentine** which is covered externally by **enamel.** The **pulp cavity** in the basal plate passes into the spine where it branches considerably in the **dentine.**

THE SKELETON

The skeleton is entirely cartilaginous and consists of (i) an **axial skeleton** (the skull and vertebral column) to which should be added a **visceral skeleton** (the skeletal parts of the jaws, the hyoidean arch and the branchial arches), and (ii) an **appendicular skeleton** (the girdles and the skeletal part of the fins).

Examine a prepared disarticulated skeleton.

THE AXIAL SKELETON

The Skull

This consists of the cranium, to which is attached the visceral skeleton which supports the jaws and pharynx, and the sense capsules.

Examine a skull from which the visceral arch skeleton has been removed.

Dorsal View

The somewhat oblong **cranium** or **chondrocranium** (so-called because it is cartilaginous) is bounded in front by the two large oval **olfactory capsules** with a large dorsal hole, the **anterior fontanelle,** in the centre between and behind them. The olfactory capsules bear three **rostral cartilages** projecting from the anterior end and constituting the **rostrum** which supports the snout. The sides of the cranium project as the **supra-orbital ridges** and the **auditory capsules** are fused

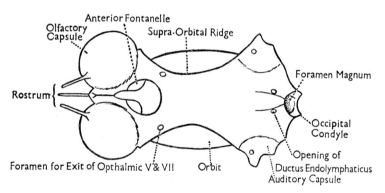

FIG. 124. **Scyliorhinus. Chondrocranium. Dorsal View.**

laterally to the posterior ends. Note the ridges which mark the positions of the anterior and posterior semi-circular canals of the membranous labyrinth ("ear"). The large hole in the hinder end of the cranium is the **foramen magnum** through which the spinal cord passes, and on either side of the foramen magnum are the rounded **occipital condyles** which articulate with the first vertebra.

The two **foramina for the exit of the ophthalmic** branches of the **Vth and VIIth nerves** will be seen just posterior and slightly lateral to the anterior fontanelle and the **opening of the ductus endolymphaticus** will be seen on each side where the auditory capsule fuses with the cranium at its inner end.

Ventral View

The base of the skull is a wide flat cartilage. Running across the posterior end will be seen a pair of **grooves for the carotid arteries.** These end in two **foramina for the carotid arteries.** Note also the **olfactory capsules** and the **olfactory openings,** considerably covered by the **nasal cartilages** and the rostrum. The **upper jaw** and **lower jaw** and the **ligaments** connecting the two halves of each will be seen. These will be described below.

Lateral View

Note the ventral rostral cartilage, which with the two dorsal rostral cartilages constitute the **rostrum,** the **olfactory capsule,** the **auditory capsule,** the **occipital condyles** and the **orbit,** bounded above by the **supra-orbital ridge** and below by the **sub-orbital ridge.** In the orbit will be seen the following **foramina:—**

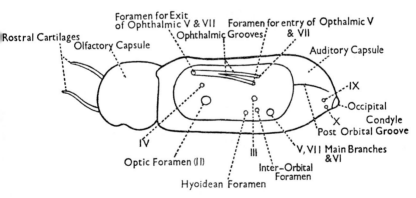

FIG. 125. **Scyliorhinus. Chondrocranium. Lateral View.**

The most posterior dorsal foramen is the **foramen for the entry of the ophthalmic branch of the VIIth nerve.** Immediately below and slightly anterior to this is the **foramen for the entry of the ophthalmic branch of the Vth.** From both these foramina, the **ophthalmic grooves** run forwards to a foramen at the anterior dorsal end of the orbit which is the common **foramen for the exit of the ophthalmic branches of the Vth and VIIth nerves** (already seen in the dorsal view). Ventral and slightly anterior to the foramen by which the ophthalmic branch of the Vth enters is the **foramen for the IIIrd nerve,** and dorsal and anterior to this, under the ophthalmic groove, is the **foramen for the IVth nerve.** Beneath this is a large foramen for the **IInd nerve,** the **optic foramen,** ventral and posterior to which is a **hyoidean foramen** for the "hyoidean" artery better known as the **efferent pseudo-branchial.** Posterior to this is the **inter-orbital foramen,** the aperture of the inter-orbital canal, behind which is a large **foramen for the main trunk of the Vth and VIIth nerves** and the **VIth nerve.** Note the **post-orbital groove** which runs back from the orbit, at the posterior end of which is the **foramen for the IXth nerve.** The **foramen for the Xth nerve** is just at the side of the foramen magnum and may not be seen in this view.

THE VISCERAL ARCH SKELETON

There are seven hoops of cartilage. The first or **Mandibular Arch** is modified to support the upper and lower jaws. The **upper jaw** is made up of the **palato-quadrate cartilage** which meets its fellow of the opposite side anteriorly. These cartilages are attached to the skull by the **ethmo-palatine ligaments** in front of the orbit and by the **pre-spiracular ligaments** in the region of the auditory capsule. The **lower jaw** is made up of a pair of **Meckel's cartilages.** Both jaws bear **labial cartilages.** This type of jaw is said to be *hyostylic.*

The second or **Hyoid Arch** is partly concealed by the jaws and consists on each side of an upper cartilage, the **hyomandibular cartilage,** and a lower one, the **cerato-hyal cartilage,** which articulates with a ventral plate of cartilage supporting the tongue, the **basi-hyal.**

The five remaining arches are the **Branchial Arches** which support the gills and are similar to each other except that the first and fifth have certain modifications.

Each arch contains a rod-like **pharyngo-branchial cartilage,** the most dorsal, an **epi-branchial cartilage,** short and plate-like, a rod-like **cerato-branchial cartilage** running inwards and forwards and a **hypo-branchial cartilage,** a short rod, running backwards and inwards. The **basi-branchial cartilage,** a long flat plate, rather pointed posteriorly, runs along the ventral side and is common to all.

In the 1st branchial arch, the hypo-branchial, is directed forwards

and it is not joined to the basi-branchial but to the basi-hyal. The
5th branchial arch has no hypo-branchial; consequently its cerato-
branchial is much wider than in the other four and directly articulates

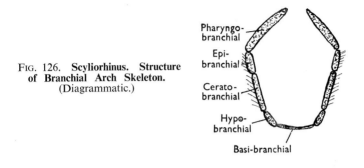

FIG. 126. **Scyliorhinus. Structure
of Branchial Arch Skeleton.**
(Diagrammatic.)

Pharyngo-
branchial

Epi-
branchial

Cerato-
branchial

Hypo-
branchial

Basi-branchial

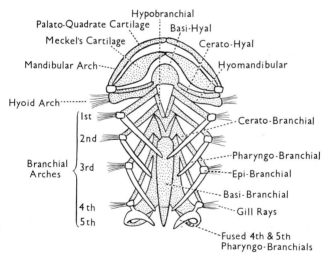

Hypobranchial
Palato-Quadrate Cartilage   Basi-Hyal
Meckel's Cartilage   Cerato-Hyal
Mandibular Arch   Hyomandibular
Hyoid Arch
Ist   Cerato-Branchial
2nd
Branchial   Pharyngo-Branchial
Arches   3rd   Epi-Branchial
                                      Basi-Branchial
4th   Gill Rays
5th
                                      Fused 4th & 5th
                                      Pharyngo-Branchials

FIG. 127. **Scyliorhinus. Visceral Arches. Dorsal View.**

with the basi-branchial. The 4th and 5th pharyngo-branchials are
fused. On the 2nd, 3rd and 4th branchial arches are the **extra-
branchials,** ventral and external to the arches, often missing from
prepared skeletons. The 5th cerato-branchials bear notches for the
Cuverian ducts.

The **gill rays (branchial rays)** are borne on the posterior edges of the epi-branchials and cerato-branchials of the first four branchial arches and on the hyomandibular and cerato-hyals of the hyoid arch.

Draw a dorsal view of the Visceral Arch skeleton.

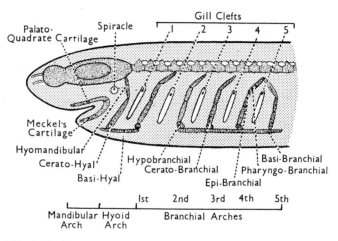

FIG. 128. **Scyliorhinus. Relation of Visceral Arches to Gill Clefts. Diagrammatic.**

The Vertebral Column

This consists of about 130 similar vertebræ.

(1) *Examine a few vertebræ from the trunk region in lateral view.*

Note the body or **centrum,** bearing the **vertebral neural plates** on the dorsal side with the **intervertebral neural plates** between them. Between the vertebral neural plates and the intervertebral neural plates are the rounded **neural spines.** The notches projecting from the sides of the centra in this region are called **transverse processes*:** they have slender **ribs*** attached to them.

(2) *Examine the anterior end of a vertebra from the trunk region.*

Note that the **centrum** is concave at both ends and therefore said to be *amphicælous* and that it contains **notochordal tissue** in the centre. The centrum is surmounted on the dorsal side by the **neural arch** composed of the two **neural processes** in contact with the centrum,

* These are not homologous with those of higher vertebrates.

above which are the **vertebral neural plates** with the **neural spine** between them on the dorsal side. The two **transverse processes** are produced laterally from the lower region of the centrum and bear very short **ribs.**

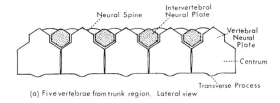

(a) Five vertebrae from trunk region. Lateral view

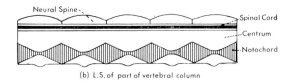

(b) L.S. of part of vertebral column

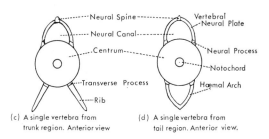

(c) A single vertebra from trunk region. Anterior view

(d) A single vertebra from tail region. Anterior view.

FIG. 129. **Scyliorhinus. Vertebral Column.**

(3) *Examine the anterior end of a vertebra from the tail region.*

Note the structures in (2) and that the transverse processes, or rather **hæmal processes,** join to form the **hæmal arch,** in which the caudal artery and vein are situated.

(4) *Cut a portion of the vertebral column vertically in the mid-longitudinal (sagittal) plane.*

Note the structures in (2) above except the transverse processes and ribs, observing the widening of the **notochordal tissue** between each pair of centra, and noting, inside the neural arch, the **spinal cord.**

THE APPENDICULAR SKELETON

The Pectoral Girdle

This is an incomplete hoop of cartilage, complete on the ventral side only, which is known as the **coracoid portion;** it has a concave **pericardial depression** on the dorsal surface for the ventricle of the heart. The curved pointed processes which arch round to the dorsal side constitute the **scapular portion.** Where the two portions join are three smooth **glenoid facets** for the articulation of the pectoral fin.

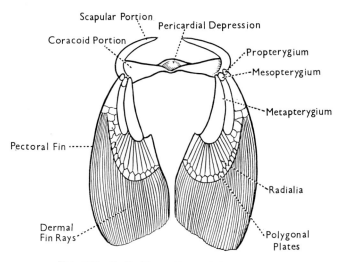

FIG. 130. Scyliorhinus. Pectoral Girdle and Fin.

The Pectoral Fins

Each articulates with the glenoid facet at the bend of the hoop, and is composed of three basal cartilages, the small anterior **propterygium,** the longer middle **mesopterygium,** and the still longer posterior **metapterygium.** Radiating out from these three cartilages, are the **radialia** (or **cartilaginous fin rays**) which bear at their distal ends a few rows of very small cartilages, **polygonal plates.** The rest of the fin is supported by the elastic **dermal fin rays (dermotrichia).**

The Pelvic Girdle

This consists of a more or less straight bar of cartilage, the middle part of which is called the **ischio-pubic portion,** the outer parts being known as the **iliac portions.** The **acetabular facets** for the attachment of the pelvic fins are on the iliac portions.

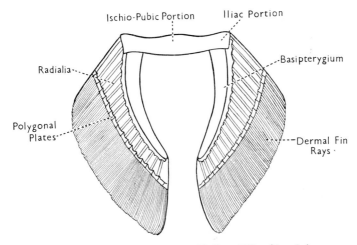

FIG. 131. **Scyliorhinus. Pelvic Girdle and Fin. (Female.)**

The Pelvic Fins

These articulate with the acetabular facets near each end of the bar, and each is supported by single long curved **basi-pterygium,** bearing on its outer edge a number of long **radialia** and then a few tiny **polygonal plates** as in the pectoral fin, the rest of the fin being supported by **dermal fin rays (dermotrichia).** In the **male,** each **clasper** is composed of a rod of cartilage, grooved on its inner border and attached to the posterior end of the basi-pterygium.

The Dorsal, Ventral and Caudal Fins

Each consists of a row of rod-like **basal cartilages** and a row of **radialia** bearing **polygonal plates** at their distal ends. The cartilaginous parts are sometimes absent in the ventral and caudal fins.

THE ARTICULATED SKELETON

Examine the articulated skeleton to see the relationship of the various parts.

INTERNAL ANATOMY
THE MUSCLES

Cut out a large rectangle of skin from the side of the fish in the tail region.

The muscles are arranged in zig-zag shaped segments, **myotomes,** separated by septa of connective tissue called **myocommata.** This shows metameric segmentation.

THE ALIMENTARY SYSTEM

(1) *Place the dogfish, ventral side upwards, on a dissecting board and fix awls through the basal parts of the pectoral and pelvic fins. Make a median, but not too deep, incision from the level of the pectoral fins to the cloaca cutting through the skin and body-wall, and pin back the flaps on either side with awls. Transverse cuts may be made at both ends to facilitate the deflection of the body wall.* (Preserved specimens if not injected are often partially opened to admit formalin.)

The **Abdominal** or **Peritoneal Cavity,** which is part of the **Cœlom,** contains the following digestive viscera.

It will be necessary to deflect the lobes of the liver outwards and the stomach to the animal's left in order to bring all the structures into view.

The lower end of the **œsophagus** leads from the pharynx to the **stomach,** a large U-shaped sac lying between the **left** and **right lobes** of the **liver.** The left lobe is somewhat divided at the proximal end forming what is sometimes called the **median lobe.** The liver is attached to the anterior abdominal wall by a **suspensory** or **falciform ligament.** The large side of the stomach is the **cardiac portion,** the narrow side the **pyloric portion** and this leads by the **pylorus** into the **intestine.*** Find the **gall-bladder,** embedded in the median lobe of the liver and the **bile duct** between the liver and the stomach, which leads from the gall-bladder to the dorsal side of the intestine (this part sometimes being called the duodenum). The **pancreas** is a whitish body lying between the pyloric portion of the stomach and the intestine, which it enters by the **pancreatic duct.** It will be seen that the intestine gradually gets wider. Continuous with it is the narrow **rectum** into which the small dorsally placed **rectal gland** opens. The rectum opens into the **cloaca.** The alimentary canal is suspended by **mesentery,** that supporting the stomach being called the **mesogaster.**

---

* The terms duodenum and ileum into which the intestine are sometimes differentiated are not aptly applicable in the dogfish as there is insufficient demarcation.

Note also the **spleen,** a large reddish body attached to the posterior end of the stomach with a branch running up alongside the pyloric portion, though it is not part of the alimentary system.

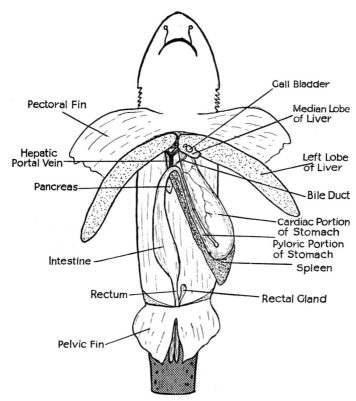

Fig. 132. **Scyliorhinus. Alimentary Canal. Liver Displaced Outwards and Stomach to Animal's Left.**

(2) *Cut lengthwise along the sides of the intestine, remove the ventral wall carefully and wash out the contents.*

Examine the so-called **spiral valve** in the intestine.

(3) *Now examine the* **blood vessels associated with the alimentary canal.**

Deflect the alimentary canal as far as possible outwards to the animal's left and turn it over in order to expose the dorsal side without damaging any of the supporting tissues. Find the following arteries which take oxygenated blood to the organs.

The short **cœliac artery** arises from the **dorsal aorta** near the anterior end of the stomach and divides into (i) the **hepatic artery** going straight to the liver, (ii) the **gastric artery** to the stomach, and (iii) the **intestino-pyloric artery** to the anterior part of the intestine and the pancreas.

The **anterior mesenteric artery** arises about 2·5 cm or so behind the cœliac and goes to the rest of the intestine and the rectum. It will be seen on the posterior side of the intestine.

The **lieno-gastric artery** arises immediately behind the anterior mesenteric and the two arteries cross: it runs to the stomach and spleen.

The **posterior mesenteric artery** arises about 4 cm. behind the lieno-gastric and goes to the rectal gland.

Now find the following veins which carry deoxygenated blood from the organs.

Find the large **hepatic portal vien** which will be found to enter the liver slightly right of the middle line. Before doing so it divides into three branches, one to each lobe. *It is best to trace this vessel backwards.* It runs alongside the bile-duct and pancreas and is formed by the union of veins from the intestine, stomach, spleen and pancreas.

The **hepatic sinuses,** one from each lobe of the liver, will be seen at the anterior end of the liver. They will be examined again later when studying the venous system.

THE VASCULAR SYSTEM

The animal has a single circulation, *i.e.*, the blood is pumped by the heart to the gills and straight on to the rest of the body and so back to the heart.

It is necessary first to remove the alimentary canal and then to expose the heart.

Open the pericardial cavity, which is the anterior part of the cœlom, by cutting through the pectoral girdle and removing the ventral body wall and the central part of the coracoid portion of the girdle.

The **pericardial cavity** is triangular in shape, the apex being directed forwards. It is almost completely filled by the heart. *Pass a seeker into the* **pericardio-peritoneal canal,** which lies dorsal to the sinus venosus (see below), and puts the pericardial and peritoneal cavities into communication.

By carefully opening the pericardium the heart will be exposed.

The Heart—External Ventral View

This consists of four chambers, and is bent dorso-ventrally into an S-shape.

Note the **sinus venosus,** a roughly triangular sac at the posterior end with the apex directed forwards. *This will be rendered more easily visible if the ventral chamber is pressed slightly forwards.* The apex of the sinus leads into the **auricle,** a large thin-walled triangular chamber on the dorsal side, part of which can be seen on each side of the ventral chamber. *To see this more fully, press the ventral chamber to one side.* The auricle leads into the prominent thick-walled ventral chamber, the **ventricle,** continuous with which is a straight tube, the **conus arteriosus,** leading in an anterior direction. Dissection of the heart must be left until the afferent branchial arteries have been examined.

The Venous System*

This is made up mostly of a number of dilated cavities called sinuses and if required, should be studied before dissecting the arterial systems. Apart from the hepatic sinuses, which enter the sinus venosus direct, all the main sinuses enter two Cuverian sinuses (or ducti Cuveri) which lead transversely into the corners of the sinus venosus. *These should therefore be opened in order to find the entrances of these sinuses; this will be simplified if seekers are inserted into their openings. Wash the blood out of the sinus venosus and ducti Cuveri after opening them.*

Trace the sinuses away from the sinus venosus (but remember that the blood in them flows to the heart).

The two short, narrow **Cuvierian sinuses,** or **ducti Cuvieri,** enter the sinus venosus laterally from the dorsal side and each receives (i) at its dorsal end the **anterior cardinal sinus,** a large space between the dorsal body wall and the gill pouches, (ii) the **posterior cardinal sinus,** lying alongside its fellow on the dorsal body wall of the abdominal cavity almost in the middle line and formed posteriorly by a number of **efferent renal veins** from the kidneys, (iii) at about its middle, the small **inferior jugular sinus** from the floor of the pharynx, at its anterior end (iv) the **subclavian vein** from the pectoral fin and neighbouring parts and (v) the **lateral abdominal vein** from the ventral body wall.

*Expose and trace the **anterior cardinal sinus** forwards on one side by making an incision from a point just dorsal to the spiracle, straight back dorsal to the gill clefts to a point just beyond the last one.*

---

* If this system is dissected a fresh specimen will be required for subsequent dissections.

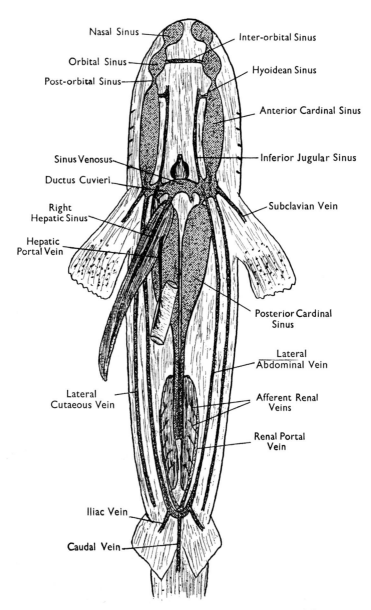

Nasal Sinus
Orbital Sinus
Post-orbital Sinus
Sinus Venosus
Ductus Cuvieri
Right
Hepatic Sinus
Hepatic
Portal Vein
Lateral
Cutaeous Vein
Iliac Vein
Caudal Vein

Inter-orbital Sinus
Hyoidean Sinus
Anterior Cardinal Sinus
Inferior Jugular Sinus
Subclavian Vein
Posterior Cardinal
Sinus
Lateral
Abdominal Vein
Afferent Renal
Veins
Renal Portal
Vein

FIG. 133. **Scyliorhinus. Venous System. Semi-diagrammatic.**

Note that this sinus communicates with the large **orbital sinus** at the back of the eye by a narrow tube, the **post-orbital sinus.** *Insert a seeker into it.* The two orbital sinuses communicate with each other by a transversely running **inter-orbital sinus** (the foramen was seen in the lateral view of the skull). The **nasal sinus** on the posterior side of the olfactory organ also communicates with the orbital sinus. The **hyoidean sinus** from the floor of the mouth enters the anterior cardinal sinus just anterior to the first gill.

Trace the posterior cardinal sinuses backwards: they are very wide sacs which narrow down as they pass backwards and pass between the kidneys, where they receive a number of **efferent renal veins.** *Cut open the posterior cardinal sinus and wash away the blood.* A number of small holes will be seen where the efferent renal veins enter it from the kidney.

Completely cut open these sinuses ventrally and wash out the blood. Note that they freely communicate with each other somewhat irregularly. The **lateral abdominal veins** are formed by the anastomosis of the two **iliac veins** from the pelvic fins. There are also two **lateral cutaneous veins** immediately beneath the lateral line and running superficially from the tail. *Examine the cut transverse edges of the body wall.* The veins will be seen immediately beneath the lateral line. They enter the posterior cardinal sinuses near their anterior ends.

The two Hepatic-Sinuses open directly into the sinus venosus on its posterior side near the centre. Into them lead the **hepatic veins** running from the two lobes of the liver. The two sinuses are incompletely separated from each other.

The Hepatic Portal System has already been seen in the dissection of the alimentary system.

The Renal Portal System consists of the two **renal portal veins** which run along the outer edge of each kidney on the dorsal side. which they enter along the entire length by **afferent renal veins.** They are formed by the bifurcation of a median **caudal vein** from the tail at the posterior end of the kidneys. The caudal vein lies in the hæmal arches of the vetebral column. This system will be examined later when dissecting the urino-genital organs.

The Arterial System

The conus arteriosus continues forwards into the **ventral aorta,** from which arise the **afferent branchial arteries,** taking blood to the gills. The re-oxygenated blood from the gills is carried in the **efferent branchial arteries** which form the **dorsal aorta** from which arise arteries taking blood to the organs of the body.

These three parts of the system will be studied separately.

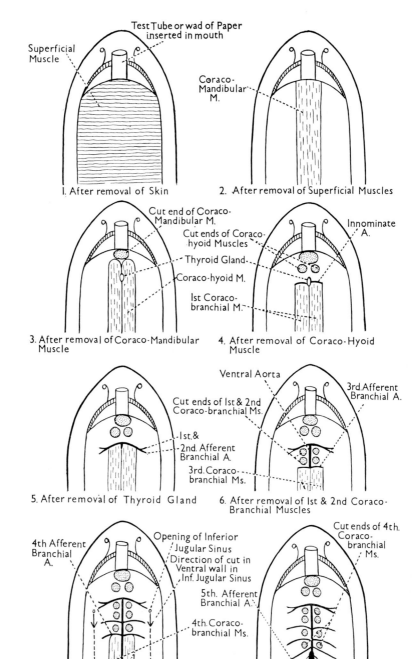

FIG. 134. **Scyliorhinus. Stages in Dissection of Afferent Branchial Arteries.**

The Afferent Branchial Arteries

First insert a test-tube into the mouth to expand it and fix the animal on the dissecting board, ventral side up with the head towards you. Cut through the skin only in the mid-ventral line from the heart to the mouth. Holding the cut edges with forceps, carefully remove the skin only on both sides of the cut on both sides as far as the pectoral girdle. This exposes the superficial muscle layer. *Carefully remove this thin layer only by making a very shallow longitudinal incision in the mid-line, following this up by gently peeling off the muscle. Observe the long median coraco-mandibular muscle now exposed. Remove this by dissection from the pectoral girdle to the lower jaw to which it is attached and paired median longitudinal muscles will be seen.* These are the coraco-hyoids which are attached to the basi-hyal cartilage. *Carefully separate them from the pectoral girdle and remove them, severing them at their anterior ends.* The thyroid gland will now be visible just behind the severed ends of the coraco-hyoids, as a pear-shaped structure in the mid-ventral line. *Remove the thyroid gland*

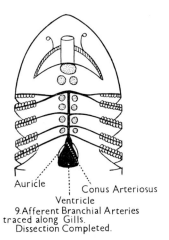

FIG. 135. **Scyliorhinus. Final Stage in Dissection of Afferent Branchial Arteries.**

Auricle

Conus Arteriosus

Ventricle

9.Afferent Branchial Arteries traced along Gills.

Dissection Completed.

carefully and a pair of **innominate arteries** will be seen below it, forming a 'T' with the anterior end of the **ventral aorta.** These divide into the **1st and 2nd afferent branchial arteries.** Behind the innominate arteries on each side of the ventral aorta is a coraco-branchial muscle running between the coracoid and hypo-branchial cartilages. *These muscles must be severed and removed as for the coraco-hyoids.* They are followed, posteriorly, by another pair *which must also be severed and removed.* The **3rd pair of afferent branchial arteries** will then be seen, followed by another pair of coraco-branchial muscles. *Sever and remove these as before.* The **4th pair of afferent branchial arteries** are posterior to these. *Remove the coraco-branchial muscles which will follow* and the **5th** (and last) **pair of afferent branchial arteries** will be found arising close to the previous pair.

Now trace these afferent branchial arteries to the gills, which are enclosed in membranous gill pouches, by careful dissection as follows:— Insert a seeker into the opening of the inferior jugular sinus, which will be found where the 2nd afferent branchial artery disappears from view. Carefully cut along the upper (ventral) wall of this sinus with small scissors, keeping the inner blade as horizontal as possible, as far as its posterior end where it enters the ductus Cuvieri. The remaining afferent branchial arteries can then be seen on the dorsal wall of the sinus and can be carefully traced to and along the gills.

The innominates divide into (i) the 1st afferents running along the outer border of the hyoid arch to supply the hemibranch borne on that arch and (ii) the 2nd afferents supplying the gills of the first branchial arch.

The pairs which follow supply the holobranches on the remaining arches. Each afferent branchial artery runs in the septum between two gill pouches. *Gently trim the gills as necessary to expose the arteries.*

Finally, carefully pare off the cartilage of the ventral side of the pectoral girdle and so expose and study the ventral view of the heart unless previously done. (See p. 166.)

Then remove the heart by cutting through the anterior end of the conus arteriosus and the posterior end of the sinus venosus. Turn it on its side and observe the S-shaped form.

The Heart—Internal Structure

To study its internal structure *cut along the mid ventral line of the conus and ventricle. Remove the ventral wall of the ventricle but pin down the walls of the conus. Wash out the blood from both.*

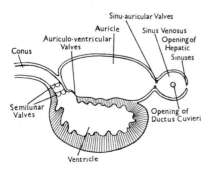

FIG. 136. **Scyliorhinus. V. S. Heart.**

Six **semi-lunar valves** in two rows of three will be seen at the entrance to the **conus arteriosus** and the **auriculo-ventricular valves** in the dorsal ventricular wall, slightly to the left of the mid-line. Note that the **walls of the ventricle** are thick and muscular while that of the **auricle** is thin. *Cut open the* **sinus venosus.** At the anterior end is the centrally placed **sinu-auricular aperture** provided with **sinu-auricular valves** and which leads into the auricle.

The Efferent Branchial Arteries

Insert the scissors into the mouth and cut back horizontally along one side as far as the last gill arch, cutting through the visceral arches. Now cut straight across to the other side behind where the heart was situated. Turn the flap over to the other side and pin it down with an awl. Do not remove the lower jaw or you will cut through a branch of the Vth nerve which must be seen later. Clean the mouth and carefully remove the mucous membrane covering the roof of the mouth but avoid removing or cutting the arteries which are now exposed. If preferred the arteries may be dissected on one side (the cut side) only. This will avoid cutting nerves which will be required later.

Four pairs of **epibranchial arteries** run inwards and backwards from the inner (dorsal) ends of the first four gill arches along the edges of the pharyngo-branchial cartilages, parallel with each other on either side. They join the **dorsal aorta** in pairs in the mid-line.

Carefully remove the pharyngo-branchial cartilage between these arteries so as to expose them fully.

Four **efferent branchial arteries** arise from the outer ends of the epibranchial arteries and loop round the first four gill-clefts. There is a half-loop on the anterior border of the fifth, the loops being connected with each other about half way along by **connecting canals.** They lie in the septa between the gill pouches.

The following arteries should also be traced:—

Arteries arising from the Dorsal Aorta

The two **roots of the carotid arteries** arise from the inner ends of the 1st efferent loops where the 1st epibranchial artery originates. Each is joined by a branch from the anterior end of the dorsal aorta, more or less opposite the hinder edge of the orbit, and immediately divides into (i) an **orbital** (or **stapedial artery**) (formerly thought to be and wrongly called the external carotid), which runs across the floor of the orbit forwards and outwards to the anterior part of the head, and (ii) an **internal carotid artery** which loops inwards and enters the cavity of the cranium, where it crosses the corresponding artery from the other side and unites with the "hyoidean" artery of

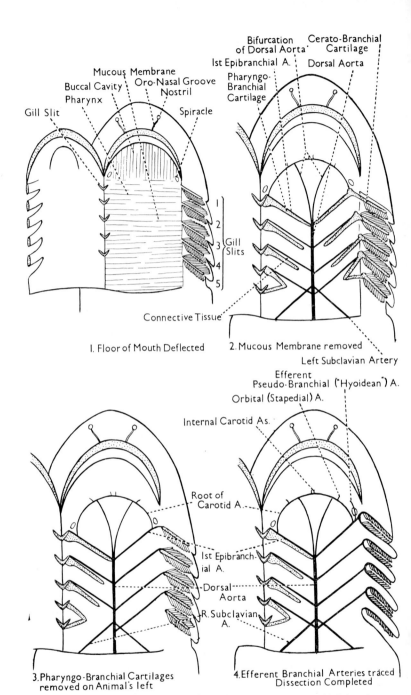

FIG. 137. **Scyliorhinus. Stages in Dissection of Efferent Branchial Arteries.**

the opposite side. There is no true external carotid artery in the dogfish. These arteries are embedded in the cartilage of the roof of the mouth.

The so-called **"hyoidean"** (**efferent pseudo-branchial**) **artery** arises from the middle of the anterior side of the 1st loop on each side but almost immediately disappears from view. *It is necessary to remove the cerato-hyal cartilage to see this.* It goes to the anterior wall of the spiracle then crosses the floor of the orbit below the eye (where it will be seen in a later dissection) and enters the cavity of the cranium. *A little more of this artery can now be traced if the edge of the spiracle and the cartilage underneath which it passes is removed.*

Trace the further course of the so-called hyoidean artery from the side of the head. Place the fish on the dissecting board, dorsal side upwards. Remove the skin so as to expose the hinder part of the mandibular arch, the hyoid arch and part of the orbital cavity. The **"hyoidean"** (**efferent pseudo-branchial**) **artery** will again be seen arising from the first efferent branchial artery. It then crosses the hyomandibular cartilage under the hyomandibular division of nerve VII, running forward to the orbit (outside the spiracle) and then turns inwards to cross the floor of the latter cavity. It will be seen again later when examining the orbit.

Now return the fish to its original position, ventral side up. Trace the dorsal aorta by careful dissection between the kidneys and find the following arteries which arise from it. Only the roots of these vessels will be seen and most of them have already been seen in this dissection of the alimentary system.

The two **subclavian arteries** arise between the 3rd and 4th epibranchial arteries on the dorsal side of the aorta and run backwards and outwards to the pectoral fins.

The **cœliac artery** is a single vessel which arises from the aorta just behind the 4th pair of epibranchial arteries and runs in the mesentery dorsal to the stomach, where it divides into branches, to the liver (hepatic), stomach (gastric) and to the anterior end of the intestine and pancreas (intestinal-pyloric).

The **anterior mesenteric artery** arises about 2·5 cm. or so behind the cœliac and goes to the intestine and rectum, with branches to the genital organs.

The **lieno-gastric artery** arises immediately behind the anterior mesenteric. It runs to the stomach and spleen. This and the anterior mesenteric cross one another as previously seen.

The **posterior mesenteric artery** arises about 4 cm. behind the lieno-gastric and runs backwards in the mesentery to the rectal gland.

The **renal arteries** and the **iliac arteries** will be seen when the urino-genital system is dissected. The continuation of the aorta as the **caudal artery** into the tail will also be seen.

THE RESPIRATORY SYSTEM

The **gills** are borne on the sides of the **gill arches** (or visceral arches) between which are the **branchial-clefts** consisting of the **gill pouches,** cavities where the gills are situated on either side except in the fifth which lacks a gill on its posterior side. These decrease in size from first to last, each having an **internal pharyngeal opening** and an external **gill slit** or **cleft;** and there are five on each side.

The gill pouches are separated from each other by **inter-branchial septa.** The cartilaginous gill rays already seen in the visceral arch skeleton pass through the septa into the gills.

The first visceral arch is modified, as already seen, to form the jaws and is called the mandibular arch. The second, the **hyoid arch,** lies between the first gill-cleft and the spiracle. It bears a gill on its posterior side, known as a **hemibranch.** There are then five **branchial arches,** the first four of which bear hemibranches on each side (each pair of hemibranches on a branchial arch being called a **holobranch**), the fifth bearing no gill as already stated.

Each spiracle contains a vestigial gill devoid of respiratory function and known as a **pseudobranch.**

(1) *Place the fish with its head towards you. If not already done, cut longitudinally through the centre of the ventral wall of one of the gill pouches as far as the gill slit. Pin back the two halves of the wall in order to expose the interior of the pouch.*

(2) *Remove and examine a* **gill.**

Each consists of a comb-like series of vascular structures, called **gill filaments** or **lamellæ.** The blood vessels in connection with the gills have already been seen.

(3) *Cut through and expose the spiracle.*

It contains the non-vascular, vestigial gill, or **pseudobranch,** on its anterior wall.

THE URINO-GENITAL SYSTEM

This dissection must be performed on an animal from which the alimentary canal and liver have been removed.

Before commencing the dissection of the male, remove the skin only on the ventral side of the pelvic region. This exposes a sac called the **siphon.**

Then cut through the centre of the pelvic girdle and through the muscle so as to expose the cloaca. Cut through this in the mid-ventral line. Fix awls through the pelvic fins.

Male

Note the two large, soft **testes** suspended from the body wall by the **mesorchium** and united at the posterior ends by a membrane (unless this has been cut or torn). The testes open at their anterior

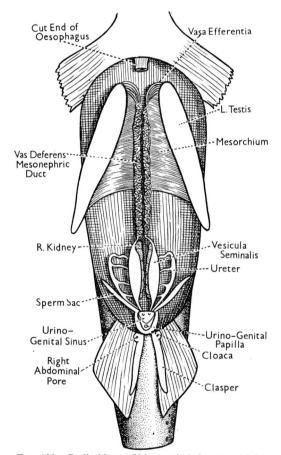

Fig. 138. **Scyliorhinus. Urino-genital Organs. Male.**

ends by a number of tiny ducts, the **vasa efferentia,** which pass through the mesorchium into the long and very sinuous **mesonephric (or Wolffian) duct,** which serves as a **vas deferens.**

Deflect the testes to one side. Carefully remove the peritoneum covering the ventral side of the kidneys.

Note the functional **kidney** on each side (the posterior part of the mesonephros), on the ventral side of which is the mesonephric or Wolffian duct which, as already stated, is the vas deferens. Towards the posterior end of the mesonephros, the vas deferens swells out to form the **vesicula seminalis,** which opens into the **urino-genital sinus** in front of the **cloaca,** into which it opens with its fellow, by the **urino-genital papilla.** *Carefully release the vesicula seminalis from the kidney.*

The two **sperm sacs** are blindly ending pouches originating from the urino-genital sinus. They are attached to the vesiculæ seminales. *Separate them carefully from these structures.* The **ureters** are formed from about five thin tubules arising trom the posterior (functional) part of the mesonephros and open into the urino-genital sinus. *It will be necessary to dissect from the vas deferens back to the cloaca to follow the course of the ureters.*

Gently dissect one kidney from the body wall at its outer edge and deflect it inwards.

The **renal arteries** from the dorsal aorta go direct to the kidneys and the efferent renal veins from the kidney enter the **posterior cardinal sinus** between the kidneys. *Open up the sinus* and observe the holes which are the **openings of the efferent renal veins.** Note the **spermatic artery** and the **spermatic vein** in the mesorchium and the **iliac arteries** which arise from the dorsal aorta and go to the pelvic fins. Note also the so-called **claspers,** in the longitudinal groove of which the sperm passes from the cloaca. Two channels from the siphon lead to the grooves on the claspers.

Find the **renal portal veins** on the inside of the dorsal edges of the kidneys and formed by the bifurcation of the caudal vein in the hæmal arch. *To see this it is necessary to remove the muscle here and then carefully pare off the cartilage of the ends of the hæmal arches.*

Female

There is one **ovary** (the right), a more or less median organ containing **ova** and attached to the body wall by the **mesovarium.** *Deflect it to the side.* The two thick-walled **oviducts** (modified Mullerian ducts) join at their anterior ends immediately in front of the liver, where the œsophagus joins the stomach, and open by a single aperture, the **internal opening of the oviduct.** They pass down to the cloaca, and near the anterior end each swells to form the **oviducal** (or **shell) gland.**

The posterior ends of the oviducts open into the cloaca by a single **cloacal opening.** The anterior end of each **kidney** (or **meson-epheros**) consists solely of a few pieces of tissue, the posterior ends only being well developed and functional. From them arise the **mesonephric** (or **Wolffian**) **ducts,** which widen and then join to form the **urinary sinus.** This opens into the **cloaca** by a **urinary**

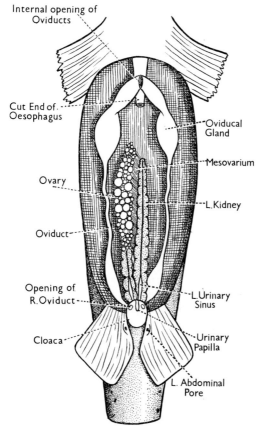

Internal opening of Oviducts

Cut End of. Oesophagus

Ovary

Oviduct

Opening of R.Oviduct

Cloaca

Oviducal Gland

Mesovarium

L.Kidney

L.Urinary Sinus

Urinary Papilla

L. Abdominal Pore

FIG. 139. **Scyliorhinus. Urino-genital Organs. Female.**

papilla. The **ureters,** composed of a series of tubules from the posterior part of the mesonephros, also open into the urinary sinus. They are difficult to identify owing to the position of the oviducts. Observe the **ovarian artery** and the **ovarian vein** in the mesovarium and *find the* **renal portal vein** as in the male.

THE NERVOUS SYSTEM

This consists of a **central nervous system** (the brain and spinal cord), a **peripheral nervous system** (the cranial and spinal nerves) and an **autonomic nervous system,** which, however, is not well marked in the dog-fish.

THE CRANIAL NERVES

I. The Olfactory Nerve

This arises from the anterior end of the olfactory bulb and supplies the olfactory organs as will be seen in examining those organs.

II. The Optic Nerve

This arises from the optic lobe, crosses its fellow on the ventral side as will be seen when examining the ventral side of the brain and, passing through a foramen on the opposite side of the cranium, enters the orbit and supplies the retina of the eye as will be seen in the examination of the orbit.

III. The Oculo-Motor Nerve

This arises from below the optic lobes on the ventral side, leaves the cranial wall behind the IInd nerve, where it branches and goes to the rectus superior, inferior and internus (or anterior) and the obliquus inferior muscles as will be seen in the orbit.

IV. The Pathetic or Trochlear Nerve

This arises between the optic lobe and the cerebellum on the dorsal side, and leaves the cranial wall by a foramen just dorsal and anterior to the IInd nerve. Its course to the obliquus superior muscle will be seen.

V. The Trigeminal Nerve

This arises from the side of the medulla and enters the orbit. Courses of its three branches will be seen in the orbit. These three branches will be traced to their destinations (together with the branches of VII), the **ophthalmic branch** to the snout, the **maxillary branch** to the upper jaw and the **mandibular branch,** bending round the angle of the jaw, to the lower jaw.

VI. The Abducens Nerve

This slender nerve arises from the ventral side of the medulla just behind the root of the VIIth nerve, which is close to the root of the Vth. It leaves the skull with the Vth nerve and, as will be seen in the orbit, supplies the rectus externus (or posterior) muscle.

VII. The Facial Nerve

This arises very close to the root of the Vth nerve and has four branches, the **ophthalmic branch** entering the orbit separately just behind and above the ophthalmic branch of the Vth nerve. The main branch enters the orbit with the Vth nerve, after which it divides into the **buccal branch** to the mouth, the **palatine branch** to the roof of the mouth, and the **hyomandibular branch to the spiracle,** ampullary canals and hyoid arch. These, too, will be seen in the orbit and they will be traced to their destinations (along with the branches of V).

VIII. The Auditory Nerve

This arises very close to the Vth and VIIth nerves, enters the auditory capsule at its inner end and goes to the membranous labyrinth. No further dissection is necessary.

IX. The Glosso-pharyngeal Nerve

This arises from the medulla just behind and ventral to the VIIIth nerve. It enters the auditory capsule on its inner side, runs backwards and outwards on its floor, and then leaves it at its posterior end and above the 1st gill cleft, divides into two branches, the **pre-trematic branch** running along the anterior edge of the 1st gill cleft, and the **post-trematic branch** running along the posterior edge.

It will be noted that this nerve is inappropriately named in the dogfish. Dissection of the nerve will accompany that of the Xth.

X. The Vagus Nerve

This is a large nerve arising from the posterior end of the medulla by several roots.

It leaves the posterior end of the skull between the auditory capsule and the cranium and runs along the inner wall of the anterior cardinal sinus when it divides into:—

The **branchial branch** running to the 2nd, 3rd, 4th and 5th branchial arches, each nerve dividing into a **pre-trematic** and **post-trematic branch.**

The **visceral** branch dividing into several branches which supply the heart, stomach and other viscera.

The **lateral** branch supplying the sense organs in the lateral line canal.

This nerve will be traced together with the IXth.

Dissect and trace the cranial nerves on one side as directed below. The roots of these nerves will be seen when examining the ventral surface of the brain.

The Eye in the Orbit

With the animal dorsal side up and the head towards you, cut away the skin surrounding one eye and dissect away the eyelids. This should be the one on the side on which the lower jaw has not been cut for the dissection of the efferent branchial arteries. The complete dissection of the nerves can then be done on this side. Do not cut away any cartilage. Pull the eye slightly outwards.

Note the spherical **eyeball** somewhat flattened on its outer surface, situated in the **orbit** and supported by six muscles.

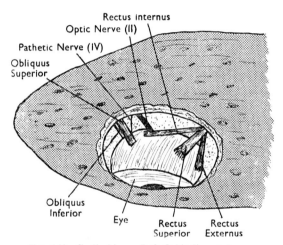

Fig. 140. **Scyliorhinus. Left Orbit dissected open.**

There are four recti muscles which have their *origins* close together in the posterior end of the wall of the orbit:—

The **rectus superior** runs forwards and outwards and is *inserted* in the dorsal side of the eyeball. It raises the eye.

The **rectus inferior** runs forwards and outwards and is *inserted* in the ventral side of the eyeball but is not seen at this stage. It lowers the eye.

The **rectus internus** (or **anterior**) runs forwards at the back of the eyeball and is *inserted* in the anterior part of the eye. It moves the eye forwards.

The **rectus externus** (or **posterior**) runs outwards and is *inserted* in the posterior part of the eye. It moves the eye backwards.

There are also two oblique muscles which have their *origin* close together in the anterior end of the wall of the orbit:—

The **obliquus superior** runs backwards and outwards and is *inserted* in the dorsal side of the eye, just in front of the rectus superior.

The **obliquus inferior** runs backwards and outwards and is *inserted* in the ventral side of the eye. *Push the eyeball backwards (not inwards) to see this insertion on the eyeball.*

They give a turning movement to the eye when used in conjunction with the recti muscles.

Note the stout **optic nerve** (cranial nerve **II**) which enters the eye at the back, and the more slender **pathetic or trochlear nerve** (cranial nerve **IV**) which enters the orbit dorsal to the optic nerve and runs to the obliquus superior muscle.

The Orbit after removal of the Eye

Cut through the eye muscles and optic nerve close to the eyeball, and remove the eye. Wash out the blood from the orbit and examine it.
Note the above muscles and nerves:—

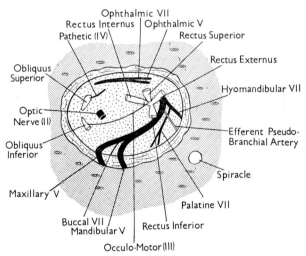

FIG. 141. **Scyliorhinus. Left Orbit after Removal of Eye.**

The **obliquus superior** is in the anterior dorsal corner of the orbit. with the **obliquus inferior** just below it. The four **recti** muscles are in the form of a cross in the posterior dorsal corner, the **rectus**

superior above, the **rectus inferior** below, the **rectus internus** (or **anterior**) running forwards and the **rectus externus** (or **posterior**) backwards.

The **optic nerve (II)** enters at the back of the orbit towards the anterior end and the **pathetic nerve (IV)** in the anterior dorsal corner runs down obliquely to the obliquus superior muscle.

Note also the following nerves:—

The **oculo-motor nerve (III)** enters behind the optic nerve, runs across the orbit, dividing into branches, one to the rectus internus, one to the rectus superior and one to the rectus inferior and obliquus inferior.

The **trigeminal nerve (V)** consists of three branches:—

(i) A broad band, the **maxillo-mandibular branch,** running across the floor of the orbit which divides near the outer edge of the orbit into (*a*) the **maxillary branch,** which passes forwards and downwards to the upper jaw, and (*b*) the **mandibular branch,** which passes downwards to the lower jaw.

(ii) A separate **ophthalmic branch** entering the orbit above the recti muscles and running forwards (with the corresponding branch of VII), to leave the orbit and its anterior end.*

The **Abducens Nerve (VI)** is slender and runs along the inferior edge of the rectus externus, which it supplies.

The **Facial Nerve (VII)** consists of four branches:—

(i) The **ophthalmic branch** enters the orbit just posterior and dorsal to the ophthalmic branch of the Vth nerve, which it then accompanies.

(ii) The **buccal branch,** a small nerve accompanies the maxillary branch of the Vth and may be seen between this branch and the mandibular branch.

(iii) The **palatine branch** runs across the floor of the orbit behind the main branch of the Vth. It is crossed by the **efferent pseudo-branchial (hyoidean) artery,** which should be noted.

(iv) The **hyomandibular branch** is a large branch which runs outwards in the posterior wall of the orbit and goes to the hyoid arch. It will be traced out later, The bulge in the posterior region of the orbit marks the position of the spiracle.

Remove the skin from the lateral side of the head on one side as far as the pectoral fin. Care must be exercised in the region of the spiracle, gill clefts and round the eye.

---

* In Squalus this is known as the superficial ophthalmic as there is also a deep ophthalmic branch.

Dissection of the Vth and VIIth Nerves
Ophthalmic Branches of V and VII

Remove the cartilage of the upper edge of the orbit by cutting parallel with it paring the cartilage very carefully until the forward continuance of the **ophthalmic branches** *of V and VII are exposed. Now trace these nerves to the ampullary canals of the snout by removing the tissue immediately dorsal and ventral to them, exercising great care to avoid cutting the nerves.*

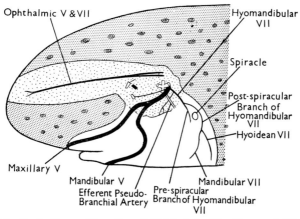

Fig. 142. **Scyliorhinus. Dissection of Ophthalmic, Maxillary and Mandibular Branches of Vth and VIIth Nerves and of Hyomandibular Branch of VIIth Nerve.**

Maxillary and Mandibular Branches of V

These branches have already been traced to the lower edge of the orbit. By careful dissection, trace each to its destination as follows:—

The **Maxillary Branch** *along the cartilage of the upper jaw.*

The **Mandibular Branch** *running backwards behind the angle of the jaw and then forwards along the cartilage of the lower jaw.*

Hyomandibular Branch of VII

This was seen in the orbit as a large nerve running posteriorly in the direction of the bulge marking the position of the spiracle. It divides into a slender **prespiracular** branch which runs anterior to the spiracle and a large **postspiracular branch** which passes posterior to the spiracle. *Trace the continuation of the latter branch backwards to the main branch by very careful* *paring away the surface only of the muscle below and behind the spiracle. It is easier than tracing it forwards. When the nerve has been identified, follow it back to the main branch in the orbit by careful dissection, cutting upwards and carefully following the nerve.* Below the spiracle the postspiracular branch will soon be seen to divide into an external **mandibular branch,** which ultimately supplies the ampullary canals on the lower jaw and a **hyoidean branch** which runs down the hyoid arch. *Find these branches, exercising the same care as before. Then find the prespiracular branch which arises from the same point on the main branch as the postspiracular branch.*

Buccal and Palatine Branches of VII

No further dissection of these branches, already seen in the orbit, is necessary.

Dissection of the IXth and Xth Nerves

The dissection should also be performed on the side opposite to that used for the examination of the venous and efferent branchial systems if the same specimen is used. Open up the anterior cardinal sinus in which part of the course of the nerves lies. It is situated dorsal and parallel to the gill slits between the body wall and the muscles of the back. It can be felt where there is less resistance to the pressure of the fingers. Make the incision just dorsal to the spiracle and cut straight back, dorsal to the gill clefts, until the posterior end of the sinus is reached. Alternatively the sinus can be found by cutting upwards from the third gill slit. You will thus pass transversley across the sinus. Having located it cut straight backwards to the posterior end of the sinus and forwards as far as the spiracle to open up the entire sinus. Wash out any blood in the sinus and remove the roof of the sinus, taking care not to cut away any nerves.

IXth Nerve

In the anterior corner of the sinus will be found a comparatively slender nerve entering from the skull. This is the IXth nerve. *Trace this by careful dissection of the 1st gill cleft.* It will be found to divide into a slender **pre-trematic branch** which runs along the anterior side of the cleft and a less slender **post-trematic branch** running along the posterior edge. *These two branches can be traced by very careful dissection and removal of such parts of the gills as is necessary, though the thinner pre-trematic is more difficult to follow for any great distance.*

Branchial Branches of X

The Xth nerve leaves the cranium and enters the sinus as a stout structure, the main stem, posterior to the IXth nerve. The **branchial branch** has four branches supplying the 2nd to 5th gill clefts and lies on the floor of the sinus. Each has a **pre-trematic** *and* **post-trematic branch.** *Trace them to the gills, following the same procedure as for those of the IXth.*

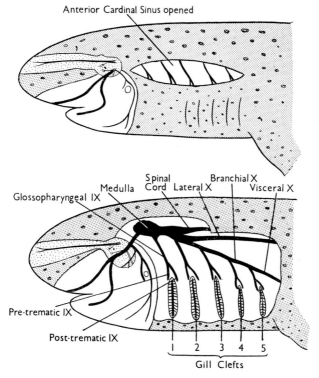

Fig. 143. **Scyliorhinus. Dissection of IXth and Xth Cranial Nerves.**

Visceral Branch of X

This arises from the main stem and closely accompanies the branchial branch to a point just before it passes to the last gill cleft; it is difficult to distinguish the two nerves anterior to this. Then it passes out of the sinus in a posterior direction to give branches to the heart, stomach and other viscera.

Trace this nerve back for a short distance by careful dissection. It is not necessary (or possible, if the alimentary system and other viscera have been removed) to trace its branches.

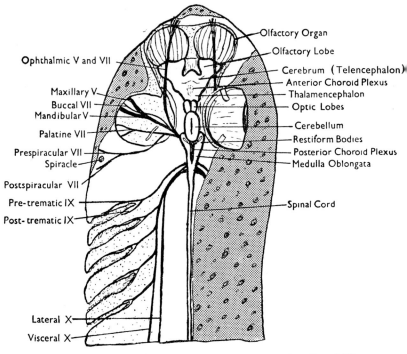

Ophthalmic V and VII

Maxillary V
Buccal VII
Mandibular V

Palatine VII

Prespiracular VII
Spiracle

Postspiracular VII
Pre-trematic IX
Post-trematic IX

Lateral X
Visceral X

Olfactory Organ
Olfactory Lobe
Cerebrum (Telencephalon)
Anterior Choroid Plexus
Thalamencephalon
Optic Lobes
Cerebellum
Restiform Bodies
Posterior Choroid Plexus
Medulla Oblongata

Spinal Cord

FIG. 144. **Scyliorhinus. Brain and Cranial Nerves in situ. Dorsal View.**

Lateral Branch of X

This branch is dorsal to the branches already seen and is more deeply seated at first, being embedded in the muscle of the body wall.

Trace it from the main stem. It soon runs inwards into the muscle. *Carefully dissect away sufficient of the muscle to expose it for a short distance. Its path to the sense organs of the lateral line need not be followed.*

Now expose the brain as directed below in order to trace the nerves back to it.

Very carefully pare off the cartilage from the roof of the cranium, starting at the posterior end, and gradually work forwards until the whole of the dorsal side of the brain is exposed. Take care not to damage the brain with your scalpel while doing this. Still being very careful, continue to remove all such cartilage as is necessary to expose the continuation of the IXth and Xth nerves to the brain. Now remove the muscle posterior to the cranium for a short distance until you come down to the vertebræ. Remove the top of the exposed vertebral column to expose the spinal cord. Continue to remove the cartilage, exercising great care, so as to expose the continuation of the IXth and Xth nerves to the brain.

THE BRAIN AND SPINAL CORD

Dorsal view

Expose the rest of the brain by carefully removing the cartilage at the anterior end of the cranium, both dorsally and laterally. To avoid damage to the ophthalmic branches of the Vth and VIIth nerves while removing the cartilage beneath them, sever them at their anterior ends and deflect them while paring off this cartilage.

In the **Fore-brain** note the **olfactory lobes,** the anterior surfaces of which are in close proximity to the large olfactory organs. These lobes arise from a rounded part of the fore-brain or **telencephalon (cerebrum),** behind which is the narrow **thalamencephalon,** from the hinder part of the thin-walled vascular roof (or **anterior choroid plexus**) on which arises the **pineal stalk** bearing the **pineal body.** This will almost certainly have been removed with the roof of the cranium. The **Mid-brain** consists of the rounded **optic lobes** which are partly hidden by the long flattened **cerebellum** belonging to the **Hind-brain,** continuous with which is the thin-roofed **medulla oblongata** covered by a triangular **posterior choroid plexus,** and bearing on its sides the wing-like laterally placed **restiform bodies.**

Press the brain carefully to one side and *look for the inner ends of the ten cranial nerves.* III and VI will not be seen in this view.

Note that the **spinal cord** is continuous with the medulla oblongata and bears a groove, the **dorsal fissure,** on the dorsal side.

Ventral View

Cut through the olfactory nerves which run from the olfactory lobes to the olfactory organs in front of the brain, and through the posterior end of the medulla where it joins the spinal cord or, better, a little further back. Then, carefully lifting the brain with forceps and working forwards, cut through the roots of the cranial nerves in succession.

Finally gently remove the brain from the cranial cavity and examine its ventral surface.

Note the **olfactory lobes, telencephalon, thalamencephalon** and **optic lobes,** already seen in the dorsal view. At the posterior end of the telencephalon the optic nerves (**II**) cross over one another forming the **optic chiasma,** behind which is the **infundibulum,** the sides of which consist of two thick lobes, the **lobi inferiores,** and between which is the **hypophysis** or **pituitary body,** the end of which will

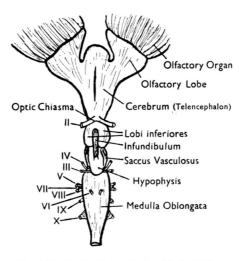

FIG. 145. **Scyliorhinus. Brain. Ventral View.**

probably have been left in the cranium. The unpaired vascular outgrowth beside and behind the infundibulum is the **saccus vasculosus.** Behind this is the **medulla oblongata.**

Look for the roots of the cranial nerves. Their positions have been given on pp. 180 & 181. IV will probably not be seen. To see III, *lift the infundibulum.*

There is also a series of three slender nerves arising from the lower side of the medulla below the roots of the Xth nerve which unite immediately outside the skull, forming a thin trunk which joins the first few spinal nerves. These slender nerves are called the **occipital nerves.** The fused nerve formed by the occipital nerves is probably more or less homologous with the hypoglossal nerve of terrestrial vertebrates, and can be seen by a dissection from the side. It runs with the vagus for a short distance, then leaves it to pass downwards and forwards round the Cuvierian duct to supply the hypo-branchial muscles.

Longitudinal Section

Cut a sagittal (median longitudinal) section of the brain, preferably after hardening in alcohol for a few days.

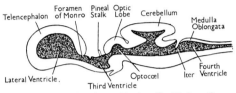

FIG. 146. **Scyliorhinus. Brain. Sagittal section.**

Note one of the **lateral ventricles** in the **telencephalon** (these two ventricles are separated by a septum), each leads into the **third ventricle** in the **thalamencephalon** by a **foramen of Monro.** The third ventricle leads by the narrow **iter** (or **aqueduct of Sylvius**) into the **fourth ventricle** in the **medulla.** Each **optic lobe** contains an **optocœl** and each **olfactory lobe** a **rhinocœl.** There is also a cavity in the **cerebellum.**

THE SPINAL NERVES

Carefully press the spinal cord to the side to show the roots of the spinal nerves.

Note that a pair of **spinal nerves** arises between each pair of vertebræ on the sides. Each nerve arises by a **dorsal (afferent** or **sensory) root** bearing a **ganglion** and a **ventral (efferent** or **motor) root** which join outside the vertebral column to form the (mixed) spinal nerve.

THE SENSE ORGANS

The Eye

Remove the remains of the muscles from the eyeball and cut a longitudinal section of the eye by cutting round it, thus dividing it into inner and outer parts.

Note that the eyeball is divided into an **anterior chamber** and a **posterior chamber** by a solid spherical **lens.** The anterior chamber contains a watery fluid, the **aqueous humour,** and the **posterior chamber (vitreous body)** contains a gelatinous fluid, the **vitreous humour.** The lens is kept in position by a circular band containing muscle fibres and known as the **ciliary body** to which it is joined by the **suspensory ligament.**

The wall of the eye consists of three coats:—

(i) The opaque white outer coat is the **sclerotic** which encloses the eye except where the optic nerve enters and in the front which is the transparent **cornea.**

(ii) The black middle coat, or **choroid,** is pigmented and vascular and is lined by a silvery-looking membrane, the **tapetum** (peculiar to the dogfish). It lines the sclerotic but not the cornea. In front is the contractile pigmented **iris,** which is perforated in the centre by an aperture, the **pupil.**

(iii) Lining the choroid is the **retina,** a membrane containing cells sensitive to light, which are continuous with the fibres of the optic nerve.

The Membranous Labyrinth

This is a very delicate and time-consuming dissection. If required the following procedure should be adopted.

Cut away the skin from the posterior part of the skull on one side and scrape the latter clean. The side used for tracing the Vth and VIIth nerves to the brain will have been damaged. Therefore when this dissection has to be done use the other side. Find the ridge running inwards and backwards from the posterior edge of the orbit and the ridge running outwards and backwards to the back of the skull. These mark the positions of the two vertical semicircular canals (see below). This is the dorsal side of the auditory capsule.

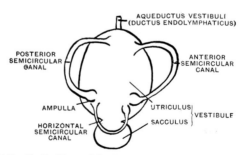

FIG. 147. Scyliorhinus. Membranous Labyrinth dissected out of Auditory Capsule.

Note the **external opening of the ductus endolymphaticus** (or **aqueductus vestibuli),** a small hole where the auditory capsule fuses with the cranium on its inner dorsal surface just behind the level of the eyes and already seen in the dorsal view of the skull.

Cut away the cartilage very carefully so as to expose the membranous labyrinth. If care is not exercised it will be damaged. The semicircular canals are similar in appearance to the surrounding cartilage and are very easily cut or broken.

Note the **ductus endolymphaticus** or **aqueductus vestibuli** which leads from the exterior to the **vestibule**. The latter is composed of:

A dorsal swelling, the **utriculus** and a ventral swelling, the **sacculus.** From the utriculus arise three **semi-circular canals,** each widening at its anterior end into an **ampulla.** The **anterior canal** is vertical and runs from front to back. The **posterior canal** is also vertical but behind the anterior canal, and the **horizontal canal** runs laterally.

The **Auditory Nerve** (Cranial Nerve **VIII**) enters the auditory capsule at the anterior end of its inner wall and gives off branches to the various parts of the membranous labyrinth.

The entire membranous labyrinth can be removed from the auditory capsule but it is very difficult to avoid damaging it and it takes a long time.

The Olfactory Organs

Cut away the skin from the dorsal surface of the snout and the cartilage of the olfactory capsules (if not already done).

Note that the **olfactory organs** are large spherical sacs, the lining of which is thrown into folds. *Cut one open to see this.* It is covered by a sensory epithelium which is supplied with twigs from the **olfactory nerve** (Cranial nerve **I**), which enters each organ from the posterior side.

TRANSVERSE SECTIONS

Now that the whole animal has been dissected, *prepared transverse sections cut through the branchial and visceral regions of the trunk and through the tail should be examined and drawn.* The structures visible will depend, of course, on the exact regions in which the sections have been made.

Branchial Region

Note (i) the **vertebra** with its **centrum** containing **notochordal tissue,** and **neural arch** enclosing the **spinal cord** (ii) the **dorsal aorta** ventral to the vertebra, (iii) the segmented **dorsal muscles (myotomes)** of the body wall, (iv) the **auricle** and **ventricle** of the **heart** enclosed in the **pericardial cavity** on the ventral side, (v) the **pharynx** across the body dorsal to the heart, (vi) the **gill pouches** laterally, (vii) the **gills,** (viii) the **coracoid portion of the pectoral girdle** ventrally, (ix) the **ventral muscles** of the body wall, (x) the **pharyngo-branchial cartilages,**

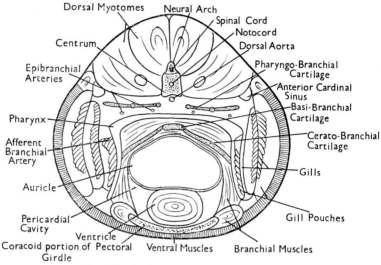

FIG. 148. Scyliorhinus. T.S. Branchial Region.

(xi) the **basi-branchial** and **cerato-branchial cartilages.** (xii) Look for **afferent, efferent** and **epi-branchial arteries** and (xiii) the **anterior cardinal sinus.**

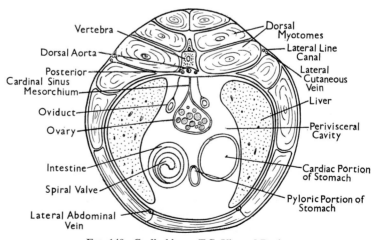

FIG. 149. Scylirohinus. T.S. Visceral Region.

Visceral Region

Note (i) the **vertebra**, (ii) the **dorsal muscles** and (iii) **dorsal aorta** as before. Also (iv) the **posterior cardinal sinus**, (v) the **perivisceral cavity** and parts of such of the following organs as are present; (vi) the **liver**, (vii) the **cardiac** and **pyloric portions** of the **stomach**, (viii) the **intestine** (look for the **spiral valve**), (ix) **testes** with **mesorchium** or **ovary** with **mesovarium** and **oviduct**, (x) **kidney**, (xi) **lateral** and **ventral muscles**, (xii) **lateral line canal**, (xiii) **lateral cutaneous vein**, (xiv) **lateral abdominal vein**.

Caudal Region

Note the (i) **vertebra** with the **neural arch** on the dorsal side and the **hæmal arch** on the ventral side, the latter containing the **caudal artery** above and the **caudal vein** below; (ii) the **myotomes** and **myocommata** of the muscles completely surrounding the vertebra.

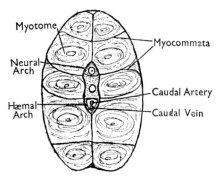

FIG. 150. Scyliorhinus. T.S. Caudal Region.

CLASS OSTEICHTHYES*

Sub-class Teleostei. Fish with the embryonic cartilaginous skeleton wholly or partially replaced by bone. Operculum and swim-bladder present. Exoskeleton of bony scales. Poikilothermic. Mouth terminal. Homocercal tail. All the common bony fish.

PHOXINUS

THE MINNOW

Examine a specimen of a **minnow** *or other typical bony fish.*

The body is streamlined, somewhat flattened laterally and in some fish quite wide dorso-ventrally. The body is covered with **bony scales**,

* This Class was formerly known as **Pisces**.

overlapping like tiles on a roof and is darker above than it is beneath. A **lateral line** is seen running along the sides of the body, which is divided into **head, trunk** and **tail**. The **mouth** is at the anterior end and dorsal to it are the **external nostrils**. The **eyes** are laterally placed and the gill-slits are hidden beneath an **operculum** which is free along its posterior edge. One, or as many as three, vertically placed **dorsal fins** and one or more **ventral fins** are present. The paired **pectoral fins** and **pelvic fins** are often closer together than is the case in the dogfish. The **caudal fin** is bilobed and symmetrical and the tail is said to be *homocercal*.

The colour varies considerably in different fish and in the minnow may be grey or yellow and is brightest in the male during the breeding season. (Internally, apart from the bony skeleton, these fish differ from the *Elasmobranchii* in the presence of an air-bladder, known also as the swim-bladder, which lies ventral to the vertebral column and which leads in most cases into the oesophagus. This helps to maintain balance and may also serve a respiratory function.)

CLASS AMPHIBIA

Larvæ aquatic. Adults usually terrestrial but return to water to lay eggs and often live in damp surroundings. All freshwater. Gills present in larvæ. Metamorphosis from larva into adult. Heart 3-chambered. Poikilothermic.

SUB-CLASSES

ANURA   URODELA

SUB-CLASS ANURA

Adult tailless. Long hind limbs adapted for swimming and jumping with webbed digits. Skin lacks scales. *Frogs and toads.*

RANA TEMPORARIA

THE COMMON FROG

OBSERVATIONS ON LIVING ANIMALS

Observe living frogs in a vivarium and out of doors.
Note how they sit, jump and swim, and examine the breathing movement of the floor of the mouth.

EXTERNAL ANATOMY

Note that the body is composed of **head** and **trunk,** and that there is no neck or tail. The skin is green or yellowish with black pigment spots, **melanophores,** and is slimy and loose. The ventral side is lighter in colour and the 'throat' of the male is white.

Put two frogs of similar colour on light and dark backgrounds respectively. Examine their colours a few hours later and draw your conclusions.

Head

On the bluntly triangular head note the two dorso-laterally placed prominent **eyes,** which have slightly movable **upper lids** and translucent **nictitating membranes** below, which can be lifted up over the eye. There are no lower lids. Behind and slightly lower than the eye on each side is the **tympanic membrane** or ear drum, well camouflaged in a pigmented area of skin. Well forward, anterior to the eyes are two small openings, the **external nares** or nostrils. The **mouth** is crescentic and large and is at the anterior end of the head.

Trunk and Limbs

On the trunk are two pairs of limbs. The **fore-limb** is short, and is composed of four parts, the **brachium** (upper arm), **ante-brachium** (forearm), the **carpus** (wrist) and the **manus** (hand). The **hind limb** is much longer and also consists of four parts, the **femur** (thigh), the **crus** (shank), the unusually long **tarsus** (ankle) and the **pes** (foot). On the **dorsal side** of the trunk towards the posterior end, note the prominent ridge; this is the **sacral prominence.** Between the hind limbs on the posterior end of the trunk is the **cloacal aperture,** the common opening of the intestine and the urino-genital duct.

*Draw a **lateral view** of the animal, showing the structures mentioned above.*

Manus

*Examine the **palmar surface** (lower surface) of the **manus** of a **male** and of a **female** frog.*

There are four short **digits,** numbered 2, 3, 4, 5 (No. 4 is the longest), the pollex, or thumb, being absent. The digits are not webbed.

In the **male,** the second digit bears a **nuptial pad,** prominent and brown in the breeding season. This is absent in the **female.**

Pes

*Examine the **plantar surface** (lower surface) of the **pes.** There are five long **digits** (No. 4 is the longest), which are **webbed.**

THE SKELETON

The skeleton is composed of the **axial skeleton,** the **visceral skeleton** and the **appendicular skeleton** as in the dogfish.

Cartilage in the skeleton of the tadpole is partly ossified when the larva develops into a frog and bones thus formed are called **cartilage bones.** Some bones, however, are formed by ossification of the enveloping membranes of the cartilage and these are known as **membrane bones.**

Examine a prepared disarticulated skeleton.

THE AXIAL SKELETON

The Skull

This consists of the chondrocranium, the sense capsules and the jaws, The skull is partly ossified and most of the bones are membrane bones though a few are cartilage bones as will be seen below.

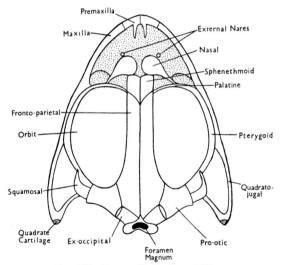

FIG. 151. **Rana. Skull. Dorsal View.**

Dorsal view

The **chondrocranium** is in the centre with the **orbit** on either side. The two **fronto-parietals** (membrane-bones) cover the chondrocranium dorsally. They partially cover the **sphenethmoid** (a cartilage-bone), visible beyond their anterior ends, and their posterior ends widen outwards to the auditory capsules. The sphenethmoid actually forms the anterior end of the cranium. Behind the fronto-parietals are the two **ex-occipitals** (cartilage-bones) forming a ring, the cavity of which is the **foramen magnum;** it is through this that the spinal cord passes.

The **Auditory Capsules** are mainly cartilaginous and are fused with the sides of the posterior ends of the cranium. Part is ossified as the **pro-otics** (cartilage-bones), which constitute part of the roof, anterior end and part of the floor.

The **Nasal or Olfactory Capsules** are at the anterior end of the skull. They are mostly cartilaginous, though in the dorsal view will be seen the two **nasal bones** (membrane-bones), roughly triangular in shape with the apices directed laterally. The **external nares** will be seen anterior to these in the cartilage.

The **Upper Jaw** is composed of cartilage covered by the **maxillæ,** forming the greater part of the upper jaw, the **premaxillæ,** small bones continuous with the maxillæ and meeting one another in the middle line in the front of the skull, and the **pterygoids,** λ-shaped bones which stretch from the auditory capsules to the maxilla on each side and form the outer border of the orbits. The jaw is supported by the **palatine,** a transversely placed bone joining the sphenethmoid to the pterygoid, the **quadrato-jugal,** joined to the hinder end of the maxilla on each side, and which runs backwards and joins one of the cross pieces of the **squamosal bone,** which is shaped somewhat like a hammer or an inverted T. Joining the quadrato-jugal and the pterygoid posteriorly is the **quadrate cartilage** but only the tip could be visible in this view.

Ventral View

The floor of the chondrocranium is formed by the **sphenethmoid** anteriorly and supported by the **parasphenoid,** shaped rather like a dagger with the blade pointing forwards, the cross pieces of the handle reaching to the auditory capsules. The two **ex-occipitals** on this surface enclose the **foramen magnum** at the posterior end of the skull and bear convex processes called **occipital condyles** for articulation with the first vertebra. The bases of the **pro-otics** in the auditory capsules can be seen. The base of the nasal capsules is composed of cartilage strengthened by the two **vomer** bones which are roughly triangular or rather triradiate and which bear the **vomerine teeth** on their posterior edges. The **internal nares** are situated on the outer edge of the vomers.

Note that the **pre-maxilla** and **maxilla** on each side bear **maxillary**

teeth. The under surfaces of the **palatines, pterygoids** and **quadrato-jugals** will also be seen. The **quadrate cartilage** will be seen where the pterygoid and quadrato-jugal join posteriorly on each side.

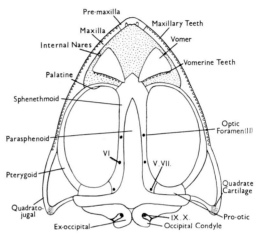

FIG. 152. **Rana. Skull. Ventral View.**

Note the following **foramina** for the cranial nerves:—

II (optic) in the cartilage on either side of the cranium.

VI (abducens) behind the foramen for II near the pro-otic.

V (trigeminal) and **VII** (facial) behind the foramina for VI, almost in the angle between the 'blade' and the 'handle' of the parasphenoid, close to the pro-otic.

IX (glossopharyngeal) and **X** (vagus) in the pit surrounded by the ex-occipitals.

The **Lower Jaw** or **Mandible** is an arch composed of two rods of cartilage, Meckel's Cartilage, fused in the centre in front and partially enclosed by membrane-bone and cartilage-bone. It is called the **dentary** although it bears no teeth. Articulation with the upper jaw is effected by the posterior part of the **angulo-splenial** bone which stretches along the inside and underside of the mandible.

The type of suspension is thus different from that in the dogfish and is said to be *autostylic*.

The Hyoid Apparatus

This is a cartilage in the floor of the mouth and is roughly rect-angular in shape, the **body** bearing at its corners processes called

anterior cornua and the **posterior cornua,** partly ossified. It cannot be examined until the dissections are completed. (See p. 227.)

The Vertebral Column

This consists of nine vertebræ, bony rings forming a canal for the protection of the spinal cord and, continuous with the last vertebra, the tapering urostyle.

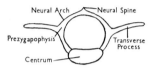

FIG. 153. **Rana. A Typical Vertebra. Anterior View.**

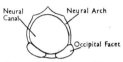

FIG. 154. **Rana. Atlas. Anterior View.**

Each **vertebra** is a bony ring, the lateral and dorsal sides of which form the **neural arch,** while the thicker base is the **centrum,** concave anteriorly and convex posteriorly except in the eighth and ninth, and therefore said to be *procœlous.* The enclosed cavity is the **neural canal.** Projecting from the top of the neural arch is the **neural spine** and, except in the first vertebra, the **transverse processes** arise from the sides. These are directed forwards in the second vertebra, outwards in the third and backwards in the rest. The vertebræ articulate with one another by processes called zygapophyses on the anterior and posterior edges of the neural arches. These are called the **prezygapophyses** and **postzygapophyses** respectively. The articular surfaces on the prezygapophyses face upwards while those on the postzygapophyses face downwards.

FIG. 155. **Rana. IXth Vertebra.**

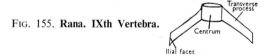

The first vertebra or **atlas,** which has a very much reduced centrum and no transverse processes, articulates anteriorly by its concave occipital facets with the occipital condyles of the skull. It lacks prezygapophyses.

The **eighth vertebra** has concave surfaces at both ends of its centrum and is therefore *amphicœlous*.

The **ninth** (or **sacral**) **vertebra** has large backwardly directed transverse processes, the ends of which, the **ilial facets,** articulate with the pelvic girdle. The centrum, the anterior surface of which is convex, has no postzygapophyses, of course, but it has two convex surfaces for articulation with the urostyle.

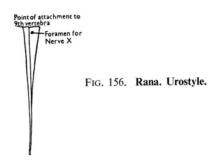

Point of attachment to
9th vertebra

Foramen for
Nerve X

Fig. 156. **Rana. Urostyle.**

The **urostyle,** though really composed of fused vertebræ, is unsegmented and tapers towards the hinder end. Anteriorly are two concave **surfaces for articulation with the ninth vertebra.** A ridge runs along its dorsal surface. There are two **foramina,** through which the **Xth** spinal nerves pass, on the sides.

THE VISCERAL SKELETON

This is composed of the cartilaginous parts of the jaws (mandibular arch skeleton), the hyoid and the columella auris of the ear (hyoid arch skeleton).

THE APPENDICULAR SKELETON

The Pectoral Girdle

This consists of two half-hoops of cartilage and bone, joined ventrally but free dorsally, and serves for the articulation of the fore-limbs. On the dorsal side above the shoulder-joint is a blade-like bone, the **scapula,** and dorsal to this a partially ossified cartilage, the **suprascapula.** On the ventral side below the shoulder joint is the **coracoid,** which widens out towards the inner end where the **epicoracoid,** a narrow strip of calcified cartilage, joins its fellow. In front of the epicoracoid is another strip of cartilage, the **precoracoid.**

The slender bones overlying the precoracoids are the **clavicles,** the only membrane-bones in the pectoral girdle, and the oval space between the precoracoid and the coracoid is the **coracoid fontanelle.** The **glenoid cavity,** into which the head of the fore-limb fits, is formed by the scapula and coracoid.

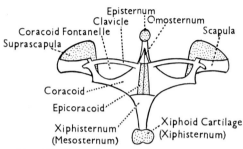

FIG. 157. **Rana. Pectoral Girdle. Ventral View.**

The central part of the girdle is composed of (*a*) the **omosternum** directed forwards and bearing at its tip a small circular plate of cartilage, the **episternum,** and (*b*) the **xiphisternum** (or **mesosternum**) directed backwards and bearing a larger plate of cartilage at its tip, **xiphoid cartilage** (or **xiphisternum**).

The Pelvic Girdle

This is also composed of two half-hoops, but they are joined at the posterior end.

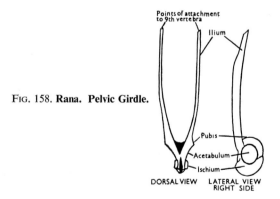

FIG. 158. **Rana. Pelvic Girdle.**

The long slender bone is the **ilium,** which articulates with the transverse process of the ninth vertebra in front and which widens out behind. Posterior to this is the **ischium,** and on the ventral side, a calcified cartilage, the **pubis.** The cavity formed by these three bones is the **acetabulum,** into which the head of the hind-limb fits.

The Fore-Limb

The skeleton of the upper arm consists of a single bone, the **humerus.** It bears a rounded swelling at its upper end, the **head** (which fits into the glenoid cavity), and a more irregular process, the **trochlea,** at its lower end. Note the **deltoid ridge** along the inner side for the insertion of the deltoid muscle.

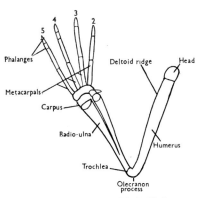

FIG. 159. **Rana. Fore Limb Skeleton.**

The skeleton of the forearm is composed of the **radio-ulna,** which corresponds to the radius and ulna of higher vertebrate animals, and which shows a groove tunning longitudinally almost to the proximal end. This bone articulates with the trochlea of the humerus by hollows in its proximal end where the ulnar portion projects backwards forming the **olecranon process.**

The skeleton of the wrist is made up of six **carpals,** in two rows of three. These may be difficult to identify individually. The proximal row consists of the **radiale, intermedium** and **ulnare,** and those in the distal row are simply called the **distal carpals.**

The palm of the manus is composed of five **metacarpals,** though only the 2nd, 3rd, 4th and 5th bear **phalanges,** which form the digits. The 2nd and 3rd have two and the 4th and 5th three.

The Hind Limb

The bones of the hind-limb correspond with those of the fore-limb, but are much longer.

The thigh-bone, or **femur,** bears a rounded **head** (which fits into the acetabulum) at its proximal end, and a **condyle** at its distal end.

The **tibio-fibula** corresponds to the tibia and fibula of higher vertebrate animals. The grooves are less conspicuous than those in the radio-ulna. Its proximal end articulates with the condyle of the femur.

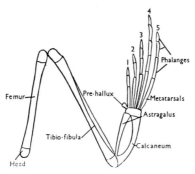

Fɪɢ. 160. **Rana. Hind Limb Skeleton.**

The ankle, like the wrist, is composed of two rows of small bones, **tarsals,** but there are only two bones in each row, (*a*) the long **astragalus** (tibiale) and **calcaneum** (fibulare), in the proximal row, separate from each other except at the ends, and (*b*) the two small **distal tarsals** in the distal row.

Six metatarsals are found in the sole of the pes, one being very small and corresponding to an extra toe, the **prehallux,** or **calcar,** which is not visible externally. The other five metatarsals bear **phalanges,** forming the digits. The 1st, the big toe or **hallux,** has two; so has the second; the 3rd has three; the 4th four and the 5th three.

THE ARTICULATED SKELETON

Examine an articulated skeleton in order to see the relationship of the various parts.

INTERNAL ANATOMY

It is an obvious advantage to carry out the dissections on the large edible frog, *Rana esculenta,* if available, but the smaller common frog is perfectly satisfactory.

THE MUSCULAR BODY WALL

Fix the frog, ventral side uppermost, to the bottom of a dissecting dish by placing pins obliquely through each manus and pes, stretching the limbs well. Hold up the skin with fine forceps in the mid-ventral line and cut through it from cloaca to mouth. Then, holding the skin with the forceps, cut through the connective tissue which joins it to the muscular body wall beneath, taking care to avoid cutting the musculo-cutaneous veins at the armpits.

The spaces between the **connective tissue** are the **subcutaneous lymph sacs.**

Make transverse cuts through the skin, level with the fore-limbs and pin it back on either side.

Note the **musculo-cutaneous** veins on the inside of the skin, the **rectus abdominis muscle** running along the lower part of the trunk, divided longitudinally in the centre by the **linea alba,** a white line of connective tissue, and transversely by **tendinous intersections.** The large fan-shaped muscle radiating from the forearm to the centre on each side is the **pectoralis muscle** and the small triangular piece of cartilage in the centre above the rectus abdominis muscle is the **xiphoid cartilage** (or **xiphisternum).** Note the **mylo-hyoid muscle,** running from the mid-line to the edges of the lower jaw. The bluish **anterior abdominal vein** will be seen through the lower part of the body wall in the mid-ventral line.

Make a median incision in the mylo-hyoid muscle and deflect the flaps to each side. The **hypoglossal nerves** (1st spinal) will be seen running up (from the spinal cord) to the mouth.

Cut longitudinally through the skin of the thigh and shank and pin it back or cut it away. Note the large spindle-shaped **gastrocnemius muscle** of the calf. It has its origin in the lower end of the femur bone and its insertion in the sole of the foot, the insertion being effected by the long **tendo achillis.** (This muscle is used in experiments in muscular contraction by electrical stimuli in the study of physiology.)

THE ALIMENTARY SYSTEM

Buccal Cavity and Pharynx

Open the mouth wide.

Note the ridge of **maxillary teeth** on the **upper jaw** and the group of teeth, **vomerine teeth,** on the roof of the mouth. There are no teeth on the **lower jaw.** Pull out the **tongue.** Note that it is fixed in front and that the tip is forked. The **internal nares** can be seen on the front of the roof of the mouth. At the back of the **buccal** cavity, *i.e.*, in the **pharynx,** is the entrance to the **œsophagus,** which leads to the stomach, and below it the slit-like **glottis** leading to the lungs. The downward projections of the large prominent eyes can be seen and, just behind them on each side, is the entrance to the **Eustachian tube** which leads to the tympanic membrane.

Pierce one of the tympanic membranes with a mounted needle and pass a bristle through one of the Eustachian tubes into the pharynx in order to see where it enters it. Pass another bristle through one of the external nares into the buccal cavity.

Dissection

Ligature the anterior abdominal vein (to prevent loss of blood) in two places, as follows: with a small scalpel, or the small scissors, make an

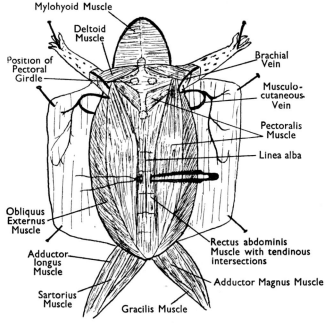

FIG. 161. Rana. **Muscular Body Wall and Method of Ligaturing the Anterior Abdominal Vein.**

incision on either side of the vein. Pull a loop of thread underneath it with the small forceps and then cut the loop at the bend (see Fig. 161). Tie each of the pieces of thread thus separated in a double knot round the vein. Cut across the vein between the ligatures, free it and deflect it out of the way.

Now expose the internal organs or **viscera** *by continuing your incisions as far back as possible and as far forward as the pectoral girdle. Cut through the coracoid and clavicle of the pectoral girdle close to the humerus on each side with the large scissors, taking care not to cut the blood vessels underneath or those going to the forelimb. Remove the freed central part of the girdle and gently pull the forelimbs farther out. Pin back the body wall and then cover the animal with water. In the* **cœlom** *or body cavity note the following, deflecting organs as necessary:—*

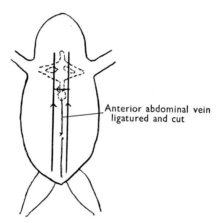

Anterior abdominal vein ligatured and cut

FIG. 162. **Rana. Method of Opening the Body Cavity.**

The large reddish-brown **liver** consists of two large lobes, the left one being subdivided into two. A small dark green sac, the **gall-bladder,** lies between them. *Deflect the liver forwards.* The **stomach** is a muscular sac on the left side of the animal into which the **œsophagus** leads from the pharynx. *Deflect the organs to the animal's right to see it.* Continuous with the stomach is the first part of the intestine (sometimes called the **duodenum**), and between them is a slight constriction, the **pylorus.** In the mesentery tissue between the stomach and the duodenum lies a whitish gland, the **pancreas.** *Squeeze the gall-bladder with forceps and thus inject bile into the* **bile duct** which comes from the gall-bladder and traverses the pancreas. The **hepato-pancreatic duct** leads from the pancreas into the duodenum. The intestine continues as a coiled tube, the **small intestine** (sometimes called the **ileum**) between the coils of which is the vascular membrane, the **mesentery.** The small intestine leads into a short wider tube, the **rectum,** which opens into the **cloaca.**

Note also the following viscera:

The **heart** in its **pericardium** is a conical reddish organ lying more or less in the centre at about the level of the fore-limbs (it may still be beating). The **lungs** lie one on each side of the heart, and the **urinary bladder** is a thin-walled bilobed sac in the posterior end of the cœlom, which opens ventrally into the cloaca.

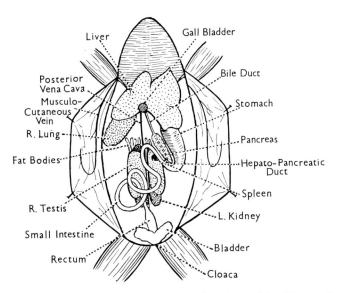

Fig. 163. **Rana. Alimentary Canal. Ventral Dissection. Male. Liver Deflected Forwards.**

Deflect the viscera slightly to the left of the animal.

In the *male* note the two yellowish oval bodies, the **testes,** and dorsal to them the reddish **kidneys** will be seen. In the *female,* the **ovaries** (containing a mass of **ova,** in the breeding season) and the coiled **oviducts** which may be distended with ova in the breeding season, will be seen. The ovaries and oviducts may considerably obscure the other viscera if they are filled with eggs, In both sexes the **corpora adiposa,** or fat bodies, yellow finger-like processes are situated at the anterior ends of the kidneys and the **posterior vena cava,** a large blue vein, runs between the kidneys up to the heart. Note also the **anterior abdominal vein,** already ligatured and cut, and the **spleen,** a small red spherical body in the mesentery near the beginning of the rectum.

Look for branches of the **mesenteric artery** and **vein** which supply the intestine lying in the mesentery.

THE VASCULAR SYSTEM

A mounted lens on a stand will be found useful in the following dissections which are best performed on a freshly killed male animal.

The circulation of the blood in the frog is a double circulation, *i.e.,* the blood is pumped by the heart to lungs and skin and, back to the heart and then on to the rest of the body and so back to the heart again.

The Heart

In a freshly killed frog, the heart may still continue to beat for some time. If this is so, note how it contracts. This will be more visible after removal of the pericardium.

Carefully free the heart from the pericardium by cutting through the latter with the fine scissors and removing it.

Note that the heart is roughly conical with the apex directed backwards and that it is composed of a single thick-walled **ventricle** at the free end with two **auricles** anterior to it. The **truncus arteriosus** arises from the right upper side of the ventricle on the ventral side and passes between the auricles, at the upper edge of which it divides into three **aortic arches.**

Turn the heart upwards and examine the dorsal view.

Note the thin-walled **sinus venosus,** roughly triangular in shape, at the apices of which the two **anterior venæ cavæ** and the **posterior vena cava** enter. The sinus leads into the right auricle, and the **pulmonary vein,** formed by a branch from each lung, leads into the left auricle.

Deflect the lungs to the sides of the animal when the pulmonary veins will be seen on their inner surfaces.

The Venous System

It is easier to trace the veins from the heart, but remember that the blood flows in them to the heart.

Deflect the heart forwards or, better, to one side. A small pin through the tip of the ventricle will keep it in place.

ANTERIOR TO THE HEART

Find the **Anterior Venæ Cavæ** which, as already seen, lead into the **sinus venosus.**

Just after leaving the auricle, the **anterior vena cava** receives:—

(i) The **External Jugular Vein,** which is the most anterior of three veins entering at this point. It runs underneath the mylo-hyoid muscle, alongside the hypoglossal nerve (already seen), and is formed by the union of the (*a*) **mandibular vein,** a small vessel from the lower jaw and (*b*) the slender **lingual vein** from the tongue and continuous with the external jugular.

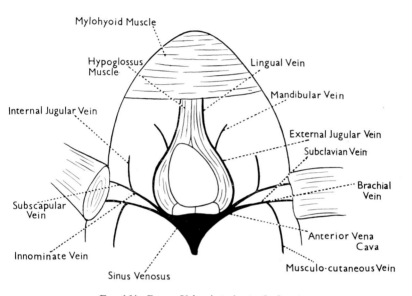

FIG. 164. **Rana. Veins Anterior to the heart.**

(ii) The **Innominate Vein** is the middle one of the three veins and leads to the angle between the shoulder and the jaw. It receives:—

(*a*) the **internal jugular vein** from the angle of the jaw;

(*b*) the **subscapular vein** which comes from the shoulder.

(iii) The **Subclavian Vein** is the posterior and largest of the three veins entering the anterior vena cava. It comes from the fore-limb, where it is formed by the union of the **brachial vein** from the arm and the **musculo-cutaneous vein** from the skin and body wall. *Gently dissect apart the muscle fibres in the limb in order to see these two veins.*

The musculo-cutaneous vein will be seen curved round on the inside of the skin which has been pinned back on each side.

POSTERIOR TO THE HEART

Find the **Posterior Vena Cava** which, as already seen, also leads into the **sinus venosus.**

At this point it receives the two **hepatic veins** from the liver (one from each lobe).

Deflect the alimentary canal to the animal's right and trace the posterior vena cava backwards.

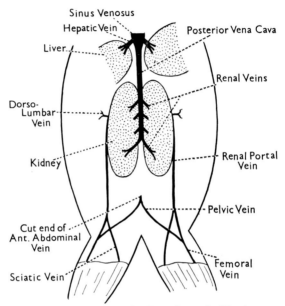

Sinus Venosus
Hepatic Vein
Liver
Dorso-Lumbar Vein
Kidney
Cut end of Ant. Abdominal Vein
Sciatic Vein
Posterior Vena Cava
Renal Veins
Renal Portal Vein
Pelvic Vein
Femoral Vein

FIG. 165. **Rana. Veins Posterior to the Heart.**

It passes between the kidneys, where it has its origin in the **renal veins,** four or five pairs of small vessels coming from the kidneys.

The **Genital Veins (Spermatic** or **Ovarian)** usually enter one of the renal veins, though they sometimes lead directly into the posterior vena cava.

The **Renal-Portal Vein** runs along the outer edge of each kidney, in which it terminates in capillaries.

Trace one of the renal-portal veins backwards on one side.

Just before entering the kidney it receives the small **dorso-lumbar vein** from the muscles of the dorsal body wall. Further back it is formed by the union of the **sciatic vein** from the inside of the thigh, and the **femoral vein** from the other side of the hind limb.

The **femoral vein** divides anteriorly into the renal portal vein (already seen) and the **pelvic vein** which runs inwards and ventralwards to join its fellow.

Find the femoral vein in the outer part of the thigh muscles immediately below the surface. Trace it backwards by carefully separating the muscles of the thigh.

Now trace the **anterior abdominal vein** *backwards.*

It is formed by the union of the two **pelvic veins.** This will usually be seen on the back of the part of the body wall where the anterior abdominal vein was ligatured.

Trace the anterior end of the severed anterior abdominal vein forwards.

It divides into two branches, one to each lobe of the liver.

The **Hepatic Portal Vein** will be found as follows:—

At the bifurcation of the anterior abdominal vein the **hepatic portal vein** will be seen entering it. It is formed by union of **gastric, intestinal** and **splenic veins** from the stomach, intestine and spleen. *Examine the mesentery again.* In it will be seen the intestinal vein. The gastric will be found on the stomach surface and the splenic coming from the spleen.

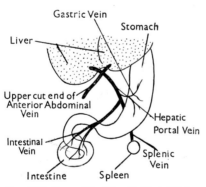

FIG. 166. **Rana. The Hepatic Portal Vein.**

The Arterial System

Ligature the anterior vena cava on one side in two places and cut between the ligatures, then carefully remove the three veins which form the anterior vena cava on that side, i.e., the external jugular, the innominate and the subclavian. (If the frog has been preserved since your last dissection, there should be little, if any, bleeding, but if it is a fresh specimen, a considerable amount of blood will be lost unless the ligature is done as directed above.) Make a small roll of paper, insert it into the mouth and push it down into the œsophagus to distend it. This will make the arteries more easily seen.

The **Aortic Arch** which as already seen arises from the **truncus arteriosus** on each side, gives rise to three arches:—

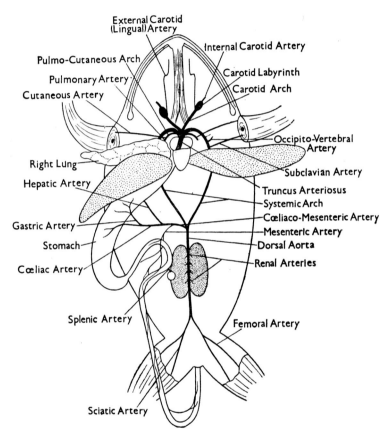

External Carotid (Lingual) Artery

Internal Carotid Artery

Pulmo-Cutaneous Arch

Pulmonary Artery

Cutaneous Artery

Carotid Labyrinth

Carotid Arch

Occipito-Vertebral Artery

Right Lung

Subclavian Artery

Hepatic Artery

Truncus Arteriosus

Systemic Arch

Cœliaco-Mesenteric Artery

Gastric Artery

Mesenteric Artery

Stomach

Dorsal Aorta

Cœliac Artery

Renal Arteries

Splenic Artery

Femoral Artery

Sciatic Artery

FIG. 167. **Rana. Arterial System.**

(i) **The Carotid arch** is the most anterior. It runs upwards and soon gives off a small **external carotid** (also called the **lingual artery**) running under the hypoglossal nerve to the tongue. Immediately after giving off the lingual artery, the carotid arch forms a labyrinth of minute vessels appearing as a swelling, called the **carotid labyrinth.** From this point the artery continues as the **internal carotid** to the orbit, though it cannot be traced without further dissection, where it continues to the brain, giving a branch, the **pharyngeal,** to the pharynx, palate, and orbit.

(ii) **The Pulmo-cutaneous arch** is the hindermost arch.

Deflect one lung to the opposite side, pin it in place, and find the arch on the exposed side.

It is a short arch which soon divides into a **pulmonary artery** running dorsally to the lung and a **cutaneous artery** which runs out towards the shoulder region on the dorsal side and then continues to supply blood to the skin of the back, along the entire length of the body.

(iii) **The Systematic arch** is the middle arch.

Deflect the alimentary canal to the animal's right and pin it in this position. Trace the arch on the exposed side.

On leaving the aorta the systemic arch on each side loops round the œsophagus to the dorsal side and runs backwards, joining its fellow just above the anterior edge of the kidneys to form the **dorsal aorta.** Find the **subclavian artery,** a large vessel which arises just after the loop of the systemic arch begins to turn downwards and which goes to the shoulder and fore-limb.

Just anterior to the origin of this artery is the **occipito-vertebral artery,** a short vessel dividing into the **occipital artery** to the head and the **vertebral artery** to the spinal cord.

The Cœliaco-mesenteric Artery arises at the junction of the systemic arches. It divides into:—

(i) the **cœliac artery,** which branches into the **gastric artery** to the stomach and the **hepatic artery** to the liver:

(ii) the **mesenteric artery,** which divides into:—

(*a*) the **anterior mesenteric artery** to the first part of the small intestine;

(*b*) the **posterior mesenteric artery** to the hinder part of the small intestine and, in some specimens, to the rectum.

(*c*) the **splenic artery** to the spleen.

The Dorsal Aorta runs in the mid-dorsal line between and dorsal to the kidneys and gives rise to:—

(i) the **renal arteries,** which enter the kidneys;

(ii) the **genital arteries (spermatic** or **ovarian),** which go to the gonads (testes or ovaries). Sometimes these arteries arise from the renal arteries;

(iii) the **rectal (or hæmorrhoidal) artery,** which goes to the hinder part of the intestine or rectum, present only in specimens in which this is not supplied by the posterior mesenteric artery.

Just posterior to the kidneys the aorta bifurcates. The two arteries thus formed are the **iliac arteries** which go to the legs.

Cut through the bony pelvic girdle with the large scissors in the mid-central line and trace the iliac artery downwards into the leg on one side.

It divides into the *sciatic and femoral arteries* on the inside and outside of the leg respectively.

THE URINO-GENITAL SYSTEM

Remove the alimentary canal by cutting through the œsophagus, the mesentery, holding the alimentary canal with forceps, and the rectum.

Male

Note the two red, flat, oval **kidneys.** They are partly hidden by the testes. Trace the kidney duct, which also serves as a genital

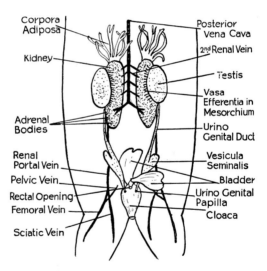

FIG. 168. **Rana. Urino-Genital Organs. Male. Bladder Deflected. Anterior Abdominal Vein not shown.**

duct and is known as the **urinogenital duct** (Wolffian duct), from the kidney to the **cloaca** which it enters by a **urino-genital papilla.** After leaving the kidney, it bears a sac, the **vesicula seminalis.** Note also

the bilobed **bladder.** On the ventral side of each kidney is the **testis.**
Ducts from this organ, the **vasa efferentia,** lie in a supporting tissue,
the **mesorchium,** and lead into the kidney.

Slit open the cloaca which will be found beneath the centre of the
pelvic girdle, already severed when dissecting the arterial system. On
its dorsal side the two urino-genital papillae will be seen side by side
in the mid line. *Then by means of a seeker or bristle find* the **opening**
of the bladder, the **rectal aperture** and the **external cloacal aperture.**

Note also the yellow digitate **corpora adiposa** at the anterior end
of the kidneys and the **inferior vena cava** between the kidneys, into
which the **renal veins** lead. Find the **femoral vein** seen already in the
muscles of the outer side of each leg, the **sciatic vein** from the inner
side of the leg. the **renal-portal vein,** formed by the union of the
sciatic and femoral veins and running along the outer edge of each
kidney and the **pelvic vein** joining its fellow from the opposite side
to form the **anterior abdominal vein** (as seen in the dissection of the
venous system).

Deflect one of the testes to one side so as to expose the ventral
surface of the kidney. Observe the light patches on the kidney: these
are the **adrenal bodies.**

Female

Note the **kidneys,** largely hidden by the ovaries, and their ducts,
the **ureters** (Wolffian ducts), each of which enters the **cloaca** by a

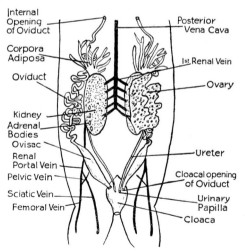

Fig. 169. **Rana. Urino-Genital Organs. Female. Right Ovary**
and Bladder Removed. Anterior Abdominal Vein not shown.

urinary papilla, the **adrenal bodies,** and the **bladder.** The lobed **ovaries** are supported by a tissue, the **mesovarium,** and may contain a large number of **ova.** The **oviducts** (Mullerian ducts) are two coiled tubes which lead to the cloaca and near the posterior ends bear swellings, the **ovisacs.** During the breeding season these may be distended with eggs. *In this case remove one of the oviducts with its ovisac. Trace one oviduct forwards and find* the funnel-shaped **internal opening of the oviduct** at the side of the œsophagus on a level with the base of the lung. *Open up the cloaca as explained for the male (p. 217). By means of a seeker or bristle find the* **oviducal apertures** *and the* **ureter openings.** *They will be found on the sides and ends of two small papillæ. Find also the* **opening of the bladder,** the **rectal aperture** and the **external cloacal aperture.** Note also the **inferior vena cava, renal, femoral, renal-portal, sciatic, pelvic** and **anterior abdominal veins** and the **corpora adiposa** as in the male.

THE NERVOUS SYSTEM

This consists of the **central nervous system,** the **peripheral nervous system** and the **autonomic nervous system,** which is more developed than in the dogfish.

The Spinal Nerves

Remove the heart, lungs, liver, kidneys and testes (or ovaries), exercising care, particularly when removing the kidneys. If a fresh animal is being used, first remove the alimentary canal, of course. Leave the systemic arch and dorsal aorta in position.

Note the bony **vertebral column** and the ten pair of **spinal nerves,** white cords which pass out between the vertebræ, being surrounded by white **calcareous concretions** as they emerge.

Trace out the following nerves:—

I. The **Hypoglossal nerve** leaves the spinal column between the 1st and 2nd vertebræ and runs forwards underneath the mylo-hyoid muscle to the tongue. *The mylo-hyoid muscle must be cut in the midline and the flaps deflected and pinned out if this has not already been done, to see the nerve, but it should already have been seen in earlier dissections.*

II, III. The 2nd nerve leaves the spinal column between the 2nd and 3rd vertebræ and is joined by the 3rd nerve, which emerges between the 3rd and 4th vertebræ. A branch from the hypoglossal joins them, thus forming the **brachial plexus.** The main trunk then continues into the arm as the **brachial nerve.**

IV, V, VI leave the spinal column between the 4th and 5th, 5th and. 6th, and 6th and 7th vertebræ respectively. They are all small and run outwards and backwards to the skin and muscles of the body wall.

VII, VIII, IX emerge from the next three intervertebral spaces and unite to form the **sciatic plexus.** Several branches arise from this, the large **sciatic nerve** to the leg being the most easily seen and being formed by the union of VIII and IX.

X. The **Coccygeal nerve** leaves the vertebral canal by a foramen in the urostyle. Branches go to the bladder, etc., and to the sciatic nerve.

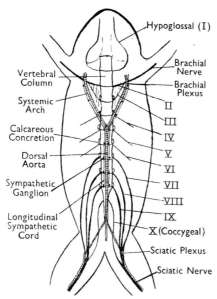

FIG. 170. **Rana. Spinal Nerves.**

The Autonomic Nervous System

Note the chain of pigmented **sympathetic ganglia** outside the vertebral column on either side of the dorsal aorta, joined by the **longitudinal cords** of the sympathetic chain. Branches, the **rami communicantes,** connect the spinal and sympathetic* nervous systems.

THE BRAIN AND SPINAL CORD

Examine again the skull and vertebral column of an articulated skeleton before you begin this dissection.

* Part of the Autonomic Nervous System.

Pin the frog, dorsal side uppermost, in the dissecting dish. Remove the skin from the dorsal side; then, carefully inserting small scissors from behind, cut through the sides of the cranium on both sides, keeping the blade of the scissors flat. Then cut through the occipito-atlantal membrane covering the space between the skull and the first vertebra and through any remaining tissues which prevent the lifting of the roof of the cranium. Deflect the roof of the cranium forwards and then remove as much as possible of the sides.

Now turn the dish round so that the head of the frog points towards you, insert the scissors into the neural canal of the first vertebra taking care not to injure the spinal cord within and carefully remove the roof (neural arch) of the vertebra. Remove the tops of the other vertebræ in succession and so expose the spinal cord.

Dorsal View

*Examine in situ from front to back, removing the pigmented membrane, the **pia mater,** as you proceed.*

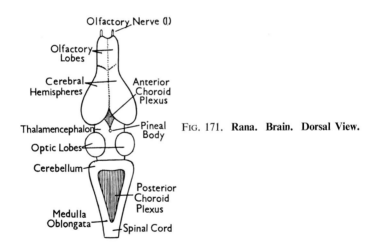

Fig. 171. **Rana. Brain. Dorsal View.**

In the **Fore-brain** note the anteriorly placed **olfactory lobes,** joined in the middle line so that they appear individed. Behind them are the large **cerebral hemispheres** separated by a median fissure. These are followed by a small portion, the **thalamencephalon,** covered by a vascular membrane, the **anterior choroid plexus.** This part of the brain bears the **pineal body** on a short stalk. (The pineal body will probably have been removed with the roof the skull so that only the stalk remains.) The side walls of the thalamencephalon are thick and are known as the **optic thalami.** The **Mid-brain** consists of two spherical **optic lobes** immediately behind the thalamencephalon. The **Hind-brain** follows first as a narrow transverse band, the **cerebellum,** and then by the tapering **medulla oblongata,** covered by the vascular **posterior choroid plexus.**

The spinal cord is continuous with the medulla oblongata. It tapers towards the posterior end, terminating in the thin **filum terminale** in the urostyle. Note that it is wider where the nerves for the fore-limb originate, the **branchial enlargement.** and again where the nerves for the hind-limb arise, the **lumbar enlargement.** Note also that the cord is somewhat flattened and that it has a fairly deep groove, the **dorsal fissure,** along its length. The hinder-most nerves run backwards in the vertebral canal for some distance before they pass out of it. This group of nerves, together with the filum terminale, is called the **cauda equina.**

Ventral View

It is best to hold the frog in the hand, resting the head along the fingers when removing the brain.

Cut through the olfactory nerves which hold the brain in position in front, then cut through the remaining nerves at the sides and base, keeping as far away from the brain as possible.

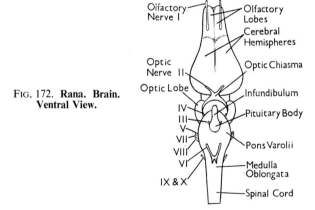

FIG. 172. **Rana. Brain. Ventral View.**

Cut along each side of the spinal cord with a small scalpel; this will sever the spinal nerves. The spinal cord can then be removed from the ventral column, the filum terminale being gently pulled out of the urostyle with forceps.

Gently turn the brain and spinal cord so as to expose the ventral surface. You may find it easier to hold the frog vertically in your hand to remove the brain and spinal cord. Everything must be done gently and on no account exercise any force.

Note the **olfactory lobes** and **olfactory nerves (I)**, **cerebral hemispheres** and the sides of the **optic lobes.** The **optic nerves (II)** cross over one another forming the **optic chiasma.** Immediately behind this is the **infundibulum,** a grooved median swelling on the floor of the thalamencephalon, attached to which is the **pituitary body** or **hypophysis.** (This may have been broken off when the brain was removed.) The **crura cerebri** join the thalamencephalon to the medulla but are hidden by the posterior end of the pituitary body.

The Spinal Cord shows a groove, the **ventral fissure,** along its length which is deeper than the dorsal fissure.

Horizontal Section

The brain must be hardened in 90 per cent. alcohol before a section is made. This takes three or four days.

With the brain preferably on its side, carefully cut off the roof of the brain with a small scalpel, inserting it into the side of the brain.

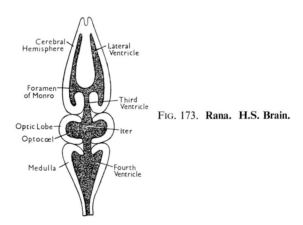

FIG. 173. **Rana. H.S. Brain.**

Note that the brain contains cavities or ventricles as follows:—

The **lateral Ventricles** are in the **cerebral hemispheres** and communicate in the centre with the **third ventricle** in the **thalamencephalon** by the **foramen of Monro.** The third ventricle communicates by the **iter** (or **aqueduct of Sylvius**) with the **fourth ventricle,** which is in the **medulla.** The **optocœls** in the optic lobes open into the iter.

THE CRANIAL NERVES

Using a fresh frog (preferably hardened in alcohol), expose the dorsal surface of the brain as before, pour in 90 per cent. alcohol and remove the skin round the eye. Trace out the cranial nerves on one side as follows:—

I. The **Olfactory Nerve** arises from the olfactory lobe and goes to the epithelium lining the nasal cavity. It is not worth while dissecting this out.

II. The **Optic Nerve** which arises from the optic lobe on the opposite side, thus forming the **optic chiasma** goes to the eyeball. *Seperate the recti muscles at the posterior end of the orbit. The optic nerve will be seen in the* **retractor bulbi muscle** *in the midst of them.*

III. The **Oculomotor Nerve** is a very small nerve arising from the ventral side of the brain between the crura cerebri and supplies four of the eye muscles, namely, the rectus superior, inferior and internus and the obliquus inferior. (These muscles have been studied in the dog-fish.) This nerve is difficult to trace.

IV. The **Pathetic or Trochlear Nerve** is another slender nerve, arising between the optic lobes and the cerebellum on the dorsal side and is difficult to trace. It supplies the obliquus superior muscle, and is too small to be worth tracing out.

VI. The **Abducens Nerve** is also very slender and difficult to trace. It arises from the ventral surface of the medulla and supplies the rectus externus muscle.

Dissection of the IXth and Xth nerves

It is best to dissect these nerves before the Vth and VIIth as the lower jaw must be removed to see the latter.

IX. The **Glossopharyngeal Nerve** arises behind the VIIIth nerve in the side of the medulla by a root common with the Xth nerve. It passes through the skull immediately behind the auditory capsule and then divides into:—

    (*a*) An **anterior branch,** running forwards and downwards, which joins the VIIth (facial) nerve.

    (*b*) A **posterior branch,** also running forwards and downwards, which goes to the base of the tongue.

X. The **Vagus Nerve.** This arises with the IXth nerve as already seen. It also leaves the skull with the IXth nerve and then, after a ganglionic swelling, gives rise to four branches as follows:—

(a) The **laryngeal branch** which runs under the hypoglossal nerve anteriorly to the larynx.

(b) The **cardiac branch** which runs to the heart.

(c) The **pulmonary branch** which runs alongside the pulmonary artery to the lung.

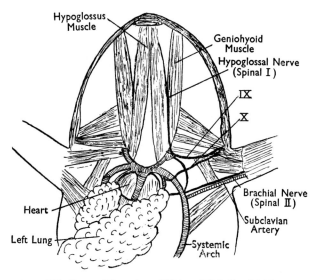

FIG. 174. **Rana. Dissection of IXth and Xth Cranial Nerves.**

(d) The **gastric branches** (usually two) which go to the stomach.

Open up the ventral side of the body to expose the viscera in the usual way.

Carefully remove the mylohyoid muscle. Find the hypoglossal nerve (Spinal I) and the geniohyoid muscle running longitudinally from the anterior part of the lower jaw and dividing posteriorly. Note also the hypoglossus muscles in the mid-line running from the anterior end of the lower jaw and dividing posteriorly. *Now find the external jugular vein.* The **IXth nerve** runs from the angle of the jaw, crosses the hypoglossal nerve alongside the external jugular vein and runs under the geniohyoid muscle to the floor of the mouth. The **Xth nerve** runs more or less parallel with the IXth with some muscle between before it divides into the four branches mentioned above.

Find the **cardiac, pulmonary** and **gastric branches.**

Cut through the external jugular, innominate and sub-clavian veins on one side where they join the anterior vena cava and remove these veins: otherwise the nerves will be hidden. Deflect the lung of this side to the opposite side of the body. The IXth and Xth nerves will then be seen.

Dissection of the Vth and VIIth Nerves

V. The **Trigeminal Nerve** is a large nerve—the largest cranial nerve—and arises from the side of the anterior end of the medulla. It bears a swelling or ganglion, the **Gasserian ganglion,** just before it enters the bone of the skull. After leaving the skull, just anterior to the auditory capsule, it divides into:—

(*a*) the **Ophthalmic Branch** which passes through the orbit between the bone of the skull and the eye, and out at the anterior end where it divides into two branches supplying the skin of the anterior part of the head. *Move the eyeball away from the inner wall of the orbit to see this branch.* It will be seen again when studying the VIIth nerve.

(*b*) the **Maxillary Branch.** This branch, which will also be seen later, runs at the back of the orbit to the upper jaw or maxilla and arises from a second branch of the Vth nerve, the maxillo-mandibular branch.
    Press the eyeball sideways to see this branch.

(*c*) the **Mandibular Branch** arises from the maxillo-mandibular branch and runs behind and at first parallel to the maxillary branch. The mandibular branch leaves the maxillary branch at the outer edge of the eye and runs to the lower jaw or mandible, in close association with the mandibular branch of the VIIth nerve but external to it, It, too, will be seen later in the dissection.

VII. The **Facial Nerve** arises from the side of the medulla immediately behind the Vth nerve, leaving the skull behind it and then divides into a **palatine branch** and a **hyomandibular branch.**

(*a*) The **Palatine Branch** runs straight forwards in the floor of the orbit from the inner posterior corner to the nasal region on the roof of the mouth. *Trace this branch as follows:—*

With the frog ventral side up, cut transversely from the angle of the jaw on one side and deflect the lower jaw to the opposite side. If the mucous membrane covering the roof of the mouth is now carefully cut away, this branch will be seen on the ventral surface of the eye.

Move the eyeball outwards.

The **ophthalmic branch of the Vth nerve** will now be seen running forwards along the inner edge of the orbit.

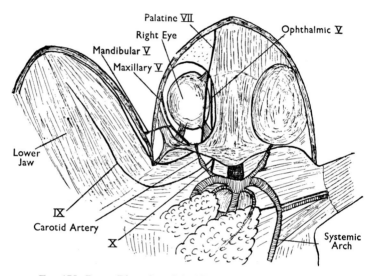

Fig. 175. Rana. **Dissection of the Vth and VIIth Crainial Nerves.**

Now find the **maxillo-mandibular branch of the Vth nerve** *at the posterior end of the orbit and trace it forwards to the point where it divides. Follow the* **maxillary branch** *in the floor of the orbit to the upper jaw. Then trace the* **mandibular branch** *across to the now deflected lower jaw. It runs along the outer edge of the jaw and will be seen if the muscles there are carefully separated.*

(*b*) The **Hyomandibular Branch of the VIIth nerve** is not easy to follow.

It runs round the anterior end of the auditory capsule where it divides into a posterior **hyoidean branch** to the muscles of the hyoid in the floor of the mouth, and a **mandibular branch** running along the inner edge of the lower jaw, in close association with the mandibular branch of the Vth.

Again examine the edge of the **lower jaw.** *Carefully separate the muscles there and find the* **mandibular** branch running along the inner edge of the jaw. After a short distance it gives off a branch which anastomoses with the mandibular branch of the Vth nerve.

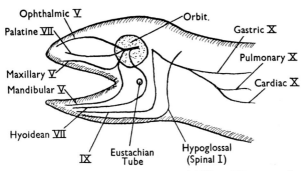

FIG. 176. **Rana. Cranial Nerves. Laterial View. Diagrammatic.**

VIII. The **Auditory Nerve** arises from the side of the medulla, immediately behind the VIIth nerve, enters the auditory capsule and supplies the internal ear.

THE HYOID APPARATUS

Now that you have completed the dissection of the nerves, you can proceed to expose the hyoid apparatus.

Remove all the nerves and any traces of muscle from the floor of the mouth, taking great care that in so doing you do not damage the hyoid and with the knowledge that there is some muscle attachment to the hyoid. The outline of the **body of the hyoid** *will become visible and should be gently cleaned up with a small scalpel. When this is completed very carefully trace out the* **anterior** *and* **posterior cornua** *from the corners.*

SUB-CLASS URODELA

Tailed throughout life. All limbs short and of equal length. Some retain gills in adult. The larva of the *Axolotyl* persists and becomes sexually mature. This is the larva of the *Amblystome* into which it can metamorphose. *Newts* and *salamanders.*

TRITON and MOLGE

NEWTS

*Examine a specimen of a **newt** externally.*

The **Common Newt,** *Triton,* is about 10 cm. in length and the **Crested Newt,** *Molge,* about 15 cm. The body is thus longer than that of the frog and it is also thinner and has a slightly flattened tail. The limbs are short and of equal length and are devoid of claws. The skin of *Triton* is of a pale brown colour dorsally and yellow ventrally and both surfaces have black pigmented spots. In Spring the colour changes to orange and the male develops a crest. *Molge* has a dark brown, rather warty upper surface and a yellow one with black spots beneath. In both the skin lacks scales and is moist.

CLASS REPTILIA

Completely adapted to a terrestrial life though some are aquatic. Skin covered with scales. Breathe by lungs. Four-chambered heart. Poikilothermic. Fertilisation internal. Lay heavily yolked eggs protected by shells. *Lizards, snakes, tortoises, turtles, crocodiles* and *alligators.*

LACERTA

THE LIZARD

*Examine the **Common Lizard** (Lacerta vivipara).*

The body is divided into head, neck, trunk and tail and is completely covered with **horny scales.** Dorsally the animal is green but on the ventral side it is brown or yellow.

The **head** is comparatively small, somewhat pyramidal in shape and slightly flattened dorso-ventrally. The **mouth** is wide and anteriorly placed, the **external nostrils** being dorsal to it. The **eyes** are dorso-lateral and have upper and lower lids and a nictitating membrane. Behind each eye is a brown circular patch which is the **tympanic membrane,** there being no external ear.

The long **trunk** is rather flattened laterally and at its posterior end is the **cloaca.** The limbs are short, the **fore-limbs** being situated at the anterior end of the trunk and the **hind limbs** at the other end. Both manus and pes have five digits which terminate in claws.

The **tail** is extremely long, almost twice the length of the rest of the body, and is cylindrical and tapers towards its extremity.

TROPIDONOTUS NATRIX

THE GRASS SNAKE

Examine the **Grass Snake.**

Snakes are, of course, limbless reptiles and the body is divided into **head** and **trunk.** The shape of this body is cylindrical and it may attain a length of 1·2 to 1·8 m. The skin is covered with **scales** which are shed periodically as a continuous slough. The colour varies in different species of snakes and the grass snake is greenish with black and yellow markings on its head. The **eyes** are dorsally situated. The **mouth** is expansible and in the buccal cavity is a long and **protrusible tongue.** The **teeth** are fused to the bone of the upper and lower jaws. The **cloaca** is at the posterior end of the trunk. (Venomous snakes have pointed poison fangs in the mouth and these have grooves which serve as ducts for the poison which is secreted by the poison gland.)

CLASS AVES

Body covered with feathers except on hind limbs in some cases in which they are covered with horny scales. Forelimbs modified into wings. Jaws encased in horny beak and devoid of teeth. Homoiothermic. Fertilisation internal. Lay large-yolked eggs in nests. *Birds.*

COLUMBA

THE PIGEON

Pigeons are varieties of the wild Rock Dove (*Columba livia*). The well-known Wood Pigeon is *C. palumbus.*

Examine a **pigeon** *externally. A plucked bird will also be required.*

The body, which is divided into head, neck, trunk and tail is completely covered with feathers except on the beak and feet. The colour is bluish-grey with a greenish blue sheen on the neck.

The **head** is spherical and bears a pointed **beak** composed of upper and lower jaws devoid of teeth and encased in horny skin. At the base of the beak are the **external nostrils,** partly hidden by the **cere,** a swollen, sensory patch of skin of waxy appearance. The **eyes** are laterally placed, each has its own field of vision and each is protected by upper and lower lids and a nictitating membrane. Posterior to the eyes, hidden in feathers known as the **ear coverts,** are the **auditory apertures.** The long **neck** can be turned through 180°. *Examine the plucked bird* and note that the **trunk** is boat-shaped. It has a distinct **keel** on the ventral side and at the posterior end is the **cloacal** aperture. The fore-limbs are modified to serve as **wings.** The skeleton of this limb is of typical vertebrate form but is devoid of digits. The wing is folded flat against the body when at rest and it cannot be straightened out completely. This is due to two folds of skin, one across the elbow, the **propatagium,** and the other across the arm-pit, the **postpatagium.** *Examine the plucked bird to see these.* The **hind-limbs** are vertically placed, the thigh being hidden beneath the skin, and the **pes** is covered with **scales** similar to those found in reptiles. There are four digits, three directed forwards and one, the **big-toe,** backwards. This enables the bird to perch. The toes end in claws. *Again examine the plucked bird.* The **tail** is short and stumpy and at its apex on the dorsal side is the **oil gland aperture.**

Feathers

There are three kinds of feathers:—

Contour feathers which include the **covert** feathers which cover the whole body (except the beak and feet) and the **down feathers** between

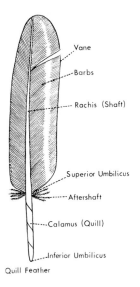

Vane

Barbs

Rachis (Shaft)

Superior Umbilicus

Aftershaft

Calamus (Quill)

Inferior Umbilicus

Quill Feather

FIG. 177. **Columba. Quill Feather.**
(*From Wallis—Practical Biology*)

them. **Quill feathers** are found on the wings and tail and **filoplumes,** minute thread-like feathers lie all over the body between the coverts. (These are left when a bird is plucked and their removal entails singeing.)

Examine a **quill feather from the wing** *or* **tail.** *Use a lens for detail.*

The feather is composed of a **shaft** or **rachis,** convex on its upper surface and with a groove on the lower, on each side of which is the flattened and expanded **vane** or **vexillum.** This is made up of **barbs** on either side of the rachis arranged like teeth on a comb, though they are obliquely placed. *Examine with a lens.* It will be seen that each barb bears branches called **barbules** which interlock by means of tiny hooks known as **barbicels.** This cohesion gives the feather a continuous surface during flight. At the lower end of the vane is a small tuft of small feathers devoid of barbicels and known as the **aftershaft.** The continuation of the rachis beyond the vane is known as the **quill** or **calamus** and at its upper end is a minute pit called the **superior umbilicus** while at its lower extremity is a hole called the **inferior umbilicus.** The quill feathers on the wing are known as **remiges** and those on the tail as **rectrices.** They differ slightly in that in the former the vane is asymmetrical, one side being wider than the other and somewhat concave on the upper surface while the rectrices are flat and have a symmetrical vane. Feathers are made of keratin.

Examine a **covert feather.**

This is similar to a quill feather but the rachis is pliable and the vane much softer.

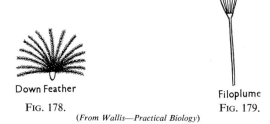

Down Feather

Fig. 178.

Filoplume

Fig. 179.

(*From Wallis—Practical Biology*)

Examine a **down feather.**

This, too, is very soft. The rachis is short and very soft barbs stretch out from it.

Examine a **filoplume.**

This is composed of a thin thread bearing a small tuft at tis tip.

Examine a **complete wing.**

The arrangement of the coverts and quill will be seen in the illustration. The **primary quills** are at the tip. These are followed by the **secondary quills** and then the **primary coverts** followed by the smaller **secondary coverts**. A small group of feathers, known as the **thumb feathers** (also called the **bastard wing**) will be found on the upper surface.

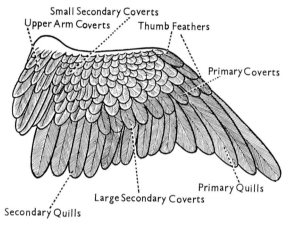

FIG. 180. **Columba. Wing.**
(From Wallis—Practical Biology)

Examine the plucked bird again.

The feathers are arranged in definite tracts known as **pterylae,** the tracts devoid of feathers being called **apterylae.** This may not be clearly visible.

CLASS MAMMALIA

The body is divided into a **head** and **trunk,** separated by a distinct **neck,** the trunk bearing **four limbs** and is covered with **hair,** or some modification of it. The skin possesses **sebaceous glands** for lubrication of the hair, and **sudorfic** or **sweat glands** for excretion and osmo-regulation. The trunk is divided into an anterior **thorax** and a posterior **abdomen,** separated internally by a muscular partition, the **diaphragm.** The bones of the skull are immovably articulated by dovetail joints called **sutures.** The long bones and vertebræ bear epiphyses and between the vertebræ are inter-vertebral discs. The **cerebrum** is large and the surface is thrown into folds or **convolutions** in the higher types. The heart is **four-chambered,** having two auricles and two ventricles, and there is a double circulation (though not in the same sense as in Amphibia). The heart is situated between the lungs, which are enclosed in sacs, the **pleura,** in the thorax. There is **no** renal-portal system. Homoiothermic. The red corpuscles are non-nucleated. The intestinal and urino-genital openings are **separate.** Except in the Mono-tremata which lay eggs, the offspring develop in a special organ in the abdomen, the **uterus,** to the wall of which they are vitally joined by an **allantoic placenta.**

The ova are **small** and with **little yolk (microlecithal)**. The newly born young are fed for a limited period on milk secreted by special glands, **mammary glands,** in the skin of the mother.

SUB-CLASSES

MONOTREMATA	METATHERIA	EUTHERIA
Primitive mammals showing clear evidence of their reptilian ancestry. Lay large eggs with large yolks. Young fed by milk secreted by modified sudorific glands in skin of female: no teats are present. Have a cloaca. *Duckbilled Platypus, Spiny anteater.*	Marsupials. Primitive mammals in which young are developed in body of female but are immature at birth. Development continues in a pouch or marsupium where the young are fed on milk secreted by modified sebaceous glands. Separate urino-genital and anal apertures. *Kangaroo, Wallaby.*	

SUB-CLASS EUTHERIA

Most highly developed mammals. Young "fully developed" in uterus of female to which they are united by an allantoic placenta. Separate urino-genital and anal apertures. Highly developed brain. *All the common higher mammals.*

ORYCTOLAGUS

THE RABBIT

Though in the same **Order, Lagomorpha,** the *rabbit,* **Oryctolagus,** belongs to a different genus from that to which the *hare,* **Lepus,** belongs. Rabbits are gregarious and live in a warren of burrows. They serve as food for birds and other mammals including man but they do a great deal of damage to his crops. They breed rapidly and the young are born blind and naked. The disease *myxomatosis* introduced into this country several years ago reduced their numbers enormously but myxomatosis resistant strains have since developed. Farmers are compelled by law to keep the numbers down to avoid over population and consequent damage to crops.

Hares live in grass and bushes are and solitary animals. They also serve as food. Their young are born with fur and with their eyes open. The hind legs and external ears of hares are longer than those of the rabbit.

EXTERNAL ANATOMY

Examine a killed **rabbit.**

Note the colour of the body, which is covered with fur and is divided into head, neck, trunk, and tail.

Head

The head tapers towards the anterior end, or snout, where the **external nostrils,** will be seen as two slits on skin devoid of hair. The **mouth** below the nostrils, is bounded above by an **upper lip,** cleft in the middle, showing the *incisor teeth,* and by a **lower lip.** At the sides of the nostrils are stiff whiskers, or **vibrissæ,** which are tactile organs. The **eyes** are on the sides of the head: each has an **upper lid** and a **lower lid** both bearing **eyelashes,** and a hairless **nictitating membrane.** The **external ears,** or **pinnæ,** are large and are situated on the sides of the head at the posterior end. The entrance to the ear is called the **external auditory meatus.**

Neck and Trunk

A distinct **neck** connects the head to the trunk. The **trunk** consists of the **thorax** above, enclosed by the ribs and sternum, and the **abdomen** below, with simply a muscular body wall on the ventral side. The posterior end of the trunk bears a short **tail,** white on its under surface. The **anus** is situated at the root of the tail between the hind limbs.

In the **male,** note the **penis,** an intromittent organ in its sheath, the **prepuce.** On either side of the penis is a **scrotal sac,** which contains a testis. The region between the scrotal sacs and the anus is called the **perinæum.**

In the **female,** note the **vulva,** the slit-like urino-genital opening separated from the anus by the **perinæum,** with a small, rod-like **clitoris,** in its ventral wall. Note also the four or five pairs of **mammary glands,** bearing **teats,** on the ventral side of the abdomen.

Limbs

The **fore-limbs** of the rabbit are much shorter than the hind limbs and the **manus** bears five **digits,** whereas the **hind-limb** bears a **pes** with only four **digits,** the big toe, or hallux, being absent.

THE SKELETON

THE AXIAL SKELETON

This includes the ribs and sternum as well as the skull and vertebral column.

The skull of the rabbit is in many respects not typical of mammalia, particularly as regards the dentition. It is therefore usual to study the skull of the dog (*Canis*) in detail and then to compare with it the skull of the rabbit.

The Skull of the Dog

Cranium

The **cranium** is large and consists of three rings of bone:—

(i) A posterior **occipital ring** composed of a **supra-occipital** bone above, a **basi-occipital** below, and an **ex-occipital** on each side surrounding the **foramen magnum**. Note the **occipital condyles,** which articulate with the first vertebra (atlas) on the lower sides of the foramen magnum and the **paroccipital processes,** anterior to the occipital condyles and projecting downwards, both on the ex-occipitals.

(ii) A median **parietal ring** composed of the two **parietal bones** above, the **alisphenoids** at the sides, and the **basi-sphenoid** below. The small bone between the posterior ends of the parietals is the **inter-parietal,** forming a ridge, the **sagittal crest.**

Between the occipital and parietal rings is the **squamosal,** bearing a forwardly directed **zygomatic process** which with the jugal bone forms the **zygomatic arch** as will be seen below, and a **glenoid fossa** for the articulation of the lower jaw.

(iii) An anterior **frontal ring** composed of two **frontal bones** above, the **orbito-sphenoids** at the sides, and a **presphenoid** below. Note also the **post-orbital ridge** running from the frontals to form part of the posterior wall of the orbit.

Note the immovable joints or **sutures** between the bones of the cranium.

The anterior wall of the cranium consists of the **ethmoid** in which is the **cribriform plate.** This and the presphenoid, continue forwards as the **mesethmoid,** forming the posterior part of the **septum** of the nose, the anterior part being composed of cartliage.

Sense Capsules

The Auditory Capsule

This is formed by the **periotic bone,** divisible into (*a*) a **petrous** portion, placed internally, containing the auditory organ, and (*b*) a **mastoid portion,** situated partly on the surface between the ex-occipital and the **external auditory meatus.** This will be found on another bone, the **tympanic bone,** ventral and partly external to the periotic between the squamosal and the basisphenoid, bearing the **opening of the Eustachian tube** and protruding as a swelling, the **tympanic bulla.**

Examine the isolated **auditory ossicles,** *or bones of the ear, as follows:—*

The **malleus,** or hammer bone, the **incus,** or anvil bone, and the **stapes,** or stirrup bone. They are situated in the tympanic cavity.

Nasal Capsule

This is composed of two narrow **nasal** bones above, connected behind with the frontals, the trough-like **vomer** supporting the cartilaginous nasal septum from below and the **facial portions of the maxillæ** at the sides. The chamber is divided longitudinally by the **mesethmoid** posteriorly and by cartilage anteriorly which together form the **nasal septum,** and is bounded behind by the **cribriform plate,** as already seen. Note the **naso-turbinals,** the **maxillo-turbinals** and the **ethmo-turbinals,** three pairs of scroll-like bones on the inner sides of the chamber, and the **internal** and **external nostrils.**

Jaws and Orbit

The **Upper Jaws** are composed posteriorly of the **maxillæ,** with (*a*) the **facial** portions, each bearing **1 canine, 4 premolar** and **2 molar teeth,** and (*b*) the **palatal** portions: the last pre-molars are pointed and are known as **carnassial teeth.** Anteriorly the upper jaw is composed of the two **pre-maxillæ** joined together in the front, each of which bears the **3 incisor teeth.** The posterior part of the roof of the mouth is formed by the two **palatine** bones which surround the external nostrils and continue backwards with the vertically placed **pterygoids.** The zygomatic arch continues forwards as the **jugal** bone to the maxilla on each side.

The **Orbit** is bounded above by the frontal, and below by the jugal, anteriorly by the maxilla, and posteriorly by processes from the jugal and frontal. In the anterior wall is the **lacrimal,** in which the lacrimal canal leads to the nasal chamber. The orbito-sphenoid surrounds the optic foramen in the orbit.

The **Lower Jaw** (or **mandible**) consists of **2 dentaries** or **rami** fused anteriorly at the **mandibular symphysis,** each of which articulates posteriorly with the glenoid fossa of the squamosal by the somewhat rounded and elongated **condyle.** The **3 incisor, 1 canine, 4 premolar** and **3 molar teeth** are borne on each ramus. The first molars are **carnassial teeth** similar to those in the upper jaw. Between the condyle and the teeth is the flat upwardly directed **coronoid process,** and below the condyle is the **angle of the jaw** for muscle attachments.

From the above description and with the aid of the illustrations note the bones as follows:—

Dorsal View

The **supra-occipital, inter-parietal, parietal, squamosal, jugal, zygomatic arch, orbit, orbito-sphenoid, frontal, nasal,** anterior end of the **vomer, maxilla, pre-maxilla, incisor** and **canine teeth** (the others may not be visible in this view) and the **sutures.**

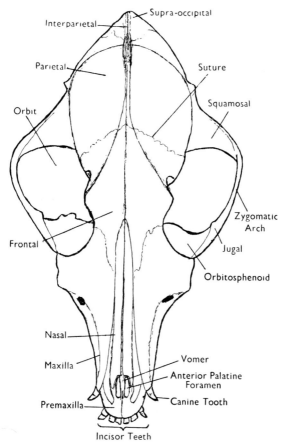

Supra-occipital

Interparietal

Parietal

Suture

Orbit

Squamosal

Frontal

Zygomatic
Arch

Jugal

Orbitosphenoid

Nasal

Vomer

Maxilla

Anterior Palatine
Foramen

Premaxilla

Canine Tooth

Incisor Teeth

FIG. 181. **Canis. Skull. Dorsal View.**

Ventral View

The **supra-occipital; ex-occipital; occipital condyles; foramen magnum; basi-occipital; condylar foramen** for the XIIth nerve anterior to the condyle; **tympanic bulla; foramen lacerum posterius;** a long foramen for the IXth, Xth and XIth nerves between the ex-occipital and the tympanic bulla; **squamosal; stylo-mastoid foramen** for the VIIth nerve outside the tympanic bulla; **basi-sphenoid; alisphenoid; foramen rotundum** for the maxillary branch of the Vth nerve in the alisphenoid; **foramen ovale** for the mandibular branch of the Vth nerve posterior to the foramen rotundum; the **Eustachian foramen** for the Eustachian tube just inside and slightly posterior to the foramen ovale; **alisphenoidal canal** for the external carotid artery between the foramen ovale and the foramen rotundum; and the **foramen lacerum medium** for the internal carotid artery next to

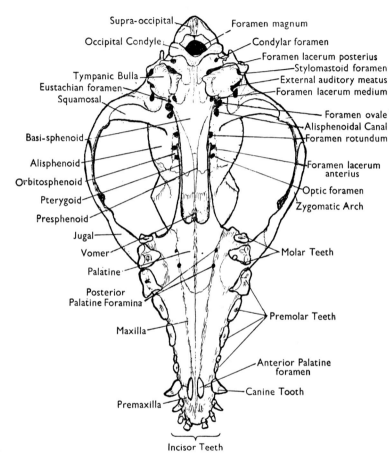

Supra-occipital
Foramen magnum
Occipital Condyle
Condylar foramen
Foramen lacerum posterius
Stylomastoid foramen
Tympanic Bulla
External auditory meatus
Eustachian foramen
Foramen lacerum medium
Squamosal
Foramen ovale
Alisphenoidal Canal
Basi-sphenoid
Foramen rotundum
Alisphenoid
Orbitosphenoid
Foramen lacerum anterius
Pterygoid
Optic foramen
Presphenoid
Zygomatic Arch
Jugal
Vomer
Molar Teeth
Palatine
Posterior Palatine Foramina
Premolar Teeth
Maxilla
Anterior Palatine foramen
Canine Tooth
Premaxilla
Incisor Teeth

FIG. 182. **Canis. Skull. Ventral View.***

the Eustachian foramen on the inside; **pterygoid; presphenoid; jugal; zygomatic arch; orbito-sphenoid; foramen lacerum anterius,** a large foramen between the alisphenoid and orbito-sphenoid for the IIIrd,

* In this and the subsequent illustrations it will be noticed that an extra pre-molar tooth is present in the upper jaw. This abnormality was present in the specimen, from which the drawing was made.

IVth, VIth and ophthalmic branch of the Vth nerve; **optic foramen** for the IInd nerve anterior to the foramen lacerum anterius in the orbito-sphenoid; **palatine; pre-maxilla;** a large aperture at the anterior end of the maxilla on the ventral side, the **anterior palatine foramen** which leads to the nasal chamber; and the **posterior palatine foramen** between the palatine and the maxilla; **incisor, canine, premolar** (including **carnassial**) and **molar teeth.**

Lateral View

Pre-maxilla; maxilla; teeth as before, **orbit; infra-orbital foramen** for the maxillary branch of the Vth nerve on the side of the maxilla; **frontal; squamosal; zygomatic arch; glenoid fossa** on the squamosal for the articulation of the lower jaw; part of the **palatine; pterygoid;**

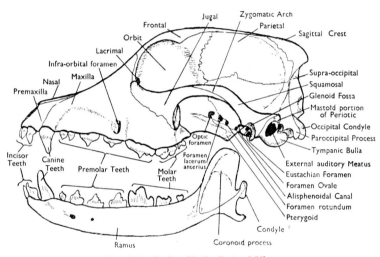

FIG. 183. **Canis. Skull. Lateral View.**

parietal; sagittal crest; supra-occipital; ex-occipital and **occipital condyle; paroccipital process; tympanic bulla; external auditory meatus; foramen ovale; foramen rotundum; foramen lacerum anterius; optic foramen; lacrimal,** a small bone in the anterior corner of the orbit bearing the **lacrimal foramen** leading to the lacrimal canal which runs to the nasal cavity.

Mandible—incisor, canine, premolar and molar (including carnassial) teeth; coronoid process; angle; condyle.

Write the **dental formula of the dog.**

Longitudinal Section of Skull

Pre-maxilla; nasal; mesethmoid; turbinals; vomer; frontal; parietal; inter-parietal; alisphenoid; squamosal; periotic; supra-occipital; ex-occipital; paroccipital process; tympanic bulla; basi-occipital; basi-sphenoid; presphenoid; orbito-sphenoid; pterygoid; palatine; maxilla; ethmoid and **cribriform plate** bearing **foramina for branches** of Ist nerve; **teeth** as in lateral view.

The **hyoid bone** (part of the visceral skeleton) will be best examined later, in position, in the dissection of the neck.

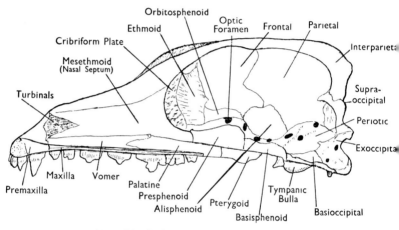

Fig. 184. **Canis. Skull. Longitudinal Section.**

The Skull of the Rabbit

Examine the skull of the rabbit and compare it with that of the dog.

Note the difference in shape of the skull and the absence of canine teeth (the rabbit is herbivorous in its diet). There are two incisors, one *behind* the other, three premolars and three molars on each side in the upper jaw, but only one pair of incisors, two premolars and three molars, in the lower jaw. The wide space between the incisors and premolars is called the **diastema.** The mastoid portion of the periotic is more easily seen on the surface, and the paroccipital process is closer to the tympanic bulla. The zygomatic arch is closer to the skull. The maxillæ take part in the formation of the palate only across the posterior end. In front of this are the two long, narrow **palatine foramina,** separated by the palatine processes of the pre-maxillæ. The posterior part of each ramus of the lower jaw is deeper than it is in the dog, and the coronoid process is hardly noticeable.

The foramen rotundum and foramen lacerum anterius are absent and are replaced by an elongated aperture, the **sphenoidal fissure.**

Write the **dental formula of the rabbit.**

The Vertebral Column of the Rabbit

The vertebræ are all built on a common plan, a ventral cylindrical part, the **centrum,** surmounted by a **neural arch** enclosing a **neural canal** and from the sides are projections for muscle attachment called **transverse processes.** Between the centra are **intervertebral discs** of fibro-cartilage and on each end of the centrum is a flat plate of bone, the **epiphysis** but this is so fused with the centrum that it may be difficult to identify as such.

The vertebral column is divided into five regions as follows:—

Cervical	. .	7 vertebræ, **atlas, axis** and 5 others.
Thoracic	. .	12 vertebræ (sometimes 13).
Lumbar	. .	7 vertebræ (6, if 13 thoracic).
Sacral	. .	4 vertebræ fused as the **sacrum.**
Caudal	. .	15 vertebræ (approximately).

Cervical Region

Atlas

This has no body or centrum, and has wide **transverse processes.** On the anterior side there are two **occipital facets** for articulation

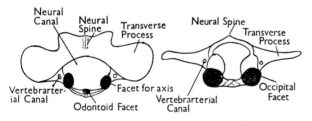

Dorsal View.   Anterior View.
FIG. 185. **Oryctolagus. Atlas.**

with the occipital condyles of the skull and on the posterior side a central **odontoid facet** for articulation with the odontoid process of axis and two lateral **facets for articulation with axis.** Above the **ligament** traversing the central cavity (often missing from a prepared skeleton) is the **neural canal;** the odontoid process of axis fits in below. The tiny hole on each side of the vertebra is the **vertebrarterial canal** for the vertebral artery. These are found in the cervical vertebræ only. The **neural spine** is an inconspicuous ridge on the dorsal side.

Draw the dorsal and anterior views. Hold a skull in position on an atlas and examine the nodding movement of the skull.

Axis

This has a wide flat **centrum,** produced forwards as the **odontoid process** (probably originally the centrum of atlas), with a **facet for articulation with atlas** on each side, small backwardly directed transverse processes called **cervical ribs, vertebrarterial canals,** and a large vertical blade-like **neural spine.** Note the **postzygapophysis,** a dorsally placed and backwardly directed process on each side of the posterior end of the **neural arch** and the **neural canal.**

Draw a lateral view. Fit an atlas and an axis together and examine the turning movement of atlas on axis

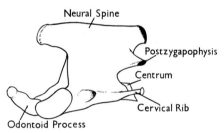

Neural Spine

Postzygapophysis

Centrum

Cervical Rib

Odontoid Process

FIG. 186. **Oryctolagus. Axis. Lateral View.**

3rd to 7th cervical vertebræ*

Each has a short broad **centrum,** a small **neural spine,** transverse processes called **cervical ribs** through which pass the **vertebrarterial canals,** two **prezygapophyses** projecting anteriorly, and two **postzygapophyses** projecting posteriorly. Note also the **neural canal.**

Draw the anterior view. Fit two neighbouring cervical vertebræ together and observe the articulation of their zygapophyses.

Thoracic Region

Each **thoracic vertebra** has a short thick **centrum,** with a **capitular demi-facet** for the articulation of the capitulum of a rib (together with a corresponding demi-facet on the next vertebra) on the upper

* See Fig. 187 p. 243.

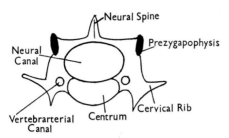

FIG. 187. Fig. 187. **Oryctolagus. A Typical Cervical Vertebra.
Anterior View.**

side at each end and the short **transverse processes** bear on their
lower surfaces a **tubercular demi-facet** for articulation with the
tuberculum of a rib.

*Take a rib of suitable size and see how articulation with a thoracic
vertebra is effected.*
(See p. 245.)

The **prezygapophyses** project upwards and outwards and the **postzygapophyses** downwards and inwards. The **neural spine** is long and
backwardly directed except in the last three or four thoracic vertebræ

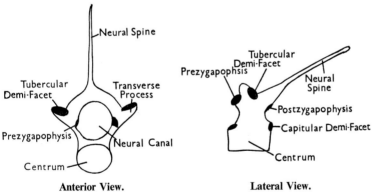

Anterior View.   Lateral View.
FIG. 188. **Oryctolagus. Thoracic Vertebra.**

which are somewhat similar to lumbar vertebræ, and have larger
centra and shorter neural spines. They bear rib facets, though
they differ from the others in that there is only one, the anterior one,
the capitular demi-facet, because the ribs articulating with them
have a capitulum only. Note also the **neural canal.**

Draw the anterior and lateral view.

Lumbar Region

Each **lumbar vertebra** has a large **centrum,** two long **transverse processes,** projecting forwards and downwards, a large flat **neural spine,** on each side of which is a **metapophysis** anteriorly, a large forwardly directed process bearing a **prezygapophysis** on its inner side. The **postzygapophyses** are backwardly directed processes on the dorsal side of the posterior end of the neural arch, which, in some cases, also bear small backwardly directed **anapophyses.** The first two or three lumbar vertebræ also bear a **hypapophysis,** a downwardly projecting process on the middle of the ventral side of the centrum.

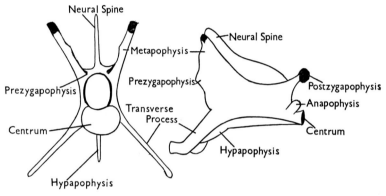

Anterior View.   Lateral View.
FIG. 189. **Oryctolagus. Lumbar Vertebra.**

Sacrum

The **sacral vertebræ** are fused together as the **sacrum.** The first, the only true sacral vertebra, is the largest and has broad so-called **transverse processes** bearing **ilial facets** for attachment to the ilium of the pelvic girdle, **prezygapophyses,** and a large **neural spine.** The other sacral vertebræ are smaller, decreasing in size from before backwards. Foramina for the exit of spinal nerves occur between the fused vertebræ.

Draw the dorsal and lateral views.

Caudal Region

The **caudal vertebræ** become smaller as they are traced backwards, gradually losing their processes and neural arches until the last few vertebræ consist merely of solid centra.

These need not be drawn.

The Ribs and Sternum

Ribs

There are twelve pairs of ribs articulated dorsally with the thoracic vertebræ, as already seen. The first seven are joined ventrally to a bone, the sternum, but the remaining five are not, the eighth and ninth being connected to the seventh and called **false ribs,** while the tenth, eleventh and twelfth are not connected with the sternum at all and are called **floating ribs.**

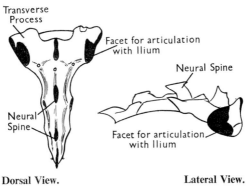

Dorsal View.   Lateral View.
Fig. 190. Oryctolagus. Sacrum.

Each **rib** is a curved, flattened, bony rod bearing at its dorsal end a process, the **capitulum,** which articulates with the demi-facets on the centra of the vertebræ, and a smaller dorsal **tuberculum,** which articulates with the demi-facets on the transverse processes. The bony shaft of the rib is called the **vertebral portion;** to this is joined the imperfectly ossified **sternal portion,** which is joined to the sternum.

Sternum

The **sternum** consists of a series of seven bony **sternebræ.** The anterior sternebra is the largest, and is called the **manubrium.** The last, the **xiphisternum,** is long and bears a flat cartilaginous plate, the **xiphoid cartilage.**

THE APPENDICULAR SKELETON

The Pectoral Girdle

This is very much reduced in the mammal and consists of two scapulæ or shoulder blades kept in position by muscles.

Examine the scapula from its dorsal side, the side bearing a ridge.

The **Scapula** consists of a triangular **blade** bearing at its apex a cavity for the articulation of the fore-limb, the **glenoid cavity.** Projecting over this is the hook-like **coracoid process.** The ridge on the blade is called the **spine:** this continues beyond the blade as the **acromion process,** at right angles to which is the **metacromion process.** In the living animal, a piece of cartilage, the **supra-scapula,** is situated on the base of the triangle and may or may not be present in the prepared skeleton.

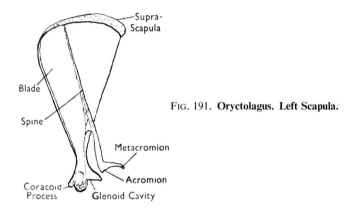

FIG. 191. **Oryctolagus. Left Scapula.**

The **clavicle** is a small membrane bone connected by ligaments to the coracoid process at one end and to the manubrium at the other. Like the hyoid, it will probably have been lost in the preparation of the skeleton. It should be seen in the articulated skeleton.

The Pelvic Girdle or Pelvis

This is strongly developed and is composed of two bones, known as the **ossa innominata,** which are joined ventrally by ligaments.

Each os innominatum is composed of a large anterior wing-like bone, the **ilium,** bearing on its inner surface a horse-shoe shaped **sacral facet** for the articulation of the Ist sacral vertebra, the **ishium,** posterior to the ilium and bearing the **ischial tuberosity** posteriorly, and the **pubis,** on the inner side, joining its fellow of the other os innominatum at the **symphysis pubis.** The ilium, ischium and pubis enclose a large hole, the **obturator foramen.** The **acetabulum** is a deep cup for the articulation of the hind-limb, formed by the ilium, ischium and pubis where they join on the other side. On its ventral side is a small bone, the **cotyloid.**

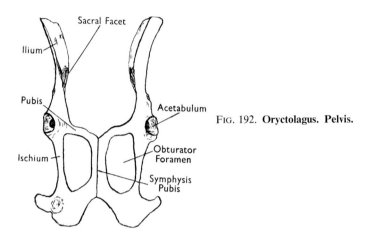

FIG. 192. **Oryctolagus. Pelvis.**

The Fore-Limb

The upper bone is the **humerus,** bearing a rounded **head** at its proximal end for articulation with the glenoid cavity of the scapula, and a groove, the **trochlea,** with ridges or **condyles** on each side, at its distal end. Just below the head, on the outer side, is a process, the **greater tuberosity,** and on the inner side the **lesser tuberosity,** both for the insertion of muscles. The groove between the tuberosities is the **bicipital groove.** The **deltoid ridge** on one edge of a lateral flattened part of the shaft below the head is for muscle attachment. At the distal end above the trochlea are two depressions, the **supratrochlear fossæ,** connected by the **supratrochlear foramen,** the upper or posterior fossa being known as the **olecranon fossa.**

The fore-arm consists of two long bones firmly attached but not fused, the **radius** and **ulna.** The ulna is longer and projects beyond a groove, the **sigmoid notch,** which fits into the trochlea of the humerus, the projection being known as the **olecranon process.** This fits into the olecranon fossa of the humerus. *Fit these bones together and observe the movement of the forearm on humerus.* The distal ends articulate with the wrist bones.

The wrist or **carpus** consists of 9 **carpals** arranged in 3 rows as follows:--in the proximal row, **radiale** and **intermedium** articulating with the radius, and **ulnare** with the ulna; in the middle, **centrale** articulating with radiale and intermedium and in the distal row, five **distal carpals,** 4 and 5 being fused.

The **manus** or hand is composed of 5 **metacarpals,** the first or **pollex** being the shortest and bearing two phlanges, while the others bear three. The distal phalanges bear pointed **claws.**

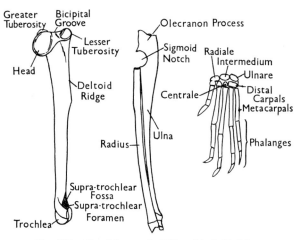

FIG. 193. **Oryctolagus. Left Fore-Limb Skeleton.**

The Hind-Limb

This is longer than the fore-limb. The upper bone is the **femur,** a long stout bone bearing a prominent rounded **head** at its proximal end, which articulates with the acetabulum of the pelvis, and at its distal end two **condyles** with a deep depression between them on the ventral side, the **inter-condylar notch.**

At the proximal end, just below the head, are three protuberances, the **greater trochanter,** the outside process on top, the **lesser trochanter** inside below the head and the **third trochanter** below the greater trochanter, all for the insertion of muscles. Continuous with the inter-condylar notch at the distal end on the dorsal side, is the **patella groove,** in which the knee-cap or **patella** glides.

The shank is composed of the **tibia** and **fibula,** separate but partially fused bones, the tibia being the larger. The tibia bears at its proximal end two prominences for articulation with the condyles of the femur, and on its anterior side a ridge, the **cnemial crest.**

The tibia fuses with the more slender fibula about half way down, and at the distal end of the fused bone will be seen the surfaces for articulation with the ankle bones.

The ankle, or **tarsus,** consists of six bones arranged in three rows as follows:—

In the proximal row, a large **calcaneum** (or fibulare), bearing a backwardly projecting **heel** and a smaller **astragalus** (fused tibiale and intermedium), both articulating with the tibia, in the middle, the **centrale,** which is, however, to one side almost under the astragalus and in the distal row, three **distal tarsals,** one being composed of two fused tarsals.

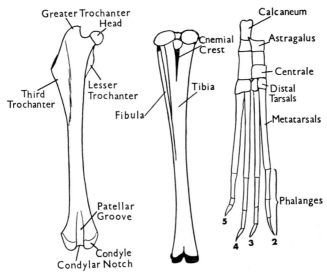

Fig. 194. **Oryctolagus. Right Hind-Limb Skeleton.**

The **pes,** or foot, is composed of four **metatarsals,** numbered 2, 3, 4, 5, No. 2 being under the centrale, and each bearing three **phalanges,** the distal ones bearing **claws.** No. 1, the hallux, is missing.

THE ARTICULATED SKELETON

Examine an articulated skeleton to see the relationship of the skull, vertebræ, ribs, girdles, and limbs and observe how to distinguish between bones on the left and right.

INTERNAL ANATOMY

Dissection is best performed on a freshly killed animal.

If the animal which is used for the general dissections is also to be used for the study of the brain, it will be necessary to take precautions at this stage to preserve the brain. The procedure is as follows:—

Remove the skin from the dorsal side of the head so as to expose the bones of the cranium. Insert the point of one blade of a pair of large scissors or small bone forceps into the cranium where the parietal and frontal bones join, and carefully remove a portion of one or both parietal bones, taking care not to push the point of the scissors into the brain. A small scalpel can be inserted in the interparietal suture and given a slight twist to facilitate the entry of the scissors.

If a trephine is available, insert this at the same point as above and remove a small circular piece of bone.

This exposure of the brain will enable the preservative fluid (usually 4 per cent. formaldehyde) in which the animal is kept between the various dissections, to penetrate into the cranial cavity.

THE MUSCULAR BODY WALL

It is advisable to wet the fur of the animal before beginning the dissection. Alternatively it can be skinned.

Place the animal, ventral side upwards, on a dissecting board. Tie the fore-limbs to the hooks in the top corners of the board or fix awls through the wrists and, stretching the body as much as possible, tie the hind-limbs to the hooks in the bottom corners of the board or fix awls through the ankles.

Pinch the skin in the centre of the abdomen and cut with the scissors, making a median incision upwards to the top of the thorax and down to the tail. Holding the skin with forceps, cut away the connective tissue between the skin and the muscular body wall as far round as possible with a scalpel and pin back the skin on both sides with awls.

Note the strip of muscle down the centre of the **abdomen,** the **rectus muscle** with the **oblique muscles** on either side.

In the **thorax,** note the **ribs,** which can be somewhat vaguely seen, the **pectoralis muscle,** a large fan-shaped muscle extending from the **sternum** (with the **xiphoid cartilage** at its lower end) to the upper-arm and the **latissimus dorsi muscle** at the side of the thorax.

Note also the **cutaneous blood-vessels** in the inner (exposed) side of the skin.

Regional Anatomy

In the mammal it is more convenient, from a practical point of view, to study the anatomy of the various regions of the body (*i.e.*, abdomen, thorax, neck and head) than to study one system at a time, though this is, of course, possible.

THE ABDOMEN

Carefully lift the wall of the abdomen with forceps and make a median incision stretching up to the xiphoid cartilage and down to the pubic symphysis. Take care not to pierce, cut or puncture any of the abdominal viscera. Now make transverse cuts in the abdominal wall just posterior to the last ribs and pin it back on either side.

THE ABDOMINAL VISCERA IN SITU

N.B. "Right" and "left" refer to the animals' right and left.

Note the shining mucous membrane, the **peritoneum,** which lines the **peritoneal cavity** (part of the cœlom) and covers the organs, and

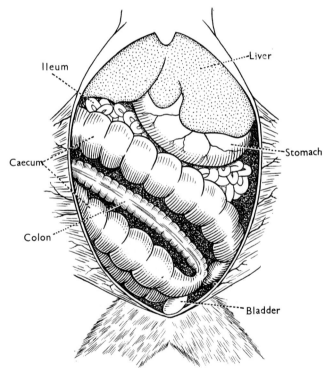

Fig. 195. **Oryctolagus. Abdominal Viscera. In situ.**

the muscular partition separating the abdomen from the thorax, the **diaphragm.** *This will be seen if the* **liver** *is gently lowered a little.* This is the large red organ behind the diaphragm and it partially covers the large white **stomach** on the left, stretching across the body. Behind the liver is part of the large brown **cæcum,** another part of which will be seen lower down. The sacculated **colon** will also be seen going across the centre of the body and parts of the **ileum** may be visible. The **urinary bladder** will be seen, particularly if it contains a quantity of urine, at the posterior end of the body cavity.

THE ABDOMINAL VISCERA AFTER
DEFLECTION OF THE INTESTINES

Deflect the liver forwards, and the cæcum and colon to the animal's right.

Find the **duodenum,** the first part of the small intestine, continuous with the stomach, to the distal loop of which a part of the rectum is attached by **mesentery.** Examination of the mesentery may reveal a number of small vesicles in it. If so, these are cysticerci (bladder-worms) of Tænia. The **rectum** contains pellets of fæces.

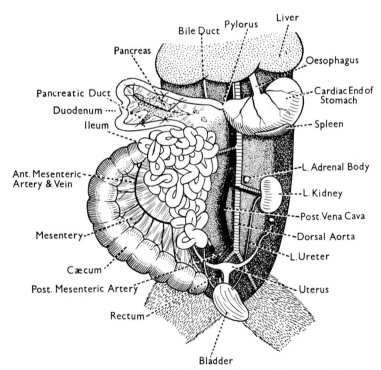

FIG. 196. **Oryctolagus. Abdominal Viscera. Intestine deflected to Right.**

Examine the **liver.** It is suspended from the diaphragm by the **falciform ligament** and is composed of five **lobes**—the **left lateral** (the larger and, in its deflected position the lower, of the two lobes on the left) **left central** (the other left lobe and, in its present position, the upper), **right central** (upper right in present position), **caudate** (partly covering the right kidney when in its normal position) and

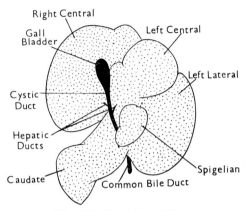

Fig. 197. **Oryctolagus. Liver.**

spigelian (in the centre and the smallest). The green **gall-bladder** lies between the right and left central lobes and is partly embedded in the right central lobe. From it arises the **cystic duct** which is joined by the **hepatic ducts** from the lobes of the liver, thus forming the **common bile duct,** which passes down on the dorsal side of the proximal loop of the duodenum, which it enters at its proximal end. Note the wide **cardiac end** and narrow **pyloric end** of the **stomach** and the constriction, the **pyloric sphincter** (or **pylorus**) at the outlet of the stomach where it joins the duodenum.

Pull the stomach down slightly in order to see the œsophagus *entering the cardiac end.*

Spread out the loops of the duodenum without cutting the mesentery.

The **pancreas** is a diffuse gland between these loops. The **pancreatic duct** leads into the distal end of the duodenum about 4 cm. beyond the bend.

The dark red body under the cardiac end of the stomach is the **spleen.**

The left **kidney,** a dark red organ, will be seen on the dorsal body wall.

Find and trace the following blood-vessels associated mostly with the alimentary canal. The veins are dark purplish red, the arteries being lighter and redder in colour and the larger ones thicker-walled:—

The **dorsal aorta** and the **posterior vena cava** both run in the mid-dorsal line on the dorsal body wall underneath the peritoneum, through which they can be seen if adequate deflection is made, the dorsal aorta being mostly dorsal to the vein.

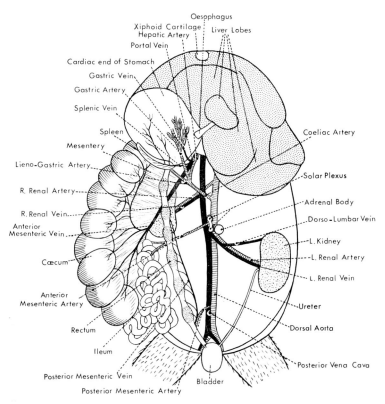

FIG. 198. **Oryctolagus. Blood vessels of alimentary canal: Solar Plexus.**

The single **anterior mesenteric artery** arises from the dorsal aorta just above the left kidney and runs in the **mesentery** supporting the intestine, to the duodenum, pancreas, ileum, cæcum and colon.

The single **cœliac artery** arises in front of the anterior mesenteric just posterior to the diaphragm and almost immediately divides into the **hepatic artery** which goes to the liver (first giving a branch to the stomach) and the **lieno-gastric artery,** which supplies the stomach (**gastric artery**) and spleen (**splenic artery**).

The **posterior mesenteric artery** is another single vessel which arises just before the aorta bifurcates posteriorly and goes to the posterior part of the rectum. It will be seen underneath the bladder.

The **duodenal vein** will be seen running alongside the **duodenal artery** (a branch of the anterior mesenteric artery) in the pancreas, where it receives branches from the duodenum itself. The vein leads into the (hepatic) portal vein* which enters the liver.

The branches of the **anterior mesenteric vein** will be seen in the mesentery. This vein also joins the portal vein at about the same point as the duodenal.

The **lieno-gastric vein** from the spleen and stomach accompanies the lieno-gastric artery but leads into the portal vein near its junction with the duodenal and anterior mesenteric veins.

The **posterior mesenteric vein** runs alongside the posterior mesenteric artery from the posterior end of the rectum on the right of the posterior vena cava to join the portal vein.

The **portal vein** is thus formed by the union of the anterior and posterior mesenteric, the lieno-gastric and the duodenal veins. It is the large vein entering the liver and can hardly be missed.

The **left renal artery** and **renal vein** will be seen running side by side to the left kidney from the dorsal aorta and to the posterior vena cava respectively.

Now find the left adrenal body and the solar plexus.

Just above the left kidney, towards the mid-line near where the renal vein joins the posterior vena cava, note the small round whitish **adrenal body,** a ductless or endocrine gland.

The **solar plexus** consists of two ganglia, one in front and one behind the point where the anterior mesenteric artery leaves the dorsal aorta, level with the left adrenal body. It may, however, be single and more central, lying on the posterior vena cava. It is white and star-shaped, is the end of the **splanchnic nerve,** which can be seen joining the ganglia, and is part of the sympathetic nervous system.

REMOVAL OF THE ALIMENTARY CANAL

Remove the alimentary canal as follows. Cut across the rectum near its lower end. Now carefully cut through the mesentery, close to the intestine, gradually unravelling the intestine. It is advisable to

---

* There is no renal portal system in the mammal.

ligature the portal vein (which lies dorsal to the stomach) in two places and cut across it between the ligatures at this stage. Take great care not to puncture the cæcum, and when you reach the duodenum try to keep the loop intact. Finally cut across the œsophagus. Now neatly lay out the alimentary canal on another dissecting board.

Measure the length.

Note the **œsophagus, stomach** (**cardiac end** and **pyloric end**), **pyloric sphincter, duodenum, ileum** ending in a swelling, the **sacculus rotundus,** where it joins the **cæcum.** This terminates in the blunt **vermiform appendix.** Note the **colon,** with its **longitudinal muscle** and **sacculations,** and the **rectum** containing pellets of **fæces.**

THE ABDOMINAL VISCERA REMAINING AFTER THE REMOVAL OF THE ALIMENTARY CANAL, IN SITU

Note the two **kidneys,** the right being more anterior than the left. These are dorsal to the peritoneum. The position of the **left adrenal body** has already been seen. The **right adrenal body** is anterior to the right kidney though usually hidden by it.

Note the **ureters,** narrow white tubes leading from the kidneys to the **urinary bladder,** and the **renal artery** and **vein** on each side leading from the **dorsal aorta** and to the **posterior vena cava** respectively. *Gently pull back the bladder slightly in order to see the entry of the two ureters.* The origins of the **cœliac** and **anterior** and **posterior mesenteric arteries** will be seen, also the bifurcation of the aorta to form the two **common iliac arteries** to the legs. Each of these gives off an **ilio-lumbar artery** to the dorsal body wall almost at once and then a small **vesical artery** to the bladder. The iliac then divides on its dorsal side into an **internal iliac artery** to the back of the pelvic cavity (consequently only its origin will be visible) and an **external iliac** or **femoral artery** to the leg.

The **internal iliac veins** from the back of the legs unite to form the posterior vena cava. They can be seen on the posterior side of the thigh muscle. Find also the **external iliac veins,** which are continuations of the **femoral veins** on the inside of the thighs, and which open into the inferior vena cava anterior to the internal iliacs. The small **vesical veins** from the bladder run near the vesical arteries and lead into the external iliac veins. The **ilio-lumbar veins** run alongside the ilio-lumbar arteries and enter the posterior vena cava, though that on the left sometimes curves forwards and runs up alongside the posterior vena cava for a short distance before joining it. The reproductive organs in the posterior part of the abdomen will be studied separately.

The Diaphragm

Now turn to the anterior end of the abdomen and examine the **diaphragm.** *It may be necessary to cut through the falciform ligament.*

Note the **central tendon** and **marginal muscular portions** of the diaphragm, in which will be seen the **phrenic artery** and **vein.**

THE KIDNEY

Examine a kidney in situ.

Note that it is of typical 'kidney', or bean shape, the concave edge or **hilus** directed inwards. It is here that the **renal artery** enters and the **renal vein** and **ureter** leave the kidney.

Remove one of the kidneys and cut a longitudinal section. Examine with a lens.

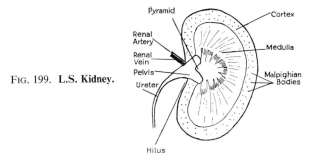

FIG. 199. **L.S. Kidney.**

Note that it consists of an outer rind, or **cortex** in which is a number of **Malpighian bodies** appearing as specks, and an inner **medulla.** The ureter widens after entering the hilus as the **pelvis** of the kidney into which the conical **pyramid** opens. The tubules of the kidney open on the pyramid.

THE REPRODUCTIVE SYSTEM

Male

At the posterior end of the body note the two **scrotal sacs** which contain the ovoid or elongated **testes.** *Open up one of the scrotal sacs lengthways* and note a mass of coiled tubes, the **epididymis,** on its inner edge, the upper part of which is the **caput epididymis** and the lower part the **cauda epididymis.** The cauda epididymis is connected with the base of the scrotal sac by a cord, the **gubernaculum.** The **vas deferens** (modified mesonephric duct) runs from the cauda epididymis into the abdomen where, after curling round the ureter, it runs to the dorsal side of the bladder and opens into the **urethra.** *Turn the penis, urethra and bladder sideways and carefully separate all attachment to the rectum which should then be freed. On the dorsal side of the bladder is a large median sac, the* **uterus masculinus** (vestige of Müllerian ducts and considered by some to be a vesicula seminalis) *and this may hide part of the vas deferens. It must therefore be carefully freed in order to expose the hidden portion.* The testis is suspended by the **spermatic cord** joined to the dorsal body wall and composed of the **spermatic artery, vein** and **nerve** and connective tissue. It passes through what is known as the **inguinal canal** from the anterior end of the testis into the cavity of the abdomen.

Carefully cut through the symphysis pubis with a scalpel and separate the two halves, or better, cut longitudinally through the centre of each os innominatum on each side of the symphysis pubis with strong scissors or small bone forceps and remove the central part of the pelvis. Deflect the bladder to the side and remove any obscuring connective tissue.

Note the **urethra** (which is a **urino-genital canal**), continuous with the neck of the bladder, and the **penis,** through which this passes. The vascular ventral wall of the penis surrounding the urethra is the **corpus spongiosum,** and the dorsal wall is composed of the paired

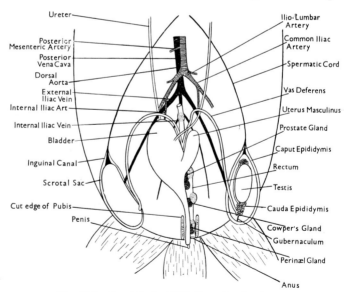

FIG. 200. **Oryctolagis. Male Reproductive Organs.**

corpora cavernosa. Both are erectile tissues. The distal end of the penis, the **glans penis,** bears the opening of the urethra, and is covered by a loose sheath, the **prepuce.**

Cut open the uterus masculinus.

It opens into the urethra by a large aperture, in front of which are the openings of the vasa deferentia. (This further justifies its being regarded as vesicula seminalis). Note the **prostrate,** a gland round the dorsal and lateral surfaces of the uterus masculinus. *Find* **Cowper's glands** posterior to the prostate but not always easily seen, and the **perineal glands** between the urethra and the rectum underneath the skin covering the perinæum.

The **spermatic artery** runs from the dorsal aorta to the testis, and the **spermatic vein** runs from the testis to the posterior vena cava.

The **rectal gland** will be seen on the dorsal side of the rectum.

Female

Carefully cut through the symphysis pubis and separate the two halves as in the male (p. 258).

Note the two oval **ovaries** on the dorsal body wall, posterior to the kidneys and the narrow convoluted oviducts (Mullerian ducts)

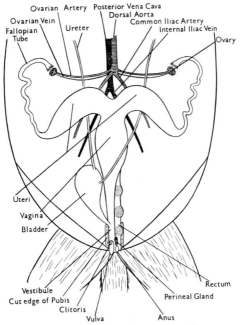

FIG. 201. **Oryctolagus. Female Reproductive Organs.**

called **Fallopian tubes** with their funnel-shaped **internal openings** close to the ovaries and in some cases partially enveloping them dorsally. These tubes lead posteriorly into the two slightly convoluted and thick-walled **uteri,** in which the embroys develop and which lead by separate openings into the **vagina,** a wide median tube which, with the **urethra** which it joins, forms the **urino-genital canal** (or **vestibule),** opening to the external **vulva.** The **perineal glands** are situated one on each side of the vestibule but Cowpers' glands are usually absent. The **clitoris** (the counterpart of the penis in the male) is a small rod-like organ in the ventral wall of the vulva.

The **ovarian artery** runs from the aorta to the ovary and the **ovarian vein** from the ovary to the posterior vena cava, though it may open into the ilio-lumbar vein on the left.

Note the **rectal gland** on the dorsal side of the rectum as in the male.

THE THORAX

Carefully remove the ventral and as much as possible of the lateral walls of the thorax, leaving the diaphragm in situ as follows:—

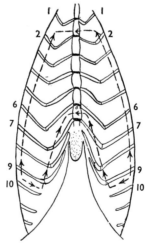

FIG. 202.   **Oryctolagus.   Method of Removal of Ventral Thoracic Wall.**

Cut through the ribs on each side from the 9th rib (inclusive) upwards to the 1st (exclusive), as low as possible. Then cut transversely in the intercostal space between the 9th and 10th ribs and continue to cut *upwards through the ribs up to the 6th sternebra above the xiphi-sternum on both sides. Cut through this sternebra, and, lifting up the ventral wall of the thorax and cutting through the **pleura** in the mid-ventral line which divides the thorax into two **pleural cavities,** continue the lateral cut on either side through the manubrium (1st stenrebra), between the 1st and 2nd ribs, Remove the ventral thoracic wall and retain it for the time being.*

THE THORACIC VISCERA IN SITU

Note the **cut surfaces of the ribs,** the **diaphragm** and the remains of the **pleura** lining the thorax on the sides. The **heart** is centrally placed with its apex directed to the left, and is enclosed in the thin **pericardium** between the two pleural cavities. On each side are the pinkish **lungs** enclosed in the **pleural cavities.** (The **pleura,** lining the pleural chambers, will have been removed with the central thoracic wall, as already seen except at the sides where it has already been observed.) The cavity between the pleural cavities and which encloses the roots of the trachea and the chief blood vessels leaving the heart and part of the œsophagus, is the **mediastinum.** It reaches from the diaphragm to the heart, enclosing this organ in its peri-cardium. The pericardial and pleural cavities and the mediastinum are all parts of the cœlom. On the lateral edges of the mediastinum will be seen the white **phrenic nerves** going to the diaphragm. The pink organ above the heart is the **thymus gland,** an endocrine gland; it is large in a young rabbit. The **trachea** divides into two **bronchi** which enter the lungs.

THE HEART AND GREAT VESSELS

Carefully dissect away the thymus gland and free the heart from the pericardium. Exercise great care in the removal of the pericardium from the base of the heart (anterior end) where the great vessels are situated. Deflect the heart as necessary as so to expose the blood vessels dorsal to it.

Note that the **heart** consists of thin-walled **right** and **left auricles** anteriorly and thick-walled **right** and **left ventricles** posteriorly.

The **pulmonary artery** is a large vessel leading from the right ventricle to the lungs. It bends towards the left and divides into two branches, one to each lung. It is crossed by the **left anterior vena cava,** which runs on the dorsal side to the right auricle. The **right anterior vena cava** will also be seen entering the right auricle.

The **pulmonary veins** come from the lungs and join, the single vein thus formed entering the left auricle. The single **posterior vena cava** passes up from the abdomen, through the diaphragm, and, running dorsal to the heart, enters the right auricle. *Now find* the **aortic arch,** which leaves the left ventricle, bends over to the left and, running dorsal to the heart, passes through the diaphragm into the abdomen as the dorsal aorta. *Find the* **ductus arteriosus,** a ligament which runs across from the aortic arch to the pulmonary artery. It will be found where the left anterior vena cava crosses the latter. On the aortic arch, note the short **innominate artery** which gives off the **right common carotid artery** almost at once and then the **right subclavian artery** to the fore-limb. The **left common carotid artery** arises from the aortic arch just beyond the origin of the innominate artery and the **left subclavian artery** arises direct from the left side of the aortic arch. Both subclavian arteries run dorsal to the anterior venæ cavæ. (Sometimes both carotid arteries arise from the inno-minate: sometimes they all arise directly from the aortic arch, the innominate artery being absent.)

The **right** and **left subclavian veins** will be seen alongside the corresponding arteries. Each enters the anterior vena cava on its own side.

The **vertebral arteries** to the head arise from the subclavian artery just beyond its origin and enter the vertebrarterial canals of the cervical vertebræ.

The **internal mammary artery** on each side arises from the sub-clavian artery just after it has given off the vertebral artery and runs along the ventral wall of the thorax.

The **internal mammary vein** enters the anterior vena cava level with the first rib.

These two vessels will have been removed with the ventral thoracic wall. *Examine the inner side of the ventral thoracic wall which has been removed and note the internal mammary artery and vein running along each side of the sternum.*

The **azygos vein** enters the right anterior vena cava just before it enters the auricle. It is unpaired, there being none on the left side.

Look for the **œsophagus** on the dorsal side of the thoracic cavity and dorsal to the **trachea** (windpipe), with the large **vagus nerve,** a white cord, on each side.

Move the aorta sideways. Observe the chain of **sympathetic ganglia** behind it, connected together by a slender nerve cord.

Finally you should find the origin of the **left recurrent** (or **posterior**) **larynegal nerve** which arises from the left **vagus nerve** just behind the subclavian artery, loops round the ductus arteriosus and then passes forwards. *Turn to the other side and find the origin of the* **right recurrent laryngeal nerve** from the **right vagus nerve** in front of the subclavian artery, afterwards looping round it and passing forwards. Further reference will be made to these origins in the dissection of the neck.

THE HEART OF THE SHEEP

The sheep's heart is much larger than the rabbit's heart and is therefore more suitable for the study of the internal structure. It differs little from that of the rabbit.

External Structure (Ventral View)

Examine a sheep's heart with the great vessels attached.

Note the two ventricles separated on each side by a groove containing **fat,** the **longitudinal sulcus.** *Feel the walls of the ventricles.* That with thicker walls is the **left ventricle,** the other the **right ventricle.** Above them are the thinner walled **left** and **right auricles** bearing small **auricular appendages.** Note the **coronary artery** and **vein** which branch over the tissue of the heart, also the **aortic arch, anterior vena cava** (single in the sheep, the right and left vessels joining before entering the heart), **posterior vena cava, pulmonary artery** and **pulmonary vein.**

Internal Structure

(i) Right Auricle and Ventricle

Place the heart with the ventral surface uppermost and make a longitudinal incision through the right auricle and ventricle just to the right of the longitudinal sulcus. You will thus expose the cavities of the right side of the heart. Wash out any contained blood.

Note the comparatively thin but muscular **wall of the right auricle** and the thicker **wall of the right ventricle.** The **inter-auricular septum** separates the two auricles and the **inter-ventriccular septum** the two ventricles. In the inter-auricular septum is a thin oval area, the **fossa ovalis** (this was open in the embryo). Note the openings of the **anterior** and **posterior venæ cavæ** into the right auricle and the **Eustachian valve,** a fold of membrane between them (representing the remains of a sinu-auricular valve). The valve between the right auricle and ventricle is called the **tricuspid valve,** because it consists of three flaps or cusps. To these cusps are attached the **chordæ tendineæ,** tendinous cords which connect them to the **papillary muscles** in the wall of the ventricle. Note the obliquely running muscular **moderator band** in this ventricle (not present in the rabbit). The other muscular columns on the ventricle wall are called **columnæ carneæ.** The **pulmonary artery** leads out of the right ventricle.

Cut open the pulmonary artery.

Note the three **semilunar valves** just inside the entrance.

(ii) Left Auricle and Ventricle

Now make a longitudinal incision through the left auricle and ventricle and wash out any blood that may be present.

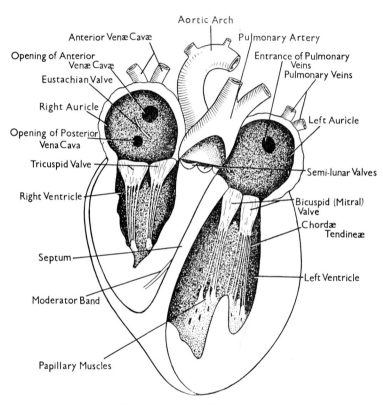

Aortic Arch

Anterior Venæ Cavæ

Pulmonary Artery

Opening of Anterior Venæ Cavæ

Entrance of Pulmonary Veins

Eustachian Valve

Pulmonary Veins

Right Auricle

Left Auricle

Opening of Posterior Vena Cava

Tricuspid Valve

Semi-lunar Valves

Right Ventricle

Bicuspid (Mitral) Valve

Chordæ Tendineæ

Septum

Left Ventricle

Moderator Band

Papillary Muscles

FIG. 203. **Ovis. Heart. Ventral Dissection.**

Examine the **wall of the left auricle** and the entrance into it of the **pulmonary vein.** The **wall of the left ventricle** is thicker than that of the right. On this side the valve between the auricle and ventricle consists of only two cusps and is called the **bicuspid (or mitral) valve.** Note the **chordæ tendineæ, papillary muscles** and **columnæ carneæ.** The **aorta** leads out of the left ventricle.

Cut open the aorta.

Note the three **semilunar valves** guarding its entrance. Just beyond this look for the openings of the **coronary artery.**

THE NECK

This is a dissection which needs patience and great care is necessary.

Extend the head and neck as much as possible and fix down the head. Cut through and dissect away the skin only of the neck from the thorax up to the lower lip, taking care not to injure the muscles. Pin back the skin on each side or remove it entirely.

Note the large **external jugular veins** on each side of the neck on the surface of the muscles and connected by a transverse **jugular anastomosis** towards their posterior ends. The **mylohyoid muscle** stretches across the **mandible** and covers the hyoid and the **mandibular muscle** runs along the inner side of each ramus.

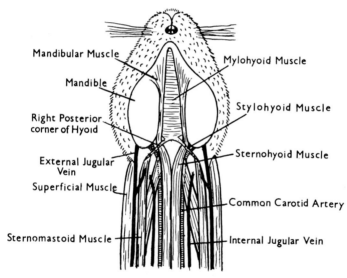

Fig. 204. **Oryctolagus. Neck Muscles.**

Now very carefully cut through the surface muscles in the mid-ventral line and pin them back on either side.

The **hyoid bone** will now be exposed. It has two **posterior cornua,** each bearing a **sternohyoid muscle.** These meet in the mid-ventral line, continue backwards and cover the trachea or windpipe at the anterior end of which is the **larynx**. *These muscles must therefore be removed.* At their tips, the posterior cornua bear the smaller **stylohyoid muscles** which pass obliquely backwards to the dorsal side. Between the cornua is the floor of the pharynx. The **anterior cornua** of the hyoid are short.

Note the **trachea**, a straight tube passing down the middle of the neck; in its wall are rings of cartilage incomplete on the dorsal side. Dorsal to it is the **œsophagus** and either side of it, dorsal to the stylohyoid muscle, is the **common carotid artery**.

Now note the **sternomastoid muscles** which are attached to the mastoid process of the skull at one end and which were attached to the sternum at the other. The **internal jugular vein** runs along the inner side of each sternohyoid muscle and alongside the trachea, opening into the external jugular vein lower down in the neck.

The trilobed **thyroid gland,** another endocrine gland, lies at the anterior end of the trachea with one lobe on either side and one on its ventral surface.

Remove the stylohyoid muscle, displace the carotid artery inwards and the sternomastoid muscle outwards and stretch and cut the connective tissue joining it to the trachea. Pin the artery by a pin alongside it and the muscle in their new positions.

Note that the **common carotid artery** divides under the stylohyoid muscle, into an **internal carotid artery** to the brain and an **external carotid artery** to the face. The **hypoglossal nerve (XIIth** cranial) crosses it and then runs outside the posterior cornu of the hyoid to the tongue. This nerve gives off a branch, the **ramus descendens,** just before it meets the internal carotid artery and crosses the carotid artery on the ventral side, supplying the sternohyoid and sternothyroid muscles.

The **Spinal Accessory Nerve (XIth** cranial) runs behind the mandible to the sternomastoid and other neck muscles.

The **Vagus Nerve (Xth** cranial) is large, leaves the skull with the XIth and then runs along the outside of and somewhat dorsal to the carotid artery, between it and the external jugular vein. *Trace it into the thorax.* In the region by the larynx it forms a ganglion just beyond which a branch known as the **anterior (or superior) laryngeal nerve** arises. It supplies the larynx and is dorsal to the sternohyoid muscles and the carotid artery which it crosses. *Displace the muscle to expose it.* The **cardiac depressor nerve** arises with the anterior laryngeal but it is very slender and not easy to see. It runs along the inside of the main branch of the vagus and supplies the heart.

The vagus gives off a third branch, the **recurrent laryngeal** (or **posterior laryngeal**) **nerve,** which arises at the posterior end of the neck,

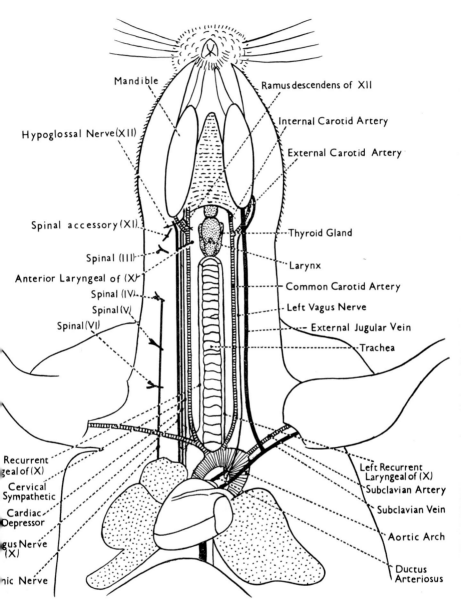

Mandible

Ramus descendens of XII

Internal Carotid Artery

Hypoglossal Nerve(XII)

External Carotid Artery

Spinal accessory (XI)

Thyroid Gland

Spinal (III)

Anterior Laryngeal of (X)

Larynx

Spinal (IV)

Common Carotid Artery

Spinal (V)

Left Vagus Nerve

Spinal (VI)

External Jugular Vein

Trachea

Recurrent
geal of (X)

Left Recurrent
Laryngeal of (X)

Cervical
Sympathetic

Subclavian Artery

Cardiac
Depressor

Subclavian Vein

gus Nerve
(X)

Aortic Arch

hic Nerve

Ductus
Arteriosus

Fig. 205. **Oryctolagus. Neck Dissection.**

The left anterior vena cava has been deflected to the animal's left and the right
external jugular vein has been omitted and the jugular anastomosis removed.
(Semi-Diagrammatic.)

runs alongside the trachea dorsal to the carotid artery and supplies the muscles of the larynx. As already seen when dissecting the thorax, on the *right* it arises just *in front of* the subclavian artery, round which it loops and passes forward alongside the trachea whereas on the *left* it arises *behind* the subclavian artery, looping round the ductus arteriosus and then passes alongside the trachea.

In the thorax, the vagus also gives other branches to the œsophagus, heart and lungs.

Now note the **anterior and posterior cervical sympathetic ganglia,** which are ganglia on the **cervical sympathetic nerve** running alongside the trachea and inside the vagus and depressor nerves, and the innermost of the three parallel nerves. The ganglia are situated one at the anterior end under the internal carotid artery and one at the posterior end near the subclavian artery.

The **Phrenic Nerve,** which has already been seen on the diaphragm, is formed from twigs of the **4th, 5th** and **6th spinal nerves** which will be found on the neck. *Trace the nerve forwards from the diaphragm.*

THE RESPIRATORY SYSTEM

Examine the respiratory organs in the neck and thorax.

At the anterior end of the **trachea** (already seen) is the sound box

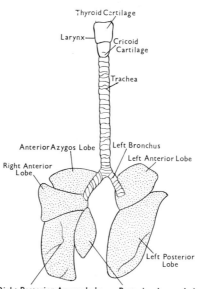

FIG. 206. **Oryctolagus. Respiratory system.**

or **larynx,** the ventral and lateral sides of which are composed of the **thyroid cartilage.** At the base is the **cricoid cartilage** separated from the thyroid cartilage by the **crico-thyroid ligament** while the dorsal side is supported by the two **arytenoid cartilages.** The trachea is composed of a series of cartilaginous rings, incomplete on the dorsal side. In the thorax it divides as already seen into two **bronchi,** one to each lung. In the lungs these branch into bronchioli, which terminate in minute air-sacs composed of alveoli.

The lungs are pink, spongy, vascular organs situated in the pleural cavities as already seen. The **left lung** is divided into an **anterior** and a **posterior lobe** and the **right lung** into an **anterior azygos lobe,** a **posterior lobe** and a **posterior azygos lobe.**

Make a drawing of the respiratory system.

THE VASCULAR SYSTEM

Now that all the blood-vessels have been seen *you should do separate dissections of the complete* **Arterial** *and* **Venous Systems.** *You should also draw connected diagrams of them.*

THE HEAD

Remove entirely the skin from the head and cut off the pinnæ close to their points of attachment to the head.

LATERAL VIEW

Note the **eye** and, immediately below it, the **infraorbital gland,** one of the four salivary glands of the mouth. It will be rendered more easily visible by carefully removing the tissue on the surface.

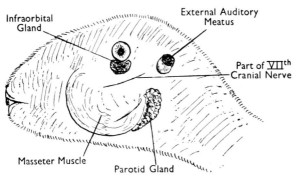

FIG. 207. **Oryctolagus. Head. Lateral View. Skin removed.**

The large **masseter muscle** covers the mandible, and posterior to it, at the angle of the jaw, is a second salivary gland (the largest), the **parotid gland,** immediately below the **external auditory meatus.** Branches of the **Facial Nerve (VIIth** Cranial) may be seen on the surface of the masseter muscle. The other salivary glands will be seen later.

The **lacrimal gland** in the posterior corner of the orbit will doubtless have been removed with the skin.

THE BUCCAL CAVITY

Cut through the muscles at the sides of the mouth and open it wide.

In the **open mouth** again note the **incisor, premolar** and **molar teeth** in the upper and lower jaws. Also observe the ridges or **rugæ of the hard palate,** the **soft palate** behind and the **pharynx** at the back. This leads ventrally into the **glottis** or opening of the trachea, with a cartilaginous flap, the **epiglottis,** on its ventral side (*seen by pressing the tongue forwards*), and dorsally into the **œsophagus.** The **posterior nasal aperture** is behind the soft palate and behind the **tonsils,** ridged depressions on either side; it leads into the **nasal chambers** in front. Note also that the **tongue,** free at its front end, has **papillæ** on its surface. These bear microscopical taste-buds.

Cut open the soft palate in the centre from front to back.

Note the **openings of the Eustachian tubes** at the sides of the narial chamber. *Insert a seeker into one of them.* It passes into the tympanic cavity.

Cut through the mylohyoid muscle which stretches across between the two mandibles in the middle line and deflect the flaps outwards (if not already done). Cut through the mandibular symphysis and separate the two half jaws slightly.

Note the **submaxillary** (or **submandibular) glands,** the third pair of salivary glands, at the angles of the jaw inside the mandible on the same level as the posterior cornua of the hyoid. The duct from each, **Wharton's duct,** *will be seen if the mandibular muscle is displaced to one side or removed.* Note also the fourth of these glands, the **sublingual gland,** which was under the tongue and will now be on the anterior inner side of the mylohyoid muscle on each side.

THE BRAIN

The brain must be examined in a freshly killed rabbit or one in which it has been preserved by treatment in accordance with the instructions on page 250, as it decomposes rapidly.

Cut through the neck between a pair of the anterior cervical vertebræ and thus sever the head completely from the body. The skin of the head having been removed, cut away the muscle and immerse the entire head in 10 per cent. hydrochloric acid to decalcify the bone. Leave it in the immersion fluid for three or four days: the bone will then be much softer and easier to cut. At the end of this period wash the head in water to remove the acid and immerse it in 70 per cent. alcohol or 4 per cent. formaldehyde (=10 per cent. formalin) for three or four days to harden the brain.

It should be noted that if a fresh animal is used, the brain should be hardened before the bone is decalcified by removal of part of the cranium by the method described on page 250, and immersion in 70 per cent. alcohol or 4 per cent. formaldehyde for three or four days.

Now remove sufficient of the cranium to enable you to remove the brain, taking care throughout the operation not to dig the scissors into the brain, by keeping the lower blade horizontal.

With the anterior end of the head facing you, insert the scissors into the hole cut in the cranium and cut through the bone back to the foramen magnum on one side. Repeat on the other side. Now turn the head right round so that it faces the opposite direction and again inserting the scissors into the hole, cut forwards round the orbit and along the outer edge of each nasal bone. Carefully remove the portions of bone thus set free. The brain will now be partially exposed. Finally remove the bone on either side of the cerebellum, the ridged part of the hind brain, so as to expose the flocculi on each side of it. The brain should now be removed from the cranium as follows:—

Cut through the spinal cord as far back as is practicable: then, lifting the free end joined to the brain, cut through the spinal nerves on each side as far from the cord as possible, working forwards, Now turn the head upside down over a dish and, holding it as close to the dish as possible, cut through the cranial nerves as far from the brain as you can, working forwards and allowing the brain to release itself from the cranium into the dish by its own weight. Finally, sever the anterior attachments of the olfactory lobes. The brain will then lie in the dish, ventral side uppermost.

It should be preserved in 70 per cent. alcohol or in formaldehyde if required on a future occasion.

Dorsal View

Note the **pia mater,** a thin vascular inner membrane which closely invests the brain. This with the **dura mater** which lines the cavity of the cranium (with which it will have been removed) and the **arachnoid** between them constitute the **meninges.** Between these membranes is the cerebro-spinal fluid.

Remove the pia mater carefully.

In the **Fore-brain** note the **cerebral hemispheres,** the two large anterior lobes with a **median fissure** between them. Their surfaces are smooth except for a few shallow grooves or **sulci,** the most prominent being the **Sylvian fissure** which begins about half-way along the outer edge of each hemisphere and runs obliquely backwards dividing the hemisphere into an anterior **frontal lobe** and a posterior **temporal lobe.**

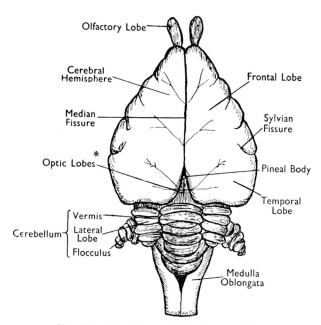

FIG. 208. **Oryctolagus. Brain. Dorsal View.**

Gently part the two hemispheres slightly.

You will see the **corpus callosum,** a transverse band of fibres which connects the two hemispheres together.

Note the small **olfactory lobes** (or **bulbs**) at the anterior end of the hemispheres.

The **Mid-Brain** consists of the **optic lobes** (or **corpora quadrigemina,** (so-called because each lobe is sub-divided into two, the two anterior

* Corpora quadrigemina

lobes receiving fibres from the retina and the two posterior lobes receiving fibres from the cochlea). They are almost completely hidden by the posterior ends of the cerebral hemispheres but *they will be visible if the cerebral hemispheres are gently separated by the fingers.*

The **pineal body** will most probably have been removed with the dura mater, but the **pineal stalk** may remain. It is situated between the hinder ends of the cerebral hemispheres on the thalamencephalon (the posterior part of the fore-brain) which is hidden by the hemispheres.

The **Hind-brain** is partly composed of the transversely ridged **cerebellum,** consisting of a central **vermis,** on each side of which is a **lateral lobe** bearing a **flocculus** and beneath and behind it is the **medulla oblongata,** continuous with which is the spinal cord.

Ventral View

Note the **olfactory lobes** bearing anteriorly the roots of the **olfactory nerves (I)** and the **cerebral hemispheres** on which will be seen the **Sylvian fissure** separating the **frontal lobe** from the **temporal**

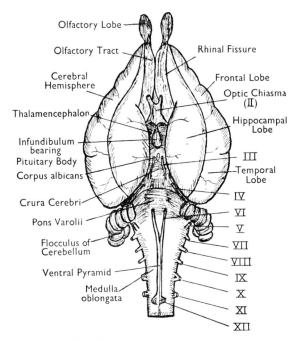

FIG. 209. **Oryctolagus. Brain. Ventral View.**

lobe. The longitudinal **rhinal fissures** mark the outer edges of the **olfactory tracts,** which are continuous with the olfactory lobes, The inner posterior **hippocampal lobes** of the cerebral hemispheres should be noted. The **optic chiasma** formed by the crossing over of the **optic nerves (II),** is situated between the hippocampal lobes at their anterior ends, and immediately behind it is the **infundibulum,** a median rounded prominence on the floor of the **thalamencephalon** bearing an endocrine gland, the **pituitary body,** which will probably have been left in the floor of the skull. A small rounded projection, the **corpus albicans** (or **corpus mammillare**) at each side of which are the roots of the **oculo-motor nerves (III),** is situated just behind the infundibulum, and behind this are the **crura cerebri,** two prominent bands of fibres which form the floor of the mid-brain. Next you will see the **pons Varolii,** a transverse band of fibres passing up on each side into the cerebellum. Note the **flocculus of the cerebellum** on each side. The **medulla oblongata** is broad in front but tapers as it runs posteriorly and shows a median longitudinal **ventral fissure,** on each side of which is a narrow strip, the **ventral pyramid** (or **pyramidal tracts).** On the ventro-lateral sides of the medulla will be seen the roots of the following nerves in order:—

Trigeminal Nerve (V).
Facial Nerve (VII).
Auditory Nerve (VIII).
Glossopharyngeal Nerve (IX).
Vagus Nerve (X), by several small roots.
Spinal Accessory Nerve (XI).

On the ventral surface of the medulla, more centrally placed, will be seen the roots of the following:—

Abducens Nerve (VI), just behind the pons.
Hypoglossal Nerve (XII) arising by several small roots posterior to the root of XI.

Horizontal Section

Place the brain, hardened in spirit, dorsal side upwards and very carefully cut a horizontal section of one of the cerebral hemispheres (the other will be required for the longitudinal section) *by inserting the point of a small scalpel with the blade horizontal not more than a quarter of an inch in the side of one of the hemispheres immediately behind the olfactory lobe. Continue to cut carefully round the lateral and posterior edges of the hemisphere and as far forwards as the olfactory lobe and remove the roof of the hemisphere. Examine with a lens.*

Note the cavity of the hemisphere, the **lateral ventricle,** towards its posterior end, a rounded thickening of the floor, the **hippocampus major,** and a further rounded thickening, the **corpus striatum,** alongside and in front of the hippocampus major. *Make a drawing at this stage.*

Now very carefully remove the floor of the ventricle by cutting round the edge of the hippocampus and wall of the hemisphere. Gently separate the two hemispheres slightly with the fingers.

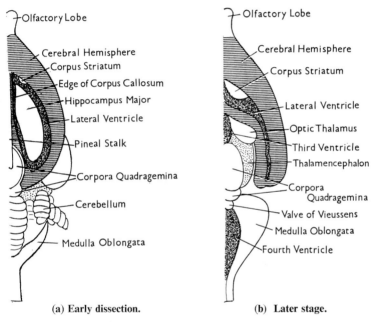

(a) **Early dissection.**   (b) **Later stage.**

FIG. 210. **Oryctolagus. H.S. Brain.**

Note the floor of the **thalamencephalon,** the posterior part of the fore-brain, behind which are the **corpora quadrigemina** with lateral thickenings, the **optic thalami,** and the **pineal stalk** anterior to them. The **anterior choroid plexus,** the vascular roof of the **third ventricle** in the thalamencephalon, will also be seen in front of the pineal stalk. Each lateral ventricle communicates with the third ventricle by a **foramen of Monro.** *Make another drawing.*

Gently remove the central vermis of the cerebellum.

Note the small cavity of the medulla, the **fourth ventricle,** which communicates with the third ventricle by the **iter.** Note also the membranous **valve of Vieussens** between the corpora quadrigemina and the vermis, and the hinder vascular part of the roof of the fourth ventricle, the **posterior choroid plexus.** *Complete your drawing.*

Longitudinal Vertical Section

Divide the brain into two halves longitudinally by a careful vertical cut. Examine the half which is complete with a lens.

Identify the **olfactory lobes,** the **cerebral hemispheres,** the edge of the **corpus callosum,** the **body of the fornix**—a thickening formed by the union in the middle line of two bands of fibres, one from the anterior border of each hippocampus. From the body of the fornix, the **anterior pillars of the fornix** run in a ventral direction to the **corpus albicans** and the **posterior pillars of the fornix** (or **fimbriæ**) run in a dorsal direction and, in fact, not posterior but anterior to the anterior pillars. Note the **lateral ventricle,** the **foramen of Monro,** the **third ventricle,** the **anterior commissure,** an oval structure ventral

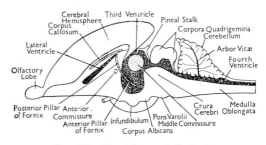

Fig. 211. **Oryctolagus. L.S. Brain.**

to the fornix, the **middle commissure** (or **massa intermedia**), a large rounded band of fibres joining the optic thalami in the thalamen-cephalon, and the **posterior commissure,** a small transverse band of fibres in the roof of the third ventricle. The **anterior choroid plexus** (or its remains), the **pineal stalk,** arising from the posterior end of the roof of the thalamencephalon, the **infundibulum,** the **corpus albicans** and the **anterior** and **posterior lobes** of the **corpra quadrigemina** with the thick **crura cerebri** forming their floor and the **pons varolii,** continuous with the crura cerebri, will also be seen. Look for the **iter** joining the third to the **fourth ventricle** in the **medulla oblongata** with the vascular **posterior choroid plexus** forming its roof. Examine the **cerebellum** and note the tree-like structure inside called the **arbor vitæ.**

You may experience a little difficulty in finding all the above structures.

RATTUS

THE RAT

Rats belong to the **Order Rodentia** to which *mice, squirrels* and *guinea pigs* also belong. The black rat is **R. rattus** and the brown rat **R. norvegicus** but the colour varies considerably. The tame rats used for dissection and experiments, whether black, white or pied, are all domesticated forms of *R. norvegicus*.

Examine a killed rat.

EXTERNAL ANATOMY

Note the colour and shape of the body and that it is covered with short hairs except on the ears and feet while a few are found on the tail. The body is divided into head, neck, trunk and tail.

Head

The head is pointed anteriorly where the **nostrils** will be seen as two slits. The black rat has a sharp snout whereas that of the brown rat is blunt. The **mouth** is below the nostrils and is bounded by two lips. At the sides of the snout are long whiskers or **vibrissae** which are tactile organs. The **eyes** are small and the external ears or **pinnae** are rounded and smaller in the brown rat than are they in the black. The entrance to the ear is called the **external auditory meatus.**

Neck and Trunk

The **neck** is short and the trunk consists of the **thorax** anteriorly and the **abdomen** posteriorly. The **tail** of the black rat is naked, slender and at least as long as the body whereas the tail of the brown rat is somewhat hairy and shorter than the body. At the root of the tail on the ventral side is the **anus.** In the *male* note the **penis,** an intromittent organ in its sheath, the **prepuce,** and the two large, **scrotal sacs** enclosed in a single fold of skin and which enclose the testes. In the *female* note the **vulva,** the slit-like genital aperture, with a small rod-like **clitoris** in its ventral wall. The genital and urinary openings are separate, unlike the rabbit, the latter opening ventral to the former. There are six pairs of **mammary glands** bearing teats, three on the thorax and three on the abdomen.

Limbs

The **fore-limbs** are shorter than the hind-limbs and the **manus** bears four distinct digits each terminating in claws while the **pes** has a short hallux or big toe and four other digits, all ending in claws.

THE SKELETON

The skeleton of the *rabbit* should be studied in preference to that of the rat, the skull of the *dog* (*Canis*) being substituted for that of the rabbit. Details for the examination of the rabbit's skeleton and the skull of the dog will be found on pp. 234–249.

The chief differences between the skeletons of the rabbit and rat are as follows:—

Dental formula of Rat $\dfrac{1033}{1033}$. Thoracic vertebræ 13. Ribs 13 pairs. The clavicle articulates with the manubrium and the acromion process of the scapula. Six lumbar vertebræ. Two sacral vertebræ are fused with two caudal to form the sacrum. Caudal vertebræ 27-30. The pes has 5 digits.

INTERNAL ANATOMY

Dissection is best performed on a freshly killed animal.

If the animal which is used for the general dissections is also to be used for the study of the brain, it will necessary to take precautions at this stage to preserve the brain. The procedure is as follows:—

Remove the skin from the dorsal side of the head so as to expose the bones of the cranium. The brain will be visible through the thin bones of the cranium. Holding the head firmly but gently, insert a small scalpel horizontally into the inter-parietal bone from behind in order to force apart the parietal bones at the sutures. Then carefully remove all these bones. The brain will then be exposed. Cover the exposed part with 4 per cent. formaldehyde. When the animal is kept in formalin between dissections, the fluid will penetrate into the cranial cavity.

THE MUSCULAR BODY WALL

Place the animal, ventral side upwards, on a dissecting board. Fix awls through the wrists and, stretching the body as much as possible, fix awls through the ankles.

Pinch the skin in the centre of the abdomen and cut with the scissors, making a median incision upwards to the top of the thorax and down to the tail. Holding the skin with forceps, cut away the connective tissue between the skin and the muscular body wall as far round as possible with a scalpel and pin back the skin on either side with awls.

Note the strip of muscle down the centre of the **abdomen,** the **rectus muscle** with the **oblique muscles** on either side.

In the **thorax,** note the **ribs,** which can be somewhat vaguely seen, the **pectoralis muscle,** a large fan-shaped muscle extending from the **sternum** (with the **xiphoid cartilage** at its lower end) to the upper-arm and the **latissimus dorsi muscle** at the side of the thorax.

Note also the **cutaneous blood-vessels** in the inner (exposed) side of the skin.

Regional Anatomy

In the mammal it is more convenient, from a practical point of view, to study the anatomy of the various regions of the body (*i.e.*, abdomen, thorax, neck and head) than to study one system at a time, though this is, of course, possible.

THE ABDOMEN

Carefully lift the wall of the abdomen with forceps and make a median incision stretching up to the xiphoid cartilage and down to the pubic symphysis. Take care not to pierce, cut or puncture any of the abdominal viscera. Now make transverse cuts in the abdominal wall just posterior to the last ribs and pin it back on either side. A great deal of fat will be found which may hide some of the structures sought. This should be carefully removed with forceps. Bleeding from capillaries will readily occur in this animal. The blood should be absorbed with cotton wool.

THE ABDOMINAL VISCERA IN SITU

N.B. "Left" and "right" refer to the animal's left and right.

Note the shining mucous membrane, the **peritoneum,** which lines the **peritoneal cavity** (part of the coelom) and covers the organs, and the muscular partition separating the abdomen from the thorax, the **diaphragm.** *This will be seen if the* **liver** *is gently lowered a little.* This is the large red organ under the diaphragm; it partially covers the large white **stomach** on the left stretching across the body, below which is part of the short **cæcum.** Part of the **colon** may also be seen and some coils of the **ileum** will be visible. The **urinary bladder** may be seen, particularly if it contains a quantity of urine, at the posterior end of the body cavity. The hook-like **vesiculæ seminales** may also be visible in this region in the male.

Two large **fat bodies** may obscure some of the organs at the posterior end of the abdomen.

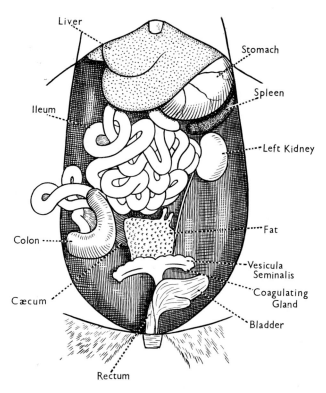

FIG. 212. **Rattus. Abdominal Viscera. (Male.)**
(From Wallis—Practical Biology)

THE ABDOMINAL VISCERA AFTER
DEFLECTION OF THE INTESTINES

Deflect the liver forwards and the viscera to the animal's left to ensure as complete an exposure of the alimentary canal as possible. Care must be exercised as the small and large intestines are coiled over one another. They should be released from each other by cutting through the **mesentery** *which supports them. Remember to remove any fat which obscures the organs.*

Find the **duodenum,** the first part of the small intestine, into which the **stomach** enters and continuous with which is the **ileum.** This leads into the small, short **cæcum** from which arises a short **colon.** The last part of the large intestine is the **rectum.**

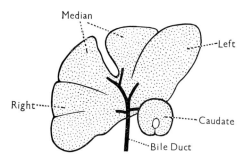

FIG. 213. **Rattus. Liver.**
(From Wallis—Practical Biology)

Examine the **liver.** It is suspended from the diaphragm by the **falciform ligament** and is composed of four **lobes**—the large sub-divided **right lobe** (the posterior one on the animal's right), the large undivided **left lobe** (the anterior one on the left), the deeply cleft **median lobe** (in the centre anteriorly) and a curved **caudate lobe** (posteriorly on the right and looping round the œsophagus). There is no gall bladder. The **bile duct** formed from the hepatic duct from each lobe enters the proximal loop of the duodenum. Note the wide translucent **cardiac end** and narrow **pyloric end** of the **stomach,** and the constriction, the **pyloric sphincter** (or **pylorus**) at the outlet of the stomach where it joins the duodenum.

Pull the stomach down slightly in order to see the **œsophagus** *entering the cardiac end.*

Spread out the loops of the duodenum without cutting the mesentery.

The **pancreas** is a large diffuse gland between these loops. The **pancreatic ducts,** (there are several), open into the bile duct along its length as the latter traverses the pancreas.

The **spleen** is a dark red body lying alongside the cardiac end of the stomach.

Now find two of the blood vessels associated with the alimentary canal. The others will not be seen until it has been removed.

The **portal vein** to the liver will be found in the mesentery alongside the colon. *Turn the stomach over to the animal's right.* The **cœliac artery,** which arises from the dorsal aorta, will be revealed running across to the stomach.

The left **kidney,** a dark red organ, will also be seen on the dorsal body wall. There is no renal portal vein.

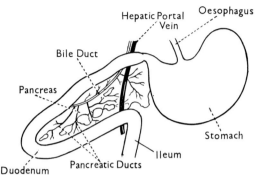

FIG. 214. **Rattus. Duodenum and Pancreas.**
(*From Wallis—Practical Biology*)

REMOVAL OF THE ALIMENTARY CANAL

Remove the alimentary canal as follows: ligature the portal vein in two places and cut across it between the ligatures. Cut across the rectum near its lower end. Now carefully cut through the mesentery, close to the intestine, gradually unravelling the intestine. Take great care not to puncture any part of the intestine, and when you reach the duodenum try to keep the loop intact. Finally cut across the œsophagus. Now neatly lay out the alimentary canal on another disecting board.

Note the **œsophagus, stomach (cardiac end** and **pyloric end), pyloric sphincter, duodenum, ileum** and **cæcum.** At the apex of the cæcum is **lymphoid tissue** which corresponds to the **appendix.** Continuous with the cæcum is the short **colon** followed by the **rectum.**

THE ABDOMINAL VISCERA REMAINING AFTER THE REMOVAL OF THE ALIMENTARY CANAL, IN SITU

Note the two **kidneys,** the right being only slightly anterior to the left: these are dorsal to the peritoneum. The **left adrenal body** is situated just above the left kidney. The **right adrenal body** is anterior to the right kidney though usually hidden by it.

Note the **ureters,** narrow white tubes leading from the kidneys to the **urinary bladder,** and the **renal artery** and **vein** on each side leading from the **dorsal aorta** and to the **posterior vena cava** respectively. The dorsal aorta and the vena cava run in the mid line, the latter being hidden by the aorta which lies ventral to it. *Gently pull back* *the bladder slightly in order to see the entry of the two ureters.* The reproductive organs in the posterior part of the abdomen will be studied separately.

The Diaphragm

Now turn to the anterior end of the abdomen and examine the **diaphragm,** which separates the abdomen from the thorax. *Cut through the falciform ligament.*

Note the **central tendon** and **marginal muscular portions** of the **diaphragm.** The œsophagus and the aorta and posterior vena cava pass through it.

THE REPRODUCTIVE SYSTEM

Male

It will probably be necessary to remove a great deal of fat before the organs and blood vessels are clearly exposed. Find the scrotal sacs and cut along the ventral surface.

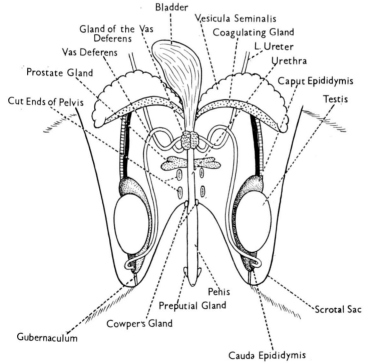

Fig. 215. **Rattus. Male. Reproductive System.**
(*From Wallis—Practical Biology*)

It has already been noted that the two **scrotal sacs** are enclosed in a single fold of skin externally. Each contains an ovoid **testis** with a large mass of coiled tubes, the **epididymis,** on its inner surface, the upper part of which is the **caput epididymis** and the lower part the **cauda epididymis,** the connection between the two parts being narrow. A cord, the **gubernaculum,** joins the cauda epididymis to the scrotal sac. Also arising from the cauda is a duct, the **vas deferens** (modified mesonephric duct) which passes into the abdomen, loops round the ureter and then passes dorsal to the bladder to enter the urethra. Note the two large hook-shaped sacculated **vesiculæ seminales** and the two **coagulating glands** alongside them.

Carefully cut longitudinally through the ischium and pubis on each side of the pelvis, i.e., on each side of the symphysis pubis, and remove the central part of the girdle. Deflect the bladder and the vesiculæ seminales and the coagulating glands to one side.

There are two **prostate glands,** *each subdivided into two, lying at the neck of the bladder. Deflect them with the structures above.*

From the neck of the bladder the **urethra** arises and passes through the erectile tissue of the **penis** to its free end, the **glans penis,** enclosed in a loose sheath, the **prepuce.** In the ventral wall of the penis is a cartilaginous process. Two **Cowper's glands** will be found alongside the urethra near the bladder. The **prostate glands** are attached to the neck of the bladder and there is a gland around the neck known as the **gland of the vas deferens.** The blood vessels associated with these organs will be seen later.

Female

Carefully cut through the pelvic girdle and separate the two halves as indicated for the male.

Note the two small, irregularly-shaped **ovaries** on the dorsal body wall lateral to the kidneys and the very short, narrow and coiled **Fallopian tubes** which arise close to the ovaries. These lead posteriorly into the thick-walled **uteri** in which the embryos develop. From the uteri a wide median tube, the **vagina,** leads to the exterior, opening into the slit-like **vulva.** Unlike the rabbit, the **vaginal opening** is separate from and dorsal to the **urethral opening.** In the ventral wall of the vulva is a small rod-shaped structure, the **clitoris** (the homologue of the male penis). The blood vessels associated with these organs will also be examined later. See Fig. 216, p. 285.

THE ABDOMINAL BLOOD VESSELS

The abdomen is now ready for examination of the blood vessels. *After removal of obscuring fat, including the large fat bodies at the*

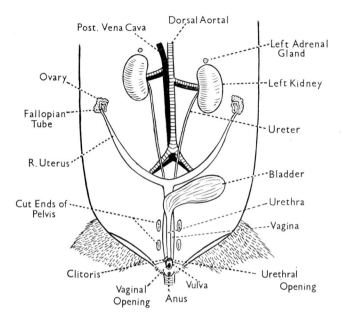

FIG. 216. Rattus. Female Reproductive System.
(From Wallis—Practical Biology)

posterior end of the abdomen, deflect out of the way any organs and structures (such as the bladder and rectum, the seminal vesicles in the male and the vagina in the female) which hide the blood vessels and pin them down. Remove any remaining mesentery which may also be obscuring the vessels and cut through the ureters near the kidneys and again near the bladder and remove the portion between. The arteries and veins will then be exposed.

It should be noted that there is considerable variation in the position and arrangement of some of the blood vessels and they may not, therefore, appear quite as described below.

Find the **dorsal aorta** and the **posterior vena cava** in the mid-line on the dorsal body wall. *Now find and identify the following arteries:—*

The single **coeliac artery** arises from the dorsal aorta just posterior to the diaphragm and divides into a **gastric artery** to the stomach (already seen), a **splenic artery** to the spleen and an **hepatic artery** to the liver. Behind it arises a single **anterior mesenteric artery** which supplies the greater part of the intestine. Next the paired **renal**

arteries leading to the kidneys will be found. These are followed by the **genital arteries (spermatic** in the male and **ovarian** in the female) but the left genital artery may arise from the renal artery on that side. A pair of **ilio-lumbar arteries** then arise from the aorta at slightly different levels, the left being slightly anterior to the right.

Towards the posterior end of the abdomen the aorta bifurcates and at this point an unpaired **posterior mesenteric artery** which supplies the rectum will be found. The bifurcations are the **common iliac arteries** which pass into the hind limbs where each divides into an **external iliac (femoral) artery** and an **internal iliac artery.** *These should*

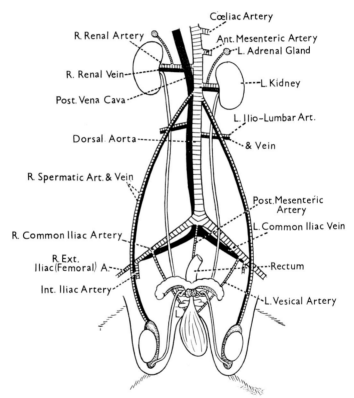

FIG. 217. **Rattus. Abdominal Blood Vessels. Male.**

be traced into the legs by careful separation of the muscles and con-nective tissue. A small **vesical artery** to the bladder arises from each common iliac. There is much variation in the branching of the iliacs and they may even be asymmetrical.

The following veins should now be found and traced:—

In some specimens two **common iliac veins** formed by the union of the **external iliac veins** (continuations of the **femoral veins**) and **internal iliac veins,** both from the legs, join to form the posterior vena cava. In other specimens it seems the vena cava is formed by the union of the external ilacs and the internal iliacs occur as a series of small veins which enter the external iliacs. There appears to be variation here and some confusion over nomenclature. On the whole, it is perhaps better to name the two veins which form the posterior vena cava the **common iliac veins.** The **ilio-lumbar, genital** and **renal veins** should now be identified alongside the corresponding arteries. The **portal vein** has already been seen in an earlier dissection; it is formed from veins from the alimentary canal. Finally the **hepatic veins** from the liver will be seen entering the posterior vena cava just below the diaphragm.

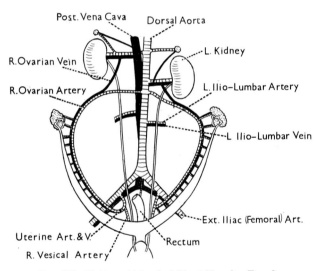

FIG. 218. **Rattus. Abdominal Blood Vessels. Female.**

THE KIDNEY

Examine a kidney in situ. Removal of surrounding fat will be necessary.

Note that is of typical "kidney" or bean shape.

The inner concave edge is known as the **hilus.** The **renal artery** and **vein** will be seen entering and leaving at this point, from which the **ureter** also emerges.

Remove one of the kidneys and cut a longitudinal section with a scalpel. Examine the cut half with a hand lens. (See Fig. 199, p. 257.)

Note that it consists of an outer rind, or **cortex,** in which is a number of **Malpighian bodies** appearing as specks, and an inner **medulla.** The ureter widens after entering the hilus as the **pelvis** of the kidney into which the conical **pyramid** opens. The tubules of the kidney open on to the pyramid.

THE THORAX

Remove the ventral wall and as much as possible of the lateral walls of the thorax. It is not as easy to leave the diaphragm in its normal position as is the case when dissecting the rabbit but the method suggested for the rabbit may be tried if desired.

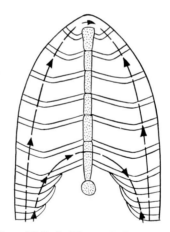

FIG. 219. **Rattus. Method of Removal of Ventral Thoracic Wall.**

Insert the scissors below the last rib on one side and well down towards the side. Take care not to damage the organs inside the thorax with the lower blade of the scissors. Cut upwards and inwards through the ribs towards the xiphoid cartilage. Continue to cut above the cartilage through the xiphisternum and then down through the ribs on the opposite side to a position corresponding to the original cut on the other side. Now cut upwards through the ribs, well down towards the outside, first on this side and then on the other, right to the top of the thorax so that the two cuts meet. Be careful at the anterior end not to cut the anterior vena cava. The ventral wall of the thorax can then be removed and the diaphragm pinned down to expose its upper surface.

THE THORACIC VISCERA IN SITU

Note the bilobed **thymus gland** which persists in the adult rat and partly covers the **heart** lying in its **pericardium.** On each side of it is a **lung** enclosed in **pleura** but the latter will most probably have been removed when opening the thoracic cavity. The **mediastinum** will also have been removed. This lies between the pleural cavities from the diaphragm to the pericardium and encloses the lower part of the trachea and œsophagus (which is dorsal to the trachea) and the main blood vessels in this part, such as the posterior vena cava. More of the **trachea** will also be visible anterior to the heart.

THE HEART AND GREAT VESSELS

Carefully remove the thymus gland and free the heart from the pericardium. Remove any fat which obscures the blood vessels and deflect the heart to the animal's right and pin it down.

The **heart** consists of thin-walled **right** and **left auricles** anteriorly and thick-walled **right** and **left ventricles** posteriorly. The right auricle will probably be at least partially hidden in its deflected position. Immediately anterior and dorsal to the heart you will see the **aortic arch** and the **innominate artery** arising from it. To the left of this the **left common carotid artery** will be found. Further to the left is the **left subclavian artery.** The **pulmonary artery** from the right ventricle of the heart runs across the lungs and between it and the aortic arch is a ligament known as the **ductus arteriosus.** The **pulmonary veins** from the lungs unite and the single vein thus formed enters the left auricle. The right and left **anterior venæ cavæ** and the **posterior vena cava** will also be found. These enter the right auricle. The two **subclavian veins** enter the anterior venæ cavæ.

The **azygos vein** from the intercostal muscles between the ribs should also be seen entering the left anterior vena cava near the heart.

THE HEART OF THE SHEEP

The sheep's heart, being much larger than that of the rat, is more suitable for the study of the internal structure. It differs little from that of the rat. Instructions for the dissection will be found on pp. 263–265.

THE NECK

Owing to the fact that the neck is short and that the nerves are very slender, this dissection needs considerable care. A lens will be found helpful for identification of the nerves and blood vessels.

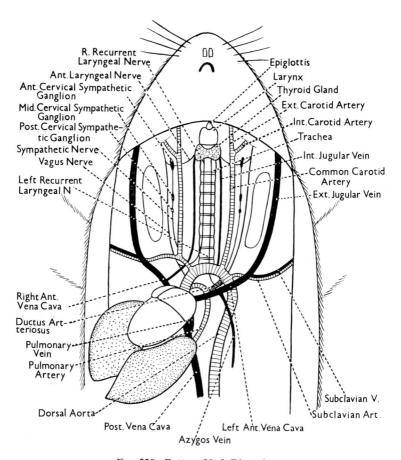

R. Recurrent Laryngeal Nerve
Ant. Laryngeal Nerve
Ant. Cervical Sympathetic Ganglion
Mid. Cervical Sympathetic Ganglion
Post. Cervical Sympathetic Ganglion
Sympathetic Nerve
Vagus Nerve
Left Recurrent Laryngeal N

Epiglottis
Larynx
Thyroid Gland
Ext. Carotid Artery
Int. Carotid Artery
Trachea
Int. Jugular Vein
Common Carotid Artery
Ext. Jugular Vein

Right Ant. Vena Cava
Ductus Arteriosus
Pulmonary Vein
Pulmonary Artery

Dorsal Aorta
Post. Vena Cava
Azygos Vein
Left Ant. Vena Cava

Subclavian V.
Subclavian Art.

Fig. 220. **Rattus. Neck Dissection.**

See that the head is well extended and pin it down. Use a small scalpel and remove fat and connective tissue and so expose the two very large **external jugular veins** *which run forwards from the anterior venæ cavæ. Cut through the sternohyoid muscles longitudinally, taking* *care not to cut beneath them. Then sever the two halves at each end and remove them. Now carefully remove the muscles on either side and thus expose the blood vessels, nerves and trachea.*

On either side of the **trachea** the slender **internal jugular veins** will be found. Note the **thyroid gland** at the anterior end of the trachea. It is composed of two lobes, one on each side of the trachea and joined on the ventral side. Anterior to this gland is the **larynx.** *Careful separation of the following blood vessels and nerves running more or less parallel with the trachea will now be necessary if they are to be identified. Find* the **common carotid artery,** the origin of which was seen earlier. It divides into an **internal carotid artery** and an **external carotid artery** at the level of the thyroid gland. Now find the **vagus nerve (X)** which runs alongside and the **recurrent (or posterior) laryngeal nerves** which lie on the surface of the trachea. *Trace them backwards to their origins from the main branch of the vagus.* The **left recurrent laryngeal nerve** arises just below the aortic arch, loops round the ductus arteriosus and then passes up the neck. The **right recurrent laryngeal nerve** arises a little further forward and then follows a similar course upwards. Near the larynx the main branch of the vagus gives off a small branch to that organ known as the **anterior laryngeal nerve.** This in turn gives off a slender **cardiac depressor nerve** to the heart: it is difficult to find.

A ganglion, the **anterior cervical sympathetic ganglion,** will be found alongside the origin of the internal carotid artery and a **sympathetic nerve** runs backwards from it. **Middle** and **posterior cervical sympathetic ganglia** will be found close to each other further back. Note the **phrenic nerve** which lies outside the external jugular vein and supplies the diaphragm. Finally *find the* **hypoglossal nerve (XII)** which runs across the carotid arteries at the anterior end, level with the larynx and then along the posterior cornua of the hyoid to the tongue.

THE RESPIRATORY SYSTEM

Remove and examine the complete respiratory organs of the neck and thorax. (See Fig. 221, p. 284.)

As already seen the **larynx** is situated at the anterior end of the **trachea.** Ventrally and laterally it is composed of the **thyroid cartilage** with the ring-like **cricoid cartilage** posterior to this. The trachea is composed of rings of cartilage, incomplete dorsally, and divides posteriorly into two **bronchi,** one to each lung. The **right lung** is divided into four lobes, the **anterior, middle, posterior** and **post-caval** lobes while the left lung is undivided.

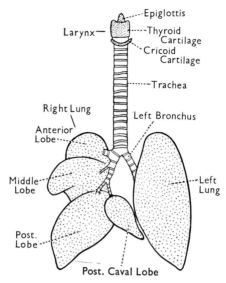

FIG. 221. **Rattus. Respiratory System.**

THE VASCULAR SYSTEM

Now that all the blood-vessels have been seen *you should do separate dissections of the complete* **Arterial** *and* **Venous Systems.** *You should also draw connected diagrams of them.*

THE HEAD

The only part of the head which will be studied is the brain.

THE BRAIN

The brain must be examined in a freshly killed rat or in which it has been preserved in accordance with the instructions on p. 278, because it decomposes rapidly. It is soft and delicate and very gentle and careful dissection is necessary. *Cut through the skin in the mid-line on the dorsal side of the head and pull the two portions apart. The roof of the cranium is thin and the brain can be seen through the bone.* (No decalcification is therefore necessary as is the case with the rabbit.) *The cranium roof must now be removed. The brain is small and, as already pointed out, great care is necessary owing to the thinness of the bone.*

Holding a very small scalpel horizontally, insert its point into the *posterior end of the interparietal bone at the back of the skull, taking care not to penetrate so as to damage the brain. This will loosen the roof of the cranium which should now be removed gradually and carefully. Now remove the rest of the bone, again very carefully, so as to expose the brain completely from end to end. The brain should now be removed from the cranium as follows:—*

Remove the surfaces of the first few vertebræ and cut through the spinal cord thus exposed. Now place the scalpel under the cranium at its posterior end and, working forwards and cutting through the nerves on the ventral side, gradually free the brain from the cranium. Great care must be exercised in the region of the cerebellum as part of it, the paraflocculi on the sides, are encased in bone. On no account touch the brain with the scalpel or other metal instruments or it will be damaged. Gently place the brain in a small dish (perti dish or large watch-glass). It should be preserved in 70 per cent. alcohol or 4 per cent. formaldehyde if required on a future occasion.

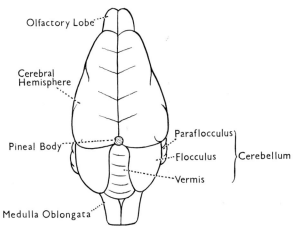

FIG. 222. **Rattus. Brain. Dorsal View.**

Dorsal View

The **Fore-Brain** is composed of two **cerebral hemispheres** with a few slight convolutions anterior to which are the proportionately large **olfactory lobes.** The posterior part of the fore-brain the thalamencephalon, is hidden in this view by the cerebral hemispheres. The rounded structure between the hemispheres at their posterior end (if it has not been removed) is the **pineal body** (possibly an endocrine gland of which little is known).

The **Mid-Brain** is composed of the corpora quadrigemina but these are not visible in the dorsal view.

The **Hind-Brain** is composed of the **medulla oblongata** with the cerebellum on its dorsal side. This consists of a central **vermis** on each side of which is a **flocculus** bearing a small **paraflocculus** externally towards its anterior end. The spinal cord is continuous with the medulla.

Ventral View

Again note the **olfactory lobes** and **cerebral hemispheres** also the **olfactory tracts** running backwards in the centre from the olfactory lobes. In the mid-line, a little further back, will be seen the **optic chiasma,** formed by the crossing over of the **optic nerves (II).** Behind this is the **pituitary body** followed by the **medulla oblongata.** Parts of the **flocculi** and the **paraflocculi** will also be visible in this view.

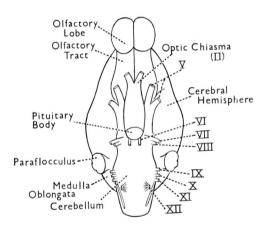

FIG. 223. **Rattus. Brain. Ventral View.**

The roots of the **trigeminal nerves (V)** run alongside the posterior part of the optic chiasma. *It is necessary to remove the roots of the trigeminal on one side to reveal the* **oculo-motor nerve (III)** and the **trochlear nerve (IV)** beneath. Immediately posterior to the trigeminal nerve, the roots of the **facial nerve (VII)** and the **auditory nerve (VIII)** will be seen and, a little further back still, those of the **glossopharyngeal nerve (IX),** the **vagus nerve (X)** and the **spinal accessory nerve**

(XI). The roots of the **abducens nerve (VI)** and of the **hypoglossal (XII)** are more ventrally situated, the former being just behind the pituitary body and the latter posterior to the spinal accessory nerve.

Longitudinal Section

This must be done on a brain hardened in alcohol. Make a longitudinal vertical cut through the centre of the brain with a razor blade and examine the cut surface. Retain the other half.

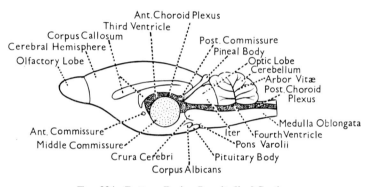

FIG. 224. **Rattus. Brain. Longitudinal Section.**

Identify the **olfactory lobes,** the **cerebral hemispheres,** the cut edge of the **corpus callosum** and the **anterior choroid plexus** immediately behind and below it. The large rounded structure ventral to this is the **middle commisure** (or **massa intermedia**). The small **anterior commissure** lies just anterior to this. The **body of the fornix** is to be found just dorsal to the anterior commissure, continuous with the corpus callosum. The **posterior commissure** lies dorsal and posterior to the middle commissure. The **corpora quadrigemina** can now be seen (it will be remembered that they were not visible in the external views of the brain). They lie between the cerebral hemispheres and the **cerebellum** which is easily identified, with its tree-like **arbor vitæ** internally, on the dorsal side of the **medulla oblongata.** The thick **crura cerebri** form the floor of the mid-brain and beneath this lies the **pons varolii.** The **fourth ventricle,** roofed over by the **posterior choroid plexus,** will be seen in the medulla and the **iter** connecting it to the **third ventricle** in the **thalamecephalon.** The **pineal body** on its long **pineal stalk** (if either is still present) and the **pituitary body** will be found in their respective places. Just anterior to the latter the **corpus albicans** will be found.

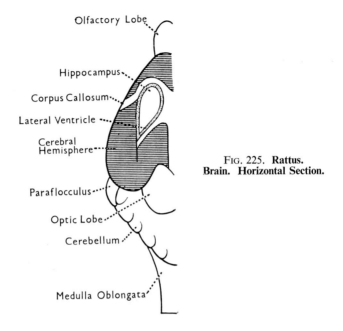

Olfactory Lobe

Hippocampus

Corpus Callosum

Lateral Ventricle

Cerebral Hemisphere

Paraflocculus

Optic Lobe

Cerebellum

Medulla Oblongata

FIG. 225. **Rattus.**
Brain. Horizontal Section.

Horizontal Section

Now take the other half of the brain and cut a horizontal section a little above half way down, again using a razor blade.

Note the **olfactory lobe** and the **cerebral hemisphere.** In the latter will be found the wide **hippocampus** and the narrow **lateral ventricle.** The **optic lobe, cerebellum** and **medulla oblongata** will also be seen.

PART III

CYTOLOGY AND HISTOLOGY

INTRODUCTORY NOTES

Cytology is the study of cell structure and **Histology** is the study of tissues.

The frog provides good material for the study of animal histology, though tissues from other animals can be used. In some cases only prepared slides should be examined. A freshly killed animal dissected dry should be used for most preparations though the preserved dogfish or frog can be used for cartilage. *Make a few permanent preparations from the animals as you dissect them. In all cases permanent prepared slides should also be examined.*

Refer to Part I for instructions on the making of slides and to the directions for drawing microscopical preparations on p. 2.

CELL STRUCTURE

Examine an electron micrograph of an animal cell. (See figs. 226 and 227, p. 298.)

The cell is enclosed in a **plasma membrane** through which oxygen, water and dissolved substances move into the cell and carbon dioxide and excretory substances out of it. This plasma membrane is considered to be composed of lipid material lined on both sides with protein and appears to be around 70 Å* in thickness. Within the plasma membrane is the **cytoplasm** in which lies the **nucleus** and various **cell inclusions.** The cytoplasm is composed of a matrix known as **hyaloplasm** containing an **endoplasmic reticulum.** The latter consists of small sacs or vesicles enclosed in membranes and on the surface of these lie granules known as **ribosomes.** These contain **R.N.A.** (*ribonucleic acid*) and proteins and are the sites of protein synthesis from amino-acids. In addition to the ribosomes are very minute **microsomes** which are mainly composed of R.N.A. of which they may form a store. Somewhat rod-shaped or rounded inclusions occur very freely; these are the **mitochondria** *and are visible under the optical microscope.* The electron microscope shows them to contain a series of membranes which divide them up into transverse sacs. They are self-replicating and the sacs contain the enzymes necessary in the various stages of respiration in the cell. The mitochondria are responsible for the transfer of the released energy. *Examine a slide showing* **mitochondria** *under the high power.* The **Golgi Body (Golgi**

*Å = Ångström = 10^{-8}çm.

297

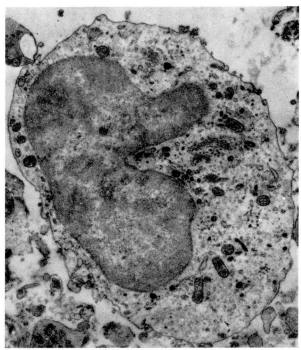

FIG. 226. Electronmicrograph of an animal cell (Baboon monocyte) × 12,000.

(*By courtesy of Dr. D. Kay and Dr. J. C. Poole*)

(*From Wallis—Human Biology*)

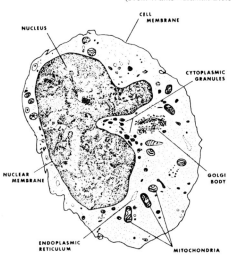

FIG. 227. Drawing of above electronmicrograph to show cell inclusions.

(*From Wallis—Human Biology*)

Apparatus), composed of a series of elongated sacs and vacuoles is a more compact cell-inclusion and though the function is uncertain it has been suggested that it is concerned with secretion as it is well developed in secretory cells but this is not by any means certain.

The most prominent structure lying in the cytoplasm is, of course, the **nucleus** enclosed in a **nuclear membrane** in which openings have been observed with the electron microscope. The nucleus is often spherical but, as will be seen, it can assume other shapes, as in the leucocytes of the blood. Within the **nucleoplasm** or **nuclear sap,** as the protoplasm in the nucleus is called, are one or more spherical bodies called **nucleoli.** They contain R.N.A. while the nucleoplasm itself contains **D.N.A.** (*deoxyribonucleic acid*) and protein. Also dispersed in the nuclear sap is the substance called **chromatin** which may be present in the form of a network in the resting nucleus but at nuclear divisions a definite number of **chromosomes** is visible and it may be that, though invisible, they exist as such in the resting state. As will be seen in our study of genetics, the number of chromosomes is constant for any one species of plant or animal. In the frog it is 12 pairs, in the rabbit 11 pairs, in *Drosophila* 4 pairs and in man 23 pairs. The chromosomes contain D.N.A. and proteins such as *histone*. It is possible that the nucleoli are concerned in the transference of nucleic acid between chromosomes and the cytoplasm.

THE STRUCTURE OF TISSUES

A **tissue** is an aggregation of cells similar in structure and function. There are five chief kinds of tissue:—

      (1) Epithelial tissue.
      (2) Connective tissue.
      (3) Blood.
      (4) Muscular tissue.
      (5) Nervous tissue.

EPITHELIAL TISSUES

Epithelium is the tissue which forms either the external covering of the body and its organs or the lining of such structures as the cœlom, the alimentary canal and the blood vessels. The latter are known as **endothelia.** Epithelium may consist of one layer of cells—**simple epithelium,** or of several layers—**compound epithelium.** The shape, structure and function of epithelial cells vary and the tissue may be classified as follows:—

Simple Epithelia	Compound Epithelia
Columnar	Transitional
Cubical	Stratified
Ciliated	
Squamous (or Pavement)	
Glandular	
Pseudo-stratified	

SIMPLE EPITHELIA

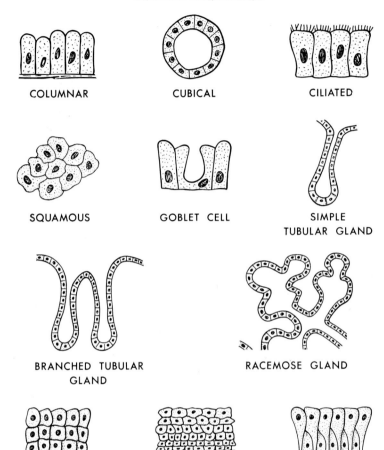

FIG. 228. **Simple Epithelia.**
(From Wallis—Human Biology)

COLUMNAR EPHITHELIUM

(i) *Take a small piece of the lining of the small intestine e.g.: of the frog. Place in 2 per cent. osmium tetroxide for a few hours and then leave in Ranvier's alcohol for twenty-four hours. Stain with picro-carmine. Mount in dilute glycerine and examine under the high power.*

Note the elongated column-like cells and their rather elongated **nuclei,** the cells standing on a **basement membrane.**

(ii) *Examine a prepared slide of* **T.S. stomach** or **intestine.**

In the mucous membrane (the innermost layer) note the single layer of elongated cells close together in columns and at right angles to the surface, which it lines, and resting on a non-cellular **basement membrane.** The **nuclei** are somewhat elongated.

CUBICAL EPITHELIUM

Examine a prepared slide of a **T.S. of the kidney.**

Note the cubical cells and their central rounded **nuclei** lining the uriniferous tubules.

This kind of tissue also occurs as the lining of sweat and other glands.

CILIATED EPITHELIUM

Examine under the high power a prepared slide of (a) the **membrane lining the roof of the mouth of the frog,** (b) *the* **trachea or bronchus of a mammal,** *or* (c) **isolated ciliated epithelial cells.**

The cells are columnar and bear on their free edges hair-like protoplasmic processes called **cilia.** Note also their **nuclei.** This tissue occurs characteristically in the bronchial passages.

SQUAMOUS (or PAVEMENT) EPITHELIUM

(i) *Examine a prepared slide of* **squamous epithelium** (*e.g. from the* **mesentery of a mammal** *which has been treated with* 1 *per cent. silver nitrate and reduced in sunlight*).

Note the blackened outlines of the thin, flat cells which fit into one another like 'crazy paving' and which have rather large **nuclei.**

(ii) *Carefully scrape the inside of your cheek with the back of a scalpel. Put the scrapings on a slide. Stain with methylene blue and examine under the high power. These squamous epithelial cells form the superficial layer of a stratified epithelium.*

When the edges of the cells are wavy as found in the peritoneum, the tissue is said to be **tessellated.**

GLANDULAR EPITHELIUM

(i) *Examine a prepared slide of the* **T.S. of the ileum of a mammal.**

Look for cup-shaped **goblet cells** formed by the infolding of the walls of the columnar epithelial cells here and there in the mucous membrane.

(ii) *Examine a prepared slide of the* **T.S. of the frog's skin.**

Note the somewhat flask-shaped **simple saccular glands** opening on to the surface. (See Fig. 237, p. 318.)

(iii) *Examine a prepared slide of the* **T.S. of the small intestine of the frog** *or* **mammal** *under the high power.*

Note the **simple tubular glands** lined by **columnar epithelial** and **secretory cells** in the mucous membrane.

(iv) *Examine a prepared slide of the* **T.S. of the stomach (pyloric end)** of a frog or mammal.

Note the **branched tubular glands** lined by columnar epithelial cells in the mucous membrane.

(v) *Examine a prepared slide of a* **T.S. of the salivary gland of a mammal.**

Observe the very much branched tubules constituting the compound saccular or **racemose gland** which are lined by **secretory cells.**

PSEUDO-STRATIFIED EPITHELIUM

Examine a prepared slide of the **T.S. of a mammalian salivary gland.** *Examine one of the larger ducts.*

Note that although the cells lining it appear to be in several layers they all, in fact, lie on a basement membrane, though some do not reach up to the free surface. It is therefore best regarded as a simple epithelium.

COMPOUND EPITHELIA

TRANSITIONAL EPITHELIUM

Examine a prepared slide of a **section of the wall of the urinary bladder of a mammal.**

Note that the tissue consists of three or four apparent layers of cells, those on the surface being large and flattened. This tissue is capable of considerable stretching, when the nunber of layers is reduced, sometimes to a single layer.

STRATIFIED EPITHELIUM

Examine again the prepared slide of the **T.S. of the skin of a frog** *or* **a mammal.**

Note that there are several layers of cells in the outer region or **epidermis.** (The dermis beneath is composed mainly of connective tissue.)

CONNECTIVE TISSUES

Connective Tissue joins various organs and structures together and its composition and histological structure varies with its function and location. In some cases it is purely skeletal. A characteristic feature is the presence of a non-cellular matrix in which the cells are embedded and which is secreted by the cells. The tissue is classified as follows:—

Fibrous Tissue	Adipose Tissue
Elastic Tissue	Cartilage
Areolar Tissue	Bone

WHITE FIBROUS TISSUE

(i) *Pull up the connective tissue between the muscles of the thigh of a frog. Remove with small scissors and mount in physiological saline (0.75 per cent.). Examine under the high power.*

Note the white wavy bundles of thin fibres crossing each other in all directions. These are non-elastic.

Irrigate with 1 per cent. acetic acid.

The fibres disappear and the **connective tissue cells** become visible.

(ii) *Remove one of the white cords* (**tendons**) *attaching the leg muscles of the frog to the bone. Mount in physiological saline and tease out the fibres from one end with a mounted needle, holding the other end with another needle. Examine under the high power.*

Note the **white fibres** in parallel bundles and the **tendon cells** between them.

Irrigate with 1 per cent. acetic acid.

Note the **nuclei** of the tendon cells. The fibres swell up into a mass.

YELLOW ELASTIC TISSUE

Examine a prepared slide of **elastic tissue from the ligamentum nuchæ of the ox or other mammal.**

Note the coarse straight fibres which branch and anastomose at intervals.

This tissue rarely occurs pure but is found in abundance in these powerful ligaments of the neck.

AREOLAR TISSUE

This is the most frequently occurring tissue found in the body.

(i) *Examine a prepared slide of the* **subcutaneous tissue of a rabbit or rat** *or* (ii) *make your own preparation of this tissue, mounting it in physiological saline* (0.9 *per cent.*).

It is composed of bundles of **white fibres** and **yellow elastic fibres,** the latter being lesser in quantity and thinner than those found in the pure tissue.

Fix in Bouin's fluid and stain with van Gieson's stain.

The white fibres are stained red and the elastic fibres dark brown.

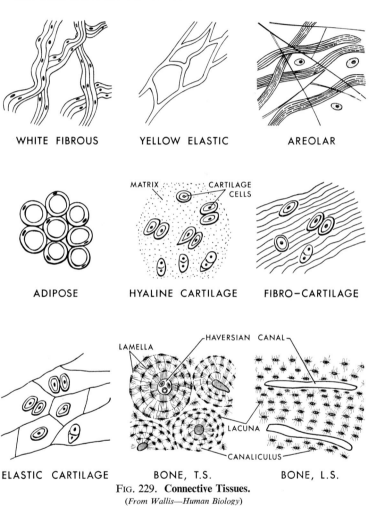

WHITE FIBROUS YELLOW ELASTIC AREOLAR

MATRIX CARTILAGE CELLS

ADIPOSE HYALINE CARTILAGE FIBRO-CARTILAGE

LAMELLA HAVERSIAN CANAL

LACUNA

CANALICULUS

ELASTIC CARTILAGE BONE, T.S. BONE, L.S.

FIG. 229. **Connective Tissues.**
(From Wallis—Human Biology)

ADIPOSE TISSUE

(i) *Examine a prepared slide of the* **mesentery of a rat or other mammal,** or (ii) *Examine a prepared slide of* **the T.S. mammalian skin.**

Note the concentration of rounded **fat cells** (in the lowest region of the dermis) *or* (iii) *Take a small piece of* **mesentery** *or* **subcutaneous connective tissue.** *Fix in Bouin's fluid, stain with Sudan III or 1 per cent. osmium tetroxide.* The **fat cells** are stained red (Sudan III) or black (osmium tetroxide).

Note the connective tissue forming a meshwork between two layers of pavement epithelium in the spaces of which lie a number of rounded **fat cells** in which the cytoplasm is reduced to lining the cells with the nuclei embedded in it, the central region being filled with fat.

HYALINE CARTILAGE

(i) *Remove one of the thin cartilages from the sternum of a frog or cut a thin section of cartilage from the chondrocranium of the dogfish. Mount in physiological saline and examine under the high power.*

Note the **cartilage cells** or **chondroblasts** with their **nuclei** lying in spaces called **lacunæ,** some occurring singly, others in groups of two or four, embedded in a clear **matrix** composed of chondrin. It occurs in the costal cartilages, in the larynx and, as articular cartilage covering the ends of bones where they form joints.

Irrigate with 1 per cent. acetic acid.

The **nuclei** become more visible.

(ii) *Make a permanent preparation of a section. Fix in formalin. Stain with Delafield's or, better, Ehrlich's hæmatoxylin. Differentiate in acid alcohol. Wash in water. Dehydrate, clear and mount in balsam.*

FIBRO-CARTILAGE

Examine a prepared slide of **fibro-cartilage.**

Note that it is a modification of hyaline cartilage in which **white fibres** are found in the matrix. It occurs in the intervertebral discs.

ELASTIC CARTILAGE

Examine a prepared slide of **elastic cartilage.**

This is a modification of hyaline cartilage, in the matrix of which **yellow elastic fibres** abound. It occurs in the epiglottis, the pinna of the ear and in the nasal septum.

BONE

Bones formed by the deposition of bone substance in connective tissue are called **membrane bones** while those formed in cartilage are called **cartilage bones.** This is a difference in origin only; histologically, they are identical. The formation of bone in these tissues is known as **ossification.**

(i) *Examine a* **T.S. of bone.**

Note the large circular **Haversian canals** (which in the living animal contain blood vessels) surrounded by concentric rings, **lamellæ** (in which calcium salts are deposited). Between each pair of lamellæ are small cavities, **lacunæ,** which contain bone cells, osteo-blasts, in the living animal. Tiny channels, **canaliculi,** run across and connect the lacunæ. The Haversian canals and their lamellæ constitute what are known as **Haversian systems.**

(ii) *Examine a* **L.S. of bone.**

Identify the structures in (i).

(iii) *Examine a section of* **membrane bone in process of ossification.** Look for (*a*) bundles of **osteogenic fibres** around which (*b*) bone cells or **osteoblasts** may be seen.

(iv) *Examine* **longitudinal sections of a long cartilage bone in process of ossification.** Look for (*a*) **unchanged cartilage,** (*b*) **cartilage undergoing calcification and disintegration,** the cartilage cells becom-ing smaller, large amœboid, multinucleate **osteoclasts** effecting this erosion, and bone cells, **osteoblasts,** which lay down the bone substance. Note that bone *replaces* the cartilage which has been destroyed.

BLOOD*

AMPHIBIAN BLOOD

To make successful blood films, it is essential that slides and coverslips are free from grease.

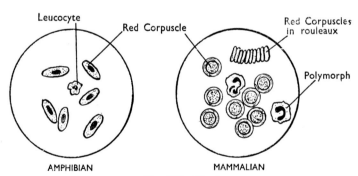

Fig. 230. **Blood.**

(i) *Take a drop of frog's blood on a cover slip. Dilute with a drop of physiological saline (0.6 per cent.) and examine under the high power.*

* Sometimes classified as a Connective Tissue, the plasma corresponding with the matrix.

Note the numerous **erythrocytes** (or **red corpuscles**), pale red or yellowish in colour, oval in shape and nucleated. Those seen from their edges show that they are biconvex.

Irrigate with 1 per cent. acetic acid.

The **nuclei** become more easily visible. Look for **leucocytes** (or **colourless corpuscles**). They are much fewer in number, and are smaller than the red. They are granular, nucleated and amœboid. The red corpuscles and leucocytes are suspended in a colourless fluid, the **plasma.**

(ii) *Put a drop of frog's blood on a coverslip and make a smear. Fix in absolute alcohol and stain for three or four minutes in alcoholic methylene blue. Wash off the excess of stain and mount.*

The **nuclei** of the **red corpuscles** and **leucocytes** will be readily seen.

MAMMALIAN BLOOD

(i) *Clean the tip of your left fore-finger with spirit and sterilise a sharp mounted needle in the flame. Quickly prick the tip of the finger. Squeeze a drop of blood on to a coverslip and invert it quickly on a slide. Make a smear and examine under the high power.*

Note the circular **erythrocytes** (**red corpuscles**), smaller than those of the frog, some occurring singly and **in rouleaux** (*i.e.*, like piles of coins). They are not nucleated, but, being biconcave discs, the central part appears light. Some **leucocytes** (**white corpuscles**) may be seen. They are larger and fewer than the erythrocytes, are nucleated and exhibit amœboid movement. There are about 600 or 700 red corpuscles to every leucocyte.

(ii) *Make another preparation as in* (i) *but dilute with physiological saline* (0.6 *per cent.*†).

The red corpuscles will be more easily examined.

(iii) *Prepare a smear of human blood, fix at once by waving rapidly in the air, and stain with methylene blue or Leishman's stain.*

The nucleated **leucocytes** will be visible. There are several kinds of leucocytes. *Search for the various types of leucocytes (see below). You are certain to find* **polymorphs** *because they are the most numerous, and probably one or two* **lymphocytes,** *but you may be unable to identify other types. The colours given below are given with Leishman's stain.*

(*a*) **Polymorphs.** The greater proportion of the leucocytes (65-70 per cent.), somewhat larger than the erythrocytes. The nucleus, stained purplish red, is composed of several lobes or is roughly

† Some prefer to use 0·75 per cent. saline.

U-shaped with the free ends lobose. Their chief function is phagocytosis.

(*b*) **Lymphocytes.** About the same size as the erythrocytes and form 20–25 per cent. of the leucocytes. The nucleus, stained blue, is round.

(*c*) **Acidophils** (or **eosinophils**). Somewhat larger than polymorphs, they form only about 3 per cent. of the leucocytes. The nucleus, stained red, may consist of two or three lobes.

(*d*) **Basiphils,** distinctly larger than polymorphs, form a mere 0·5 per cent. of the total leucocytes and have a bi-lobed nucleus which stains a purplish colour.

(*e*) **Monocytes** are the largest of the leucocytes and form about 3 per cent. of them. The nucleus, which is unstained, is usually round. These are the most actively motile of the leucocytes and their function is phagocytosis.

The **erythrocytes** will be stained an orange-pink colour.

(iv) *Examine a prepared slide showing* **blood platelets.**

Blood platelets (or **thrombocytes**) are small, rounded structures in the plasma, (colourless when unstained) about a quarter of the size of the erythrocytes. They are concerned with coagulation.

The average number of erthrocytes in man is 5,000,000 per cu. mm. (4,500,000 in woman) and their size is about 7·5μ. In the rabbit they are somewhat smaller (about 6·5μ) while in the frog they measure about 22·3μ × 15·7μ.

The average number of leucocytes in man is 7,000–8,000 per cu. mm., but the number varies at different ages, at different times and under different conditions. Thus the proportion of red corpuscles to leucocytes is about 600 or 700:1. In size they vary between 9μ and 20μ.

(v) *Prepare a blood film on a coverslip, invert on a slide and irrigate with distilled water.*

Note the effect on the erythrocytes which swell up by absorption of the water and either burst or become so distended that the hæmoglobin escapes into the plasma. This is called **hæmolysis.**

(vi) *Prepare another blood film as in* (v) *and irrigate with a hypertonic solution of sodium chloride* (0.9 *per cent.*).

Water diffuses out of the erythrocytes which shrink and become crenated.

BLOOD GROUPS

When the blood of two vertebrate animals of different species is mixed, the erythrocytes collect together in clumps and are said to **agglutinate.** This also occurs between certain animals of the same species but not in all cases. The blood of human beings falls into four **groups** and agglutination occurs only between certain groups. These groups are inherited. Agglutination is caused by the interaction of substances in the erythrocytes called **agglutinogen** (or **antigens**) and substances in the plasma called **agglutinins** (or **antibodies**). One group contains no agglutinogen in the corpuscles and this blood is therefore never agglutinated: this group is designated **O.** Of the other groups, one contains an agglutinogen **A,** another **B** and the third contains both **A** and **B.** These groups are designated **A, B** and **AB** respectively. In the plasma of Group **A** is an agglutinin **b,** in the plasma of **B** is an agglutinin **a,** in the plasma of **O** is an agglutinin **ab,** but the plasma of **AB** contains no agglutinin. The condition is inheritable.

AGGLUTINATION TAKES PLACE BETWEEN—

the plasma of **O** and the corpuscles of **A, B** and **AB**

,,　　,,　　,, **A** ,,　　,,　　　　,, ,, **B** and **AB**

,,　　,,　　,, **B** ,,　　,,　　　　,, ,, **A** and **AB**

and

NO AGGLUTINATION TAKES PLACE BETWEEN—

the plasma of **A** and **AB** and the corpuscles of **A**

,,　　,,　　,, **B** and **AB** ,, ,,　　　　,, ,, **B**

,,　　,,　　,, **AB** ,, ,,　　　　,, ,, any **Group**

,,　　,,　　,, any **Group** ,, ,,　　　　,, ,, **O**

On account of the fact that the corpuscles of **O** are not agglutinated by the plasma of any group, individuals in this group are **universal donors,** and because the plasma of **AB** cannot agglutinate the corpuscles of any group, individuals in this group are **universal recipients.**

It has been found that, regardless of group, in the case of 85 per cent. of human beings in western countries the blood is agglutinated by the serum of rabbits previously injected with the blood of the rhesus monkey. This is due to what is known as the **Rhesus factor (Rh).** Such blood is said to be **Rh positive.** The other 15 per cent. is **Rh negative.** This condition is also inheritable.

These facts form the basis of **blood transfusion,** in which agglutination must, of course, be avoided. A person's blood group is determined by mixing drops of his blood with test serums derived from people belonging to groups **A** and **B** respectively.

(vii) *Put a few drops of the blood of various pairs of individuals to each of which a few ml. of sodium citrate have been added to prevent clotting on a slide and mix them by means of a mounted needle. Examine to see whether agglutination takes place or not.*

MUSCULAR TISSUES

Muscles are composed of **fibres** which are of three kinds:—

Striated (or striped) found in the voluntary muscles, unstriated (unstriped, plain or smooth) found in the involuntary muscles, in the wall of alimentary canal, blood vessels, urinary bladder, ureters, etc. and cardiac found only in the walls of the heart.

STRIATED MUSCLE

(i) *Remove a small piece of one of the muscles of the frog's or mammal's leg. Well tease out the fibres with mounted needles on a slide and crush some of them. Mount in physiological saline (0.75 per cent.). Examine under the high power.*

Note the long cylindrical **fibres** which show alternate light and dark transverse bands (seen better when stained). Each is enclosed in a membranous sheath, the **sarcolemma,** which will be better seen where the fibres have been crushed.

Irrigate with 1 per cent, acetic acid.

The striations will be less distinct but the oval **nuclei** are now rendered more visible, scattered in the fibre, showing it is not a single cell but a synctium. Actually, each fibre is composed of a series of longitudinally running **myofibrillæ,** between which is the cytoplasmic **sarcoplasm.**

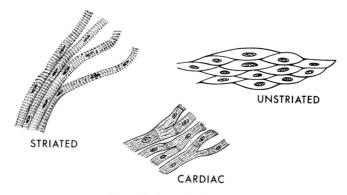

STRIATED

UNSTRIATED

CARDIAC

FIG. 231. **Muscular Tissues.**
(*From Wallis—Human Biology*)

(ii) *Remove one of the leg muscles of a cockroach. Tease out the fibres and mount in physiological saline (0.6 per cent.). Examine.*

It will be found to show a similar structure to that of the frog, though the striations are wider.

(iii) *Make a permanent preparation of teased fibres of striated muscle from the frog, the cockroach, the dogfish or mammal. Fix in formaldehyde. Stain with borax-carmine.*

UNSTRIATED MUSCLE

Spread a piece of the frog's urinary bladder on a slide with the inner surface on top. Rub off the epithelium from the inside with your finger. Fix in formaldehyde and stain with hæmatoxylin or eosin. Make a permanent mount. Examine under the high power.

Note the spindle-shaped **fibres** dove-tailing into one another, and their centrally placed **nuclei.** They are devoid of sarcolemma.

CARDIAC MUSCLE

Examine and prepare a slide of **cardiac muscle.**

Note the striated fibres with central nuclei, the striations being less distinct than in striated muscle, the fibres being devoid of any sarcolemma and being shorter than the fibres of voluntary muscle.

Also observe that the fibres branch and anastomose with neighbouring fibres.

NERVOUS TISSUES

Nervous tissue is composed of:—

Nerve cells or neurones found in the brain and spinal cord and in ganglia. Nerve fibres, which are of two kinds:—

(a) medullated or myelinated fibres found in bundles in the cranial and spinal nerves (b) non-medullated or non-myelinated fibres which occur in the terminal portions of the sympathetic nerves though they are also found in some of the peripheral nerves.

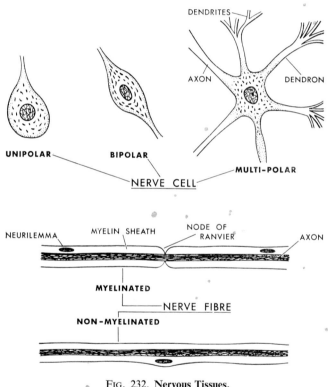

FIG. 232. **Nervous Tissues.**
(*From Wallis—Human Biology*)

NERVE CELLS (NEURONES)

Examine a prepared slide of **nerve cells from the brain or spinal cord.**

Note the large **multipolar nerve cells** or **neurones,** each composed of a **cell-body** containing a large **nucleus** in which will be seen a **nucleolus,** with small elongated bodies, **Nissl's granules,** in the cytoplasm. Some of the cell bodies bear long non-branching processes, **axons** (which become the axis cylinders of nerve fibres), and all have short processes called **dendrons,** which bear small branches called **dendrites.*** These form synapses with the dendrites of neighbouring neurones. The cells around the neurones which serve as packing tissue between them are known as **neuroglia.**

MEDULLATED NERVE FIBRES

Cut out a short piece of the sciatic nerve from the leg of a frog or mammal. Put on a slide and thoroughly tease out one end with a mounted needle. Add physiological saline (0.75 per cent.). Examine.
Note the long unbranched **nerve fibres.**
Irrigate with 1 per cent. osmium tetroxide and examine a quarter of an hour later.
Note the **axis cylinder** in the centre (a continuation of the axon of a neurone) surrounded by the fatty **medullary sheath** (stained black), outside which is the thin **neurilemma,** or primitive sheath. The medullary sheath shows interruptions at intervals, known as **nodes of Ranvier.** The **sheath nuclei** in the cytoplasm lining the neurilemma in the internodes may not be visible.

NON-MEDULLATED NERVE FIBRES

Examine a prepared slide of **non-medullated nerve fibres from the sympathetic nervous system.**
They are similar to the medullated nerve fibres except that they have no medullary sheath, they have several nuclei and they often branch and anastomose.

NERVE ENDINGS IN MUSCLE

Examine a prepared slide showing nerve endings in striated muscle.
Note the **medullated fibres** which branch two or three times, each branch running to a **muscle fibre** where they terminate in **end-plates.** In unstriated and cardiac muscle, the fibres, non-medullated at their terminations, branch and end in plexuses.

---

* The entire process (dendron and dendrite) may be called a dendrite.

MITOSIS

Under high power (or $\frac{1}{12}$ in. O.I.) *examine a slide showing* **mitosis** *in Ascaris or other suitable animal. Search for different stages of mitosis in different nuclei. You may not succeed in finding them all.*

Prophase. The **centrosome** (a small spherical structure just outside the nucleus, containing a small body, the **centriole**) divides into two

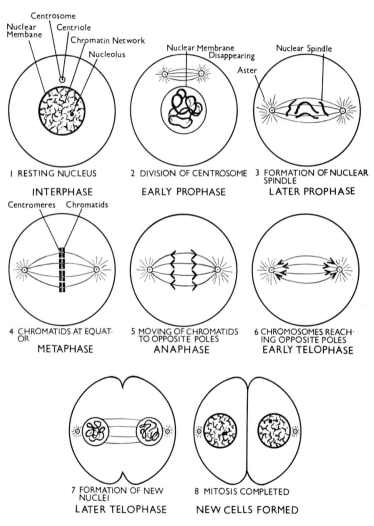

FIG. 233. **Mitosis in Animal Cells.**

and later these move to opposite poles of the nucleus, a **nuclear spindle** developing between them. Fibrils radiate from the centrosomes forming **asters** which form the poles of the spindle. The apparent chromatin network of the **resting nucleus (interphase)** actually consists of **chromosomes**† and these split longitudinally into **chromatids.** The double nature is not visible at this stage and any apparent continuous spiral is an artefact. It is due to the chromosomes being coiled round each other. Meanwhile the nuclear membrane and nucleolus disappear. The centrosomes reach opposite ends of the nucleus with the nuclear spindle between them, the mitochondria becoming arranged around the spindle.

Metaphase. The chromosomes clearly show their double structure and these chromatids arrange themselves on the equator of the spindle at points on them known as **centromeres.**

Anaphase. The chromatids begin to move towards the opposite poles of the nucleus, guided by the spindle and being pulled into V-shaped loops in the process.

Telophase. The chromatids continue to move to opposite poles and a new nucleus if formed at each pole by passing through the changes of the prophase in reverse, the chromatids becoming the new chromosomes and these become less visible. A nuclear membrane develops, the asters disappear and a single centrosome is left outside each nucleus.

Cell division follows. A constriction develops in the cytoplasm between the nuclei. This deepens and forms a cell membrane and ultimately two new cells are formed, each with its own **(diploid)** nucleus containing the original number of chromosomes.

A special form of nuclear division occurs at one stage in the formation of the gametes in which the chromosome number is halved. This is called **meiosis** and is described on pp. 326–328.

THE HISTOLOGY OF ORGANS

A microscopical examination of the structure of certain mammalian organs should now be made. They will be found to consist of more than one kind of tissue.

In the various sections examined, **blood vessels** and **lymphatics** will be seen, mostly in transverse section. The wall of an **artery** is composed of (*a*) an **external coat** of **areolar tissue,** the **Adventitia,** (*b*) a **middle coat,** the **tunica media,** of **unstriated muscle fibres** and, in the case of the larger arteries such as the aorta and the carotids,

† The shape and number of chromosomes varies with different animals but is constant for any one species as already stated on p. 299.

elastic tissue and (*c*) an **inner coat,** the **tunica interna,** composed
of **elastic tissue** lined by an **endothelium** of pavement epithelium.

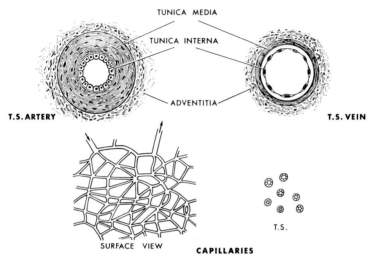

TUNICA MEDIA

TUNICA INTERNA

ADVENTITIA

T.S. ARTERY

T.S. VEIN

SURFACE VIEW   **CAPILLARIES**   T.S.

FIG. 234. **Structures of Blood Vessels.**

The wall of a **vein** is similar but thinner, there being less muscular
and elastic tissue in the middle coat and less elastic tissue in the
inner coat. The valves which occur at intervals are strengthened
infoldings of the inner coat.

The walls of **capilliaries** are thin and are composed of endothelium
only in which the cells are flattened.

The walls of **lymphatics** are in general similar to those of veins
but are much thinner.

THE SKIN

MAMMALIAN SKIN

(i) *Examine a* **T.S. of Mammalian Skin.**

The external layer, the **epidermis,** is composed of stratified epithelium, with the **stratum corneum** (or horny layer) consisting of rows of
flattened cells on the outside and, beneath it, in some parts of the
skin, a layer of granulated appearance, the **stratum granulosum.**
This is followed by a somewhat wavy layer, the **stratum Malpighii.**
There are no blood bessels in the epidermis. The stratum corneum
consists mostly of epithelial 'cells' called squames, which are devoid
of nuclei. The stratum Malpighii is the growing region and the cells

in its lower layers show mitosis. The deepest cells of this layer contain the pigment melanin, abundant in dark-skinned races. The thickness of the epidermis and particularly of the stratum corneum varies considerably in different parts of the body.

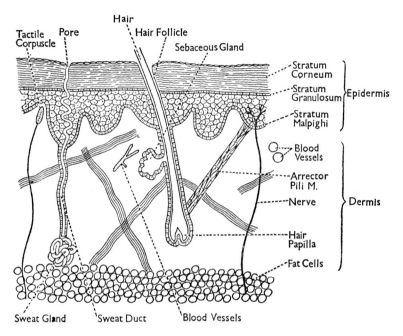

FIG. 235. **Mammalian Skin. T.S. Semi-diagrammatic.**
(From Wallis—Practical Biology)

Beneath the epidermis is the true skin or **dermis.** This is a much thicker layer composed of **connective tissue,** denser towards the outside where it projects into **papillæ** containing elastic tissue which cause the Malpighian layer to have its wavy appearance. Blood **capillaries** will be seen in the dermis and loops from them pass into the papillæ. **Nerve fibres** will also be found terminating in sense organs in some of the papillæ. These sense organs are the bulbous **tactile corpuscles,** though in certain parts such as the deeper layers of the hands and feet, the larger **Pacinian bodies** are found. Situated deep in the dermis are the **sudorific** or **sweat glands** composed of coiled tubes each provided with a duct, straight at first but spiral in the epidermis. It opens on the surface as one of the **pores** of the skin. Only parts of the ducts will be visible, of course. In the deepest layers of the dermis where the tissue is less dense is a quantity of **fat cells.**

(ii) *Examine a* **T.S. of the skin from the scalp.**

Note the **hairs,** which are epidermal in origin, each consisting of (*a*) a bulbous **root** into the base of which projects, like the bottom of a wine bottle, a small vascular **hair papilla** (this is the growing point), and (*b*) the **hair shaft.** Each hair is multicellular and hollow, and consists of an outer **cuticle** and an inner **medulla,** and is embedded in a pit-like **hair follicle** epidermal in origin. Each hair follicle is provided with (*a*) an **arrector pili** muscle near its base, composed of unstriated fibres and (*b*) one or more **sebaceous glands,** which secrete the oily sebum into the hair follicle.

Nails, claws and hoofs are are thickenings of the stratum corneum and originate from the Malpighian layer. They contain a substance known as *keratin.*

DOGFISH SKIN

Examine a slide of the **T.S. of the skin of the dogfish.**

As the mammal the skin consists of epidermis and dermis. In the lower region of the **epidermis** is the Malpighian Layer, external to which are the cells to which it gives rise, the outermost ones being flattened and dead. The dermis is composed chiefly of connective

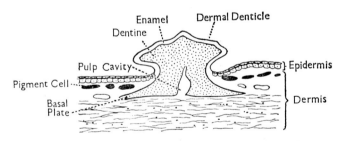

FIG. 236. **Dogfish Skin. T.S.**
(*From Wallis—Practical Biology*)

tissue containing blood vessels and nerves, and **pigment cells** are found in its upper region immediately beneath the epidermis. **Dermal denticles** are seen with their **basal plates** in the dermis and from these the backwardly directed **spines** arise and project through the epidermis. Each dermal denticle is composed mainly of **dentine** which in the **spine** is covered with **enamel** and there is a **pulp cavity** in the centre.

FROG'S SKIN

Examine a slide of the **T.S. of the frog's skin.**
The **epidermis** consists of a Malpighian layer, followed by the cells
arising from it, the outermost cells being flattened and periodically
cast off. The **dermis** is composed of **connective tissue** and bundles of

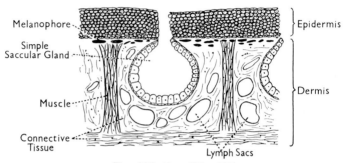

Melanophore · · ·
Simple
Saccular Gland · · ·
Muscle · · ·
Connective · · ·
Tissue
Epidermis
Dermis
Lymph Sacs

FIG. 237. **Frog Skin. T.S.**
(*From Wallis—Practical Biology*)

unstriated muscle fibres and contains blood vessels and nerves. Pigment cells or **melanophores** occur immediately below the epidermis.
Simple saccular glands, which are epidermal in origin, abound and
these secrete the watery mucus which keeps the surface of the skin
moist for purposes of respiration. Some of the glands secrete a bitter
fluid which serves to protect the animal against capture by its enemies.
Also in the dermis is a number of **lymph sacs** and it will be remembered that large sub-cutaneous lymph spaces lie between the skin
and the muscular body wall.

A TOOTH

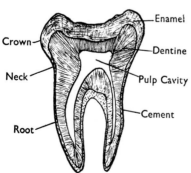

Enamel
Crown
Dentine
Neck
Pulp Cavity
Root
Cement

FIG. 238. **L.S. Mammalian Tooth.**

Examine a **L.S. of a tooth.**

Note the **crown** separated from the **root** by the **neck.** The crown is covered by **enamel** and the root by **cement,** somewhat similar to bone in that it contains lacunæ and canaliculi but deficient of Haversian canals. The greater part of the tooth is composed of **dentine** also somewhat similar to bone but devoid of Haversian canals. The **pulp cavity** in the centre contains pulp which is composed of connective tissue and contains blood vessels and nerve fibres which enter the tooth through a small hole in the root.

THE TONGUE

Examine **transverse sections of the tongue.**

Note (*a*) the **filiform papillæ,** conical in shape and edged (in man) with filaments of epithelium, (*b*) **the fungiform papillæ** amongst the filiform variety and larger in size and (*c*) the **circumvallate papillæ,** large and circular and surrounded by a narrow moat.   (These are

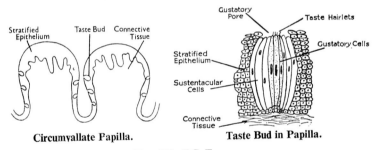

Circumvallate Papilla.   **Taste Bud in Papilla.**

FIG. 239. T.S. Tongue.

not found in the rabbit's tongue, **foliate papillæ** being present.) They form a V-shape on the tongue of man with the apex directed backwards and two small oval patches on each side at the back of the tongue in the rabbit. The fungiform and circumvallate papillæ consist of **connective tissue** enclosed in **stratified epithelium** in which are ovoid **taste-buds.** These are composed of groups of cells of two kinds, the long spindle-shaped **gustatory cells** in the centre, each with a **taste hairlet** on its free end and opening in the narrow **gustatory pore,** and the long flattened **sustentacular cells** around them which give them support.

SALIVARY GLANDS

Examine a **T.S. of one of the salivary glands.**

As already seen, it is a racemose gland. The **lobules** of which it is composed consist of small sac-like or tubular structures called **alveoli,** lined by **secretory cells,** and from which small ducts lined by cubical epithelium arise which later join others to form the main duct of the gland. The lobules are bound together by **connective tissue.** The secretory cells are mainly of two kinds—mucous cells which secrete mucin and serous cells which secrete the enzyme ptyalin. The submaxillary and sublingual glands are mixed, but the parotids contain only serous cells.

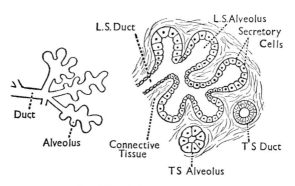

FIG. 240. **T.S. Salivary Gland.**

THE ŒSOPHAGUS

Examine a **T.S. of the œsophagus.**

It is composed of an outer coat of **connective tissue,** inside which is a **muscular layer** composed of striated fibres in the upper end of the œsophagus and unstriated fibres in the lower end. Next comes a **submucous layer** consisting of areolar tissue and finally the **mucous membrane.**

THE STOMACH

Examine **transverse sections of the wall of the stomach in the cardiac and pyloric regions.**

This consists of four coats. (i) The outer coat is the **serous coat.** (ii) The **muscular coat** is composed of an outer layer of **longitudinal fibres,** a middle layer of **circular fibres** and an inner layer of **oblique fibres.** These are all unstriated. (iii) The **sub-mucous coat** is composed of areolar tissue and contains **blood vessels** and **lymphatics.** (iv) The **mucous membrane** is lined by **columnar epithelium** containing **goblet cells** and which dips down to form the ducts of the **gastric glands.**

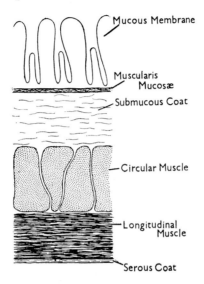

FIG. 241. **T.S. Stomach.**

These are simple or branched tubular glands and the epithelial cells secrete mucus. The **peptic cells** in these glands secrete pepsin and the ovoid **oxyntic cells** scattered amongst them secrete hydrochloric acid. The latter are not found in sections taken from the pyloric end

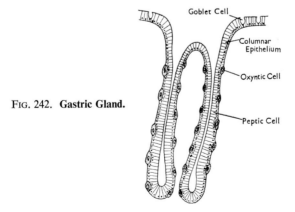

FIG. 242. **Gastric Gland.**

of the stomach where the glands have longer glandular portions but shorter ducts. The outermost region of the mucous membrane next to the sub-mucous coat consists of longitudinal and circular unstriated muscle fibres and this is known as the **muscularis mucosæ.**

THE DUODENUM

Examine a **T.S. of the duodenum** (see also Fig. 244, p. 323).

It is composed of (*a*) and outer **serous coat** (*b*) a **muscular coat** composed of two layers only, the outer being narrow and consisting or **longitudinal fibres** while the inner is thick and contains only **circular fibres** (*c*) a **submucous coat** composed of areolar tissue and containing blood vessels and lymphatics and (*d*) the **mucous membrane.** The outer region of the mucous membrane next to the submucous coat is a thin layer of muscle fibres, the external one being longitudinal and the internal circular. This is known as the **muscularis mucosæ.**

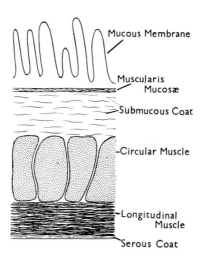

FIG. 243. **T.S. Duodenum.**

The mucous membrane is extended into the cavity of the duodenum in finger-like processes called **villi** between which are simple tubular glands, the **crypts of Lieberkühn** (or **intestinal glands**), lined by columnar epithelium and containing **goblet cells.** Each villus contains blood capillaries and a lymphatic known as a **lacteal.**

In addition small racemose glands will be found in the submucous coat; these are **Brünner's glands.** Their ducts lead either into or between the crypts of Lieberkühn.

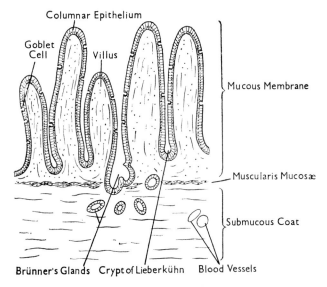

FIG. 244. **T.S. Duodenum. Mucous Membrane and Submucous Coat.**

THE ILEUM

Examine a **T.S. of the ileum.**

In general structure this is similar to the duodenum consisting of **serous, muscular** and **submucous coats** and **mucous membrane. Villi** and **crypts of Lieberkühn** are present, but there are no Brünner's glands. In addition, small nodules of lymphoid tissue (connective

FIG. 245. **T.S. Ileum.**

tissue in which the intercellular portions contain lymphocytes from the blood) called **Peyer's patches,** will be found in the mucous membrane below the crypts of Lieberkühn.

THE LARGE INTESTINE

Examine a **T.S. of the large intestine.**

Again **serous, muscular** and **submucous coats** and **mucous membrane** are present. Villi are absent, however, but **simple tubular glands** lined by columnar epithelium and containing numerous **goblet cells** are found.

THE LIVER

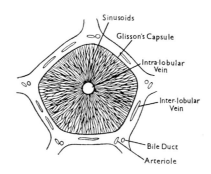

Fig. 246. **T.S. Liver, showing One Lobule.**

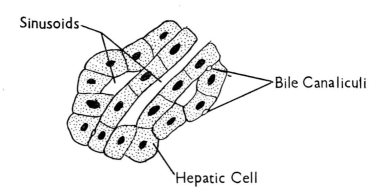

Fig. 247. **Part of Liver Lobule H.P.**

Examine a **T.S. of the liver.**

Each lobe of the liver is composed of minute many-sided **lobules** enclosed in connective tissue forming a sheath called **Glisson's capsule.** In the centre of each lobule is an **intralobular vein** which is connected by small tubules called **sinusoids** radiating round it with the **interlobular veins** in the Glisson's capsules. The interlobular veins originate from the hepatic portal vein and the intralobular veins ultimately join to form the hepatic veins. Between the sinusoids are the **hepatic cells** which secrete bile, and between these again are the minute **bile canaliculi** which eventually give rise to the hepatic ducts.

THE PANCREAS

Examine a **T.S. of the pancreas.**

This is a racemose gland somewhat similar to the salivary glands but with longer **alveoli.** It also contains isolated rounded groups of small cells, endocrine in function, which secrete insulin and which are known as the **islets of Langerhans.**

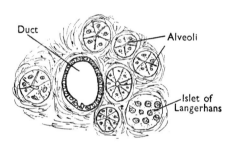

FIG. 248. **T.S. Pancreas.**

THE KIDNEY

(i) *Examine again a* **L.S. of the kidney** *with a hand lens* (refer to p. 257).

This compound tubular gland consists of an outer **cortex** and an inner **medulla** both containing the **uriniferous tubules,** which are lined for the greater part by cubical epithelium. These originate in the **Malpighian bodies** in the cortex, each Malpighian body consisting of a curved widened cup which is the beginning of the tubule and known as **Bowman's capsule.** It surrounds and encloses a tuft of blood capillaries known as a **glomerulus.** Leaving the capsule, the tubule is at first convoluted and then passes as an almost straight tube, the **descending limb,** into the medulla. Here it bends back on itself forming the **loop of Henle** and then runs, almost parallel with the descending limb, back into the cortex as the **ascending limb.** When it reaches the level of the Malpighian body, on the side opposite to that by which it left it, it again becomes convoluted and then passes across to join a straight **collecting tube** which runs down into the medulla. This is joined by other collecting tubules and then becomes wider to form the **duct of Bellini,** which opens into the **pelvis** of the kidney. The collecting tubules are bound together to form the **pyramids** of the kidney which bulge into the pelvis. In the rabbit there is only one pyramid, but in man there are several.

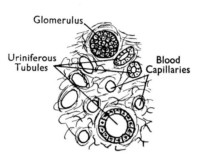

FIG. 249. **Kidney. T.S. Cortex under high power.**

(ii) *Examine a* **T.S. of the kidney cortex.**
Several **glomeruli** will be seen in their capsules, also **uriniferous tubules,** some in cross-section and some parts running longitudinally.

MEIOSIS

Maturation of the germ cells or **gametogenesis** is known as **spermatogenesis** in the male and **oogenesis** in the female.

At one stage in the formation of the gametes in both sexes (at the division of the primary spermatocytes and primary oocytes) a *reduction division* takes place in which the chromosome number is halved. This is essential in order to maintain the constancy of the number in the zygote and this method of nuclear division is **meiosis.**

At the outset, unlike mitosis, the chromosomes are thin and show no tendency to divide into chromatids. This is the **leptotene stage.** They come together in homologous pairs (one maternal and one paternal in origin) with corresponding genes in juxtaposition and each apparent chromosome is said to be **bivalent** This is the **zygotene**

stage. These bivalent chromosomes become shorter and thicker, split into two chromatids and coil round one another (**pachytene stage**). At this stage interchange of chromatic material may take place between them. This is the cause of "crossing over". The two split chromosomes partially separate (**diplotene stage**). A spindle is formed and the chromosomes arrange themselves on it (**diakinesis**)

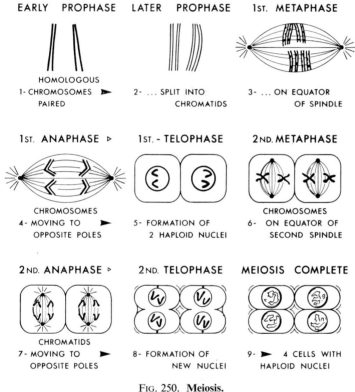

FIG. 250. **Meiosis.**
(*From Wallis—Human Biology*)

and ultimately arrange themselves on the equator of the nuclear spindle as in the metaphase of mitosis. This is the **1st metaphase.** Each is attached at two points, one for each chromosome. Next the two chromosomes, each composed of two attached chromatids, separate and move towards opposite poles (**1st anaphase**) and is followed by the **1st telophase** as in mitosis. Cytoplasmic division follows and a second nuclear division takes place in which the chromosomes pass quickly to the **2nd metaphase** which is followed by the **2nd anaphase** and **2nd telophase** stage as in mitosis.

Thus, in the first of these two divisions half of the chromosomes pass to each pole forming what corresponds with a haploid daughter nucleus and in the second the chromosomes already divided into two chromatids apiece produce four nuclei each having the **haploid** number of chromosomes.

Examine **sections showing meiosis** *under high power.*

THE TESTIS

SPERMATOGENESIS

Cells of the germinal epithelium in the seminiferous tubules divide mitotically to give rise to **spermatogonia** (*2n*) and after several such divisions these divide mitotically to give **primary spermatocytes** (*2n*).

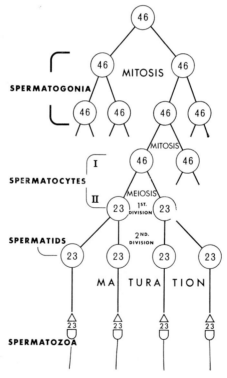

FIG. 251. **Spermatogenesis.**
(From Wallis—Human Biology)

The next division into **secondary spermatocytes** is meiotic (n). In the subsequent division **spermatids** are formed which develop into **spermatozoa,** millions of which are produced.

(i) *Examine a* **T.S. of the testis.**

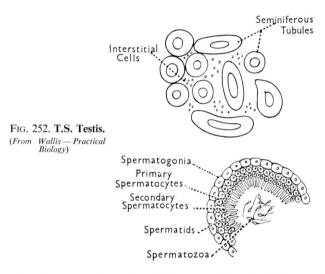

Fig. 252. **T.S. Testis.**
(*From Wallis — Practical Biology*)

Spermatogonia
Primary Spermatocytes
Secondary Spermatocytes
Spermatids
Spermatozoa

Note the large number of **seminiferous tubules** lined by **germinal epithelium** and bound together by connective tissue in which may be seen **interstitial cells** which secrete the hormones testosterone and androkinin. *If a tubule is carefully examined under the high power or, better, the* $\frac{1}{12}$ *O.I. objective, some stages on* **spermatogenesis** *may be found.* Next to the germinal epithelium are the **spermatogonia.** These divide to give rise to **primary spermatocytes** and these by meiosis to **secondary spermatocytes** which will be in the next layer. Nearer the lumen of the tubule are the small **spermatids,** formed by division of the secondary spermatocytes and, finally, in the cavity of the tubule itself, **spermatozoa** into which the spermatids develop.

(ii) *Examine a slide of* **spermatozoa** *under high power.*

Head
Middle Piece
Tail

Fig. 253. **Spermatozoa.**
(*From Wallis—Practical Biology*

Note that each consists of a **head** (ovoid and pointed in man and the monkey, but rod-like in the earthworm and frog—there is considerable variation of shape in different animals), a **middle-piece** (or **body**) and a long **tail**.

THE OVARY

OOGENESIS

This is on similar lines to spermatogenesis but differs from it in certain respects. The **oogonia** (*2n*) formed from the germinal epithelium divide mitotically to form **primary oocytes** (*2n*) and these divide meiotically to give rise to **secondary oocytes** (*n*) and what are

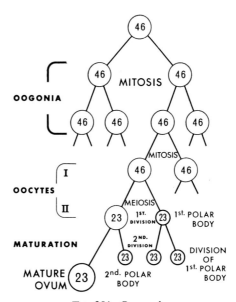

FIG. 254. **Oogenesis.**
(*From Wallis—Human Biology*)

known as the **first polar bodies,** the division being an unequal one. These divisions take place in the Graafian follicles. In the next division the **secondary oocytes** divide into an **ovum** (*n*) and a **second polar body.** This takes place after liberation from the follicle and the first polar body also divides. These polar bodies may be regarded as abortive ova, the ovum maturing at their expense, and they eventually degenerate.

(i) *Examine a* **T.S. of the ovary.**

The peritoneal covering is known as **germinal epithelium.** Inside this is mostly connective tissue constituting the **stroma** in which will be seen **Graafian follicles** in various stages of development. These arise from the germinal epithelium and those towards the outside will be small. If a larger and more mature follicle is examined it will be seen to contain two layers of stratified epithelium, one lining the cavity and the follicle, the **membrana granulosa,** the other, the **discus proligerus,** surrounding an **ovum.** Some stages in **oogenesis** *may be*

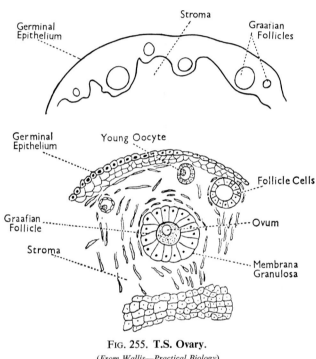

FIG. 255. **T.S. Ovary.**
(*From Wallis—Practical Biology*)

found if examined under high power. The **oogonia** arise from the **follicle cells** and divide into **primary oocytes** and each of these by meiotic division into a **secondary oocyte** and a **first polar body.** The secondary oocyte divides unequally into an **ovum** and a **second polar body,** while the first polar body also divides. However, complete maturation of the ovum usually takes place after liberation by the bursting of the follicle when the oocyte passes into the Fallopian tube. After the discharge of the oocyte, the follicle cells form a yellow body known as the **corpus luteum.**

(ii) *Examine a* **mature ovum in a follicle in a T.S. of the ovary.**

The cell is spherical and is enclosed in a thick transparent membrane, the **zona pellucida.** In the enclosed cytoplasm are **yolk-spherules** or **deutoplasm** (not in the rabbit) and a large nucleus called the **germinal vesicle** in which is a distinct nucleolus known as the **germinal spot.**

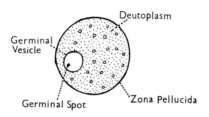

FIG. 256. **Mammalian Ovum.**

(From Wallis—Practical Biology)

MAMMARY GLAND

Examine a **T.S. of a mammary gland.**

This is a compound racemose gland and in most mammals it is a modified sweat gland. The several lobes of which the gland is composed are divided into lobules and within these lobules lie the saccular **alveoli** which are lined by columnar epithelium when milk is being secreted, and **milk globules** may then be seen in the alveoli. The T.S. should reveal the **lobules** and **alveoli.**

After secretion, however, these cells become flattened and fill up the alveoli where they may be seen. These are the conditions found in a lactating animal, otherwise the alveoli are fewer in number and smaller. **Ducts** lead from the alveoli to the surface of the teat (which contains unstriated muscle fibres) and near their outer terminations before entering the teat each swells into a small sac, the **sinus lact iferus.**

MUSCLE

Examine a **T.S. of a voluntary muscle.**

Note the bundles or **fasciculi** of **muscle fibres** surrounded and bound together by connective tissue called the **perimysium.** The tissue between the fibres is called the **endomysium** and the external covering of the muscle itself the **epimysium.**

NERVE

Examine a **T.S. of the sciatic nerve of a frog or mammal.**

Note the bundles, **fasciculi** (or **funiculi**), of **nerve fibres** bounded by connective tissue called the **perineurium.** Between the fibres in the bundles is connective tissue called the **endoneurium** while the external coat of the whole nerve is known as the **epineurium.** The **axis cylinders** of the nerve fibres will be seen in their centres.

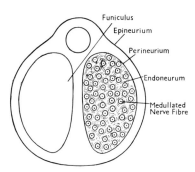

FIG. 257. **T.S. Medullated Nerve.**
(From Wallis—Human Biology)

SPINAL CORD

Examine a **T.S. of the spinal cord** *with a lens or under the low power.*

Note the outer **white matter** surrounding the central **grey matter** which is in the shape of a rough letter "H". In the centre of the cross-piece of the H is the **central canal.** The deep ventral **fissure** and the narrower **dorsal fissure** will be recognised in the white matter. The dorsal limbs of the H-shaped grey matter are known as the **dorsal horns** (or **cornua**), and the ventral limbs as the **ventral horns** (or **cornua**). From them originate the spinal nerves. Outside the spinal cord, the **dorsal root** (**sensory**) bears a **ganglion,** but the **ventral root** (**motor**) does not. Just after the ganglion, the two roots join to form the (**mixed**) **spinal nerve.**

Examine under the high power.

The white matter is composed of **medullated nerve fibres** (seen in T.S.) amongst which is a tissue composed of cells and fibres and known as **neuroglia.**

In the grey matter, **neurones** and **neuroglia** are found.

The extent to which these structures are visible will depend on the method of staining used in the preparation of the slide.

The whole spinal cord, like the brain, is invested in three membranes, the **meninges.** The thin vascular **pia mater** is the innermost and closely invests the cord, the **arachnoid,** separated from the pia mater by the subarachnoid space, is the middle membrane, while the thick **dura mater** on the outside lines the neural canal of the vertebral column. The only one likely to be present in the specimen is the pia mater.

PART IV

ELEMENTARY BIOCHEMISTRY

INTRODUCTORY NOTES

(1) The objects of the following series of experiments are (i) to give the student an insight into the nature of the chemical substances and reactions met with in the study of biology, (ii) to enable him to recognise these substances by simple tests and (iii) to help him to understand the nature of the various physiological processes met with in the study of the organism, experiments on which are given in the next part of the book.

(2) The student should work systematically through this part of the book, performing the experiments on crystalloids and colloids and on the chemical properties of organic substances and their detection in animal tissues early in his studies. The experiments in *small type* may be omitted if time does not permit all the experiments to be performed. Records can be kept in the book or part of the file to be used for Physiology.

(3) The results of some experiments have been included to enable the student to check his own observations.

(4) In all cases, use *small* quantities of the substances to be tested and add *small* quantities of reagents unless otherwise stated.

(5) Further biochemical experiments are included in Part V (Physiology).

Reference should be made to the General Directions for Practical Work in the Introduction.

BIOCHEMICAL COMPOUNDS

Biochemistry is, as the term implies, the chemistry of the living organism. It includes not only the structure and properties of the substances which constitute the organism and which the organism itself produces, but also all the chemical processes, both anabolic and katabolic, which occur in its physiology.

The most important substances which enter into the composition and metabolism of the living organism may be summarised as follows:—

(1) Inorganic compounds.
(2) Proteins.
(3) Carbohydrates.
(4) Lipides.
(5) Vitamins.
(6) Enzymes.
(7) Hormones.
(8) Other organic substances such as Pigments, Excretory substances, etc.

THE PHYSICAL PROPERTIES OF ORGANIC COMPOUNDS

Substances which dissolve in liquids forming true solutions are called **crystalloids.** Substances which form with liquids a heterogenous system intermediate between a true (molecular) solution and a suspension are called **colloids.*** A colloid **sol** thus consists of two parts: (i) the **disperse phase** composed of aggregates of molecules in constant motion (**Brownian Movement**) distributed in (ii) the **continuous phase** (or **dispersion medium**) which is, of course, liquid. If, on the addition of an electrolyte, the particles of the disperse phase coalesce or **coagulate** and are precipitated, the colloid is said to be **lyophobic.** (It is also called a **suspensoid.**) Colloids which are not precipitated in this way are said to be **lyophilic.** (They are also known as **emulsoids.**)

Water and the solvents of true solutions diffuse through membranes, but colloids are incapable of doing so. Crystalloids and colloids can therefore be separated by application of this property

---

* The size of colloidal particles varies between $1m\mu$ and 1μ ($\mu = 1$ micron = 0·001 mm. $m\mu = 1$ millimicron = 0·000001 mm and $\mu\mu = 1$ micromicron = 0·0000000001 mm.).

It should be noted that colloidal solutions of many substances such as gold and graphite can be prepared. It is therefore more accurate to speak of a substance as being in the colloid state.

in the process of **dialysis.** Membranes which allow both the solute and solvent to pass through are said to be **permeable.** But those which allow the passage of the molecules of the solvent but not those of the solute are referred to as **semi-permeable** and the diffusion of a solvent through a semi-permeable membrane is known as **osmosis.**

If a beam of light from a lantern is passed through a true solution and viewed from a position at right angles to the direction of the beam, the path of the rays cannot be traced through the solution, which appears clear. When the true solution is replaced by a colloid sol, however, the path of the beam becomes visible through the high power of a microscope owing to the scattering of the light by the particles. This is known as the **Tyndall effect** and is the principle of the **ultra-microscope.**

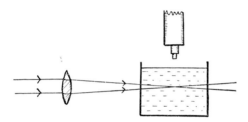

Fig. 258. **The Tyndall Effect.** Principle of the Ultra-microscope.

CRYSTALLOIDS AND COLLOIDS

EXPERIMENT 1. *Cover some crystals of potassium dichromate with a small quantity of hot water in a test-tube and shake. Examine a drop of the solution under a microscope.* It is clear and no crystals are visible until the solution cools. Potassium dichromate is a **crystalloid** and it forms a true **solution** with water. There is no expansion on solution.

EXPERIMENT 2. *Add a little hot water to a small piece of gelatine in a test tube.* The gelatine swells and disappears owing to the absorption of water, forming a **colloid sol.** *Allow the sol to cool.* It sets into a solid jelly and contraction takes place on solidifying owing to loss of water. This is a **colloid gel** and gelatine is a **reversible gel.**

BROWNIAN MOVEMENT

EXPERIMENT 3. *To a drop of Indian ink in a test tube, add water until you can see through the mixture: alternatively make a gamboge sol by rubbing a little gamboge under cold water. Filter. Place a drop of the liquid on a cavity slide and cover with a coverslip. Examine under the high power of the microscope.* Fine particles can be seen: it is a **coarse suspensoid sol.** The particles are in constant motion bombarding one another and this is known as **Brownian Movement.**

EXPERIMENT 4. *Examine a drop of aqueous Congo red sol under the microscope.* It is clear, no particles being visible, but it is a **fine suspensoid sol,** the particles being ultra-microscopic.

EXPERIMENT 5. *Examine (i) potassium dichromate solution, (ii) Indian ink sol, (iii) Congo red sol, illuminated by a Tyndall's Beam and compare the effects in the three cases.*

COAGULATION

EXPERIMENT 6. *Shake up some dried egg albumen (or white of egg) with water. It forms an opalescent sol. Divide the sol into three parts. To (i) apply heat. To (ii) add a few drops of dilute acetic acid and then heat. Compare with (i). To (iii) add a saturated solution of ammonium sulphate and note the effect.*

EXPERIMENT 7. *Make some starch sol by stirring some powdered starch with a little cold water and then adding boiling water and shaking. Divide into two parts. Allow (i) to cool. To (ii) add basic lead acetate solution. Observe results.*

FIG. 259. **Apparatus to demonstrate Osmosis.**

OSMOSIS

EXPERIMENT 8. *Cover the mouth of a thistle-funnel with a piece of parchment* or pig's bladder (a semi-permeable membrane) and tie it on securely round the rim. Using a pipette carefully run some sugar solution (which may be coloured) into the tube of the thistle-funnel until the liquid has risen a short distance up the tube. Make sure that there is no leakage round the edge of the membrane. Clamp the tube in a retort stand with the mouth of the funnel in a beaker of water. After a few minutes, mark the level of the liquid in the tube with gum-paper. Examine an hour or two later.*

DIALYSIS

EXPERIMENT 9. *Put a mixture of sodium chloride solution and starch sol into a dialysing cylinder and float this in a large quantity of distilled water. Allow it to stand for several hours with frequent changes of water. Periodically test the water for salt by adding silver nitrate solution and for starch by adding iodine solution. Draw your conclusions.*

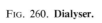 FIG. 260. Dialyser.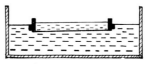

THE CHEMICAL PROPERTIES OF ORGANIC (BIOCHEMICAL) COMPOUNDS

If it is desired to ascertain the Elements present in Proteins, Carbohydrates and Fats, the following experiments should be performed with them:—

Carbon and Hydrogen

EXPERIMENT 10. *Mix about 0·5 gm. of the substance with 5 or 6 times its weight of cupric oxide previously heated in a crucible and allowed to cool in a desiccator. Place the mixture in a hard-glass test tube fitted with a delivery tube leading to a test tube of lime water. Heat the mixture. The evolution of carbon dioxide indicates the presence of* **carbon.** *If moisture appears on the cool upper part of the hard glass tube, test it with anhydrous copper sulphate.* Water indicates the presence of **hydrogen.** (It is, of course, unnecessary to test for carbon if the substance is known to be organic.)

Nitrogen

EXPERIMENT 11. (Lassaigne Test.) *Put about 0·5 gm. of the substance in a bulb ignition tube or small hard glass test tube. Add two small pellets of sodium.*

*This is not a perfect semi-permeable membrane for sugar and water as it allows a small quantity of sugar to pass through it.

Heat in the bunsen flame for a few minutes, gently at first and afterwards more strongly. Then immediately plunge the tube while hot into distilled water in a mortar, grind it up and filter.† If nitrogen is present, sodium cyanide (NaCN) will have been formed. *Test a portion of the filtrate as follows:—Add some freshly prepared ferrous sulphate solution and boil.* (Sodium ferrocyanide is thus formed.) *Acidify with conc. hydrochloric acid and add two or three drops of ferric chloride.* A deep blue coloration or precipiate indicates the presence of **nitrogen.**

EXPERIMENT 12. *Mix equal quantities of the substance to be tested and soda lime and heat in a hard glass test tube.* Ammonia, recognised by its smell and by its turning red litmus paper blue, shows the presence of **nitrogen.** A negative result, however, is not conclusive.

Sulphur

EXPERIMENT 13. *To a portion of the filtrate from Experiment* 11, *add some freshly prepared sodium nitroprusside.* A violent colour indicates the presence of **sulphur.**

EXPERIMENT 14. *To another portion of the filtrate from Experiment* 11 *add a few drops of acetic acid and then lead acetate solution.* A black precipitate indicates **sulphur.**

Phosphorus

EXPERIMENT 15. *Heat a portion of the filtrate from Experiment* 12 *with nitric acid in excess. Then add ammonium molybdate.* A yellow precipitate indicates **phosphorus.**

Halogens

EXPERIMENT 16. *Test portions of the filtrate from Experiment* 11 *as follows: —Acidify with nitric acid. Boil for a few minutes to decompose any cyanide present then add silver nitrate solution.* A white precipitate indicates **chlorine,** a yellow precipitate indicates **bromine** or **iodine** (or both). *To distinguish between bromine and iodine, add chlorine water to the acidified solution above.* Bromine and iodine will be liberated and will colour the liquid brown. *Add carbon disulphide or chloroform to this* and the **bromide** will dissolve to give a browish-red solution whereas **iodine** will give a violet solution, but if excess of chlorine water is added, the violet colour is destroyed.

Oxygen

There is no conclusive test of universal application for oxygen but if water is liberated when the dry substance is heated alone, it must obviously contain oxygen.

PROTEINS

As will have been seen in the previous experiments, proteins are compounds of carbon, oxygen, hydrogen and nitrogen. Many of them also contain sulphur and some contain phosphorus. They are of colloidal proportions and all have very complex molecules and are of high molecular weight. The protein molecule is composed

† Retain some of this filtrate to test for sulphur, phosphorus and halogens.

of a combination of **amino-acids*** by condensation. **Proteoses, peptones, peptides** and **amino-acids** are derivatives of proteins, formed in their synthesis and hydrolysis.

There are certain colour reactions given by proteins and these depend on the amino-acids present in the protein molecule. All give the "biuret" reaction (Experiment 17 below). This is due to the NH—CO group. Those containing tyrosine give the Xanthoproteic and Millon's reactions, those containing trytophane give the test of that name and those containing cystine give the cystine (sulphur) test.

In the following tests, dried albumin or white of egg can be used. *Prepare a solution of ovalbumin by shaking up a small quantity of the white powder with water.*

EXPERIMENT 17. **"Biuret" Test:** (*a*) *Add excess of sodium hydroxide solution and then a few drops of weak (1 per cent.) copper sulphate solution.* A violet colour is produced. (A similar reaction is given by biuret, the substance obtained by heating urea: hence the name of this test, but proteins do not contain biuret.) (*b*) *Perform the "biuret" test with an aqueous solution of* **peptone.** A rose pink colour is obtained.

EXPERIMENT 18. **Xanthoproteic Test:** *Add concentrated nitric acid.* A white precipitate is obtained. *Heat.* It turns yellow. *Cool under the tap and add excess of ·88 ammonium hydroxide.* It turns orange.

EXPERIMENT 19. **Millon's Test:** *Add a few drops of Millon's reagent* (this is a mixture of mercurous and mercuric nitrate in nitric acid). A white precipitate is obtained. *Boil.* It turns red.

EXPERIMENT 20. **Tryptophane Test:** *Add excess of Glacial acetic acid which has been exposed to light* (and therefore contains glyoxylic acid, on which the reaction depends). *Then, by means of a thistle funnel or by sloping the tube, carefully add some concentrated sulphuric acid.* A purple ring forms at the interface.

EXPERIMENT 21. **Arginine Test:** *Add a few drops of sodium hydroxide followed by a few drops of alcoholic α-naphthol and then a few drops of sodium hypochlorite.* A red colour is produced.

EXPERIMENT 22. **Cystine Test:** *Add a few drops of lead acetate solution.* A precipitate is obtained. *Add sufficient sodium hydroxide solution to dissolve this and then boil.* The solution turns dark brown (due to the formation of lead sulphide).

CARBOHYDRATES

It will already have been seen that carbohydrates are compounds of carbon, hydrogen and oxygen. The hydrogen and oxygen are

---

* An amino-acid is an organic acid in which a hydrogen atom attached to the carbon atom is replaced by an amino group ($-NH_2$). Amino-acetic acid or *glycine* (CH_2NH_2COOH) is the simplest. Amino-propionic acid or *alanine* is $CH_3CH.NH_2COOH$. *Tyrosine* and *cystine* are further examples.

in the same proportion as in water and the general formula is $C_m(H_2O)_n$. They are classified as follows:—

(1) **Monosaccharides:** Simple sugars of general formula $C_nH_{2n}O_n$.

(2) **Disaccharides:** Sugars formed by the condensation of two hexose (monosaccharide) groups and therefore of formula $C_{12}H_{22}O_{11}$.

$$2C_6H_{12}O_6—H_2O = C_{12}H_{22}O_{11}.$$

(3) **Polysaccharides:** Non-sugars of general formula $(C_6H_{10}O_5)_n$.

General Tests for Carbohydrates

All carbohydrates on boiling with concentrated hydrochloric acid yield substances called **furfurals** which can be identified by colour reactions such as by the addition of alcoholic thymol or **a**-naphthol.

EXPERIMENT 23. **Molisch Test:** *Add a few drops of an alcholic solution of α-napthol to a solution of a carbohydrate. Then, by means of a thistle funnel or by sloping the tube, carefully pour in a little conc. sulphuric acid.* A violet colour is formed at the junction of the liquids.

EXPERIMENT 24. **Thymol Test:** *To a small quantity of a solution of a carbohydrate (or to a solid insoluble one) add a few drops of alcoholic thymol and excess of conc. hydrochloric acid. Boil for a couple of minutes, shaking periodically.* A carmine colour is produced.

Monosaccharides (Hexoses)
Glucose

EXPERIMENT 25. **Fehling's Test:** *Mix equal quantities of Fehling's solution A and B (see Appendix I) shake and add to some glucose solution. Boil.* Note the red precipitate of cuprous oxide.*

EXPERIMENT 26†. **Barfoed's Test:** *Add about 1 ml. of Barfoed's solution and boil.* A red precipitate of cuprous oxide is formed.

EXPERIMENT 27. **Benedict's Test:** (Used instead of Fehling's reaction for detecting the presence of glucose in urine because ammonia and creatinine, invariably present in urine, dissolve cuprous oxide and so render Fehling's test less sensitive). *Add a few millilitres of Benedict's qualitative reagent to about 1 ml. of glucose solution and boil for two or three minutes.* A yellow precipitate is formed and the solution turns green.

EXPERIMENT 28. **Rapid Furfural Test:** *Add a few drops of alcoholic α-naphthol and then a few millilitres of conc. hydrochloric acid.*

---

* The reducing property of this and other sugars is due to the presence of an aldehyde (—CHO) radicle in the molecule.

† Peculiar to monosaccharides.

Boil. Observe that a violet colour is produced only after prolonged boiling.

Disaccharides*
Maltose

EXPERIMENT 29. *Carry out (a) Barfoed's and (b) Fehling's Tests with a solution of maltose.*

Lactose

EXPERIMENT 30. *Carry out (a) Barfoed's and (b) Fehling's Tests with a solution of lactose.*

EXPERIMENT 31. *To about 20 ml. of a solution of lactose in a small basin add about 5 ml. of conc. nitric acid. Evaporate to small bulk in a fume chamber. A* white precipitate of crystalline mucic acid (into which the lactose has been oxidised) separates out.

Sucrose

EXPERIMENT 32. *Perform (a) Barfoed's and (b) Fehling's Tests with a solution of sucrose.*

EXPERIMENT 33. *Hydrolyse some sucrose by boiling a solution with dilute hydrochloric or sulphuric acid for a few minutes and allowing it to stand for five minutes or so. Then neutralise by adding some caustic soda solution (test with litmus paper) and perform Fehling's Test.*

EXPERIMENT 34. *Perform the rapid Furfural Test (Experiment 28) with sucrose.* Observe the immediate formation of a violet colour.

Polysaccharides*
Starch (Amylum)

Perform the following tests with starch sol, prepared by making a paste of starch with cold water and then adding boiling water.

EXPERIMENT 35. *Perform Fehling's Test.*

EXPERIMENT 36. *Add a few drops of dilute iodine solution and observe the effect. Heat and note the result. Cool under the tap and again note the result.*

EXPERIMENT 37. *(a) Hydrolyse some starch by boiling with dilute sulphuric acid for a few minutes: neutralise by adding caustic soda solution (test with litmus paper). (b) Hydrolyse a separate portion of starch by adding a little of the enzyme diastase and leave for a quartet*

---

* Negative results will be obtained in some of the following experiments.

of an hour. Then divide each hydrolysed solution into two portions and to one add iodine solution and with the other perform Fehling's Test.

Glycogen (Animal Starch)

Make an aqueous solution of glycogen and perform the following tests:—

EXPERIMENT 38. *Perform Fehling's Test.*

EXPERIMENT 39. *Add iodine solution.* Observe that a red colour is obtained.

EXPERIMENT 40. *Hydrolyse some glycogen by boiling for a few minutes with dilute hydrochloric or sulphuric acid. Neutralise by adding caustic soda (test with litmus paper) and then carry out Fehling's Test.*

LIPIDES

Lipides are esters of higher members of a series of organic acids known as the Fatty Acids. They are compounds of carbon, hydrogen and oxygen but with a lower oxygen content than the carbohydrates. They are classified as follows:—

    (1) Simple lipides—fats, oils and waxes.
    (2) Complex lipides.
    (3) Lipide derivatives.

Fats and **oils** are esters of the polyhydric alcohol *glycerol*, $C_3H_5(OH)_3$ with higher fatty acids such as *stearic*, $C_{17}H_{35}COOH$ and *palmitic*, $C_{15}H_{31}COOH$, and with the unsaturated monobasic acid, *oleic* $C_{17}H_{33}COOH$. A fat differs from an oil in being solid whereas the latter is liquid at 20°C. On hydrolysis, a fat decomposes into glycerol and the fatty acid thus:—

$$(C_{15}H_{31}COO)_3C_3H_5 + 3H_2O = 3C_{15}H_{31}COOH + C_3H_5(OH)_3$$
$$\text{Tripalmitin} \qquad\qquad\qquad \text{Palmitic acid} \qquad \text{Glycerol}$$

$$(C_{17}H_{35}COO)_3C_3H_5 + 3H_2O = 3C_{17}H_{35}COOH + C_3H_5(OH)_3$$
$$\text{Tristearin} \qquad\qquad\qquad \text{Stearic acid} \qquad \text{Glycerol}$$

If the hydrolysis is performed with sodium hydroxide sodium stearate (soap) will be formed in place of stearic acid. This is called **saponification.**

$$(C_{17}H_{35}COO)_3C_3H_5 + 3NaOH = 3C_{17}H_{35}COONa + C_3H_5(OH)_3$$
$$\text{Sodium stearate}$$

The so-called "essential," "ethereal" or "volatile" oils such as oil of turpentine, oil of cloves, oil of lavender are not lipides though some of their reactions are similar.

Fats and Oils

EXPERIMENT 41. (*a*) *Add a little water to some olive oil in a test tube and shake.* The emulsion formed is only temporary. (*b*) *Add some caustic soda solution and shake again.* The emulsion lasts longer but eventually the oil and water separate out.

EXPERIMENT 42. *Add some ether to some olive oil in a test tube and shake. Note that the oil dissolves. Then pour some of this liquid on to a piece of filter paper and examine again when the ether has evaporated.*

EXPERIMENT 43. *Add two drops of* 1 *per cent. solution of osmium tetroxide* ("*osmic acid*") *to a few drops of olive oil in a watch glass. Leave for a few minutes when a black colour will develop.*

EXPERIMENT 44. *Shake up some olive oil with a little water and add* (*i*) *Sudan III and* (*ii*) *Sudan IV* (*Scharlach Red*) *to separate samples in test tubes. Let the tubes stand for a few minutes and you will see that the oil is stained red in both cases.*

EXPERIMENT 45. **Saponification.** (i) *Prepare an emulsion of olive oil* (*prepared as in Experiment* 41 (*b*)) *in a boiling tube. Place the tube in a beaker of boiling water and leave it there for about half an hour, shaking the tube periodically.* The emulsion separates out into an upper layer of oil and a lower layer of caustic soda, but saponification takes place at the interface and the glycerol formed passes into the lower layer.

(ii) *Pour off the liquid, carefully pouring away the unchanged oil, retaining the soap, and perform the following test for glycerol with* (*a*) *glycerol from the bottle and* (*b*) *the residual liquid;—Add some solid sodium bisulphate and heat.* Note the acrid smell of acrolein.

(iii) *Now remove excess alkali from the small piece of soap left in the original boiling tube, by washing with water. Then dissolve the soap by boiling it in water. Divide the solution into three portions,* (*a*), (*b*) *and* (*c*), *and test for soap as follows:—*
To (*a*) *add calcium chloride solution; to* (*b*) *add lead acetate solution, and to* (*c*) *add a small quantity of solid sodium chloride. Repeat this with soap solution and compare results.*

EXPERIMENT 46. Repeat Experiments 42 and 43 with some oil of turpentine, oil of cloves or other voltatile oil.

ENZYMES

The characteristics, properties and behaviour of digestive enzymes is dealt with in Part V—Physiology.

The Lenham Enzyme Kit (see Enzymes in Appendix II) contains a series of enzymes, substrates, appropriate reagents and instruction booklets and should prove useful for general work on enzymes.

TO TEST FOR THE PRESENCE OF PROTEINS, CARBO-HYDRATES AND LIPIDES IN TISSUE AND FOOD-STUFFS

EXPERIMENT 47. *Tests for proteins, carbohydrates and lipides should now be performed with various foodstuffs, and also with suitable solids or solutions provided by the laboratory.* The following scheme is suggested as suitable for quick detection of the compounds present but any other convenient method can be adopted.

ANALYTICAL TABLE
To Test a Substance for Proteins, Carbohydrates and Fats.

TEST FOR PROTEIN 1. *Add sodium hydroxide and a few drops of* 1% *copper sulphate* (Biuret Test).	VIOLET COLOUR = **Protein.**
2. If Protein present confirm as follows:— *Add* Millon's Reagent.	WHITE PRECIPITATE. *Heat* IT TURNS RED.
TEST FOR CARBOHYDRATE 1. *Add a few drops of a-naphthol. Then slowly add* *conc. sulphuric acid down the sides of the test* *tube.* (Molisch Test.)	VIOLET RING = **Carbohydrate.**

2. **If Carbohydrate present**
    Test for Monosaccharide*
    Add Barfoed's Reagent and heat.
	RED PRECIPITATE =
	Monosaccharide.
	NO PRECIPITATE =
	Disccharide or
	Polysaccharide.

If Monosachcaride present. *Confirm by adding Fehling's solution.* *Heat.*	RED PRECIPITATE = **Glucose** or **Fructose.**
If Monosaccharide absent. **Test for Disaccharide.** *Peform Fehling's Test.*	RED PRECIPITATE = **Maltose** or **Lactose.**† NO PRECIPITATE = **Sucrose** or **Polysaccharide**
If Dicaccharide present (i) **Test for Lactose** *Add conc. nitric acid. Evaporate to* *small bulk in porcelain basin in fume* *chamber.*	SEPARATION OF CRYSTALS = **Lactose.** NO SEPARATION OF CRYSTALS = **Maltose.**
(ii) **Test for Sucrose.** *Add conc. hydrochloric acid and heat.* *Confirm by boiling with dilute sulphuric* *acid for a few minutes. Neutralise with* *sodium hodroxide* (*test with litmus paper*) *and then perform Fehling's Test.*	IMMEDIATE VIOLET COLOUR = **Sucrose.** RED COLOUR = **Sucrose** (now hydrolysed to monosaccharide).
Test for Polysaccharide *Add iodine solution.*	BLUE COLOUR = **Starch.** RED COLOUR = **Glycogen.**
TEST FOR FAT *Add* 1% *Osmium tetroxide* ("*osmic acid*") or *Alkannin* or *Sudan III.*	BLACK COLOUR after a few minutes = **Fat** RED COLOUR = **Fat.**

 * It should be remembered that both monosaccharide and disaccharide and/or polysaccharide *may* be present. If so, some confusion may result but subsequent tests should obviate this to some extent.

 † Both lactose and maltose may be present but this is unlikely in simple examination tests as there is no simple specific test for maltose. If glucose is also present it will, of course, give a ppt. here, but it will have been identified already.

Milk

EXPERIMENT 48. *Examine a few drops of milk under the low and then the high power of the microscope. Note that this is an emulsion.*

EXPERIMENT 49. *To a few millilitres of milk add an equal volume of water and a few drops of glacial acetic acid. Shake thoroughly. The* **casein** *is precipitated, together with fat globules. Filter. (a) Dissolve the residue in dilute caustic soda and test for protein. (b) Boil the filtrate. The* **lactalbumin** *coagulates. Filter. (c) Make a suspension of the residue in water (it is insoluble) and test for protein. (d) Test the filtrate for* **lactose** *by applying Fehling's Test.*

VITAMINS

Vitamins are complex chemical compounds of high molecular weight which are essential to the growth and maintenance of health of the vertebrate animal, though they are required in very minute quantities only and they may be termed *accessory food factors.* They occur in natural foods, the original source of many being green plants but others occur in animal organs and products. Some are specific to certain animals. Some are fat-soluble, others are water-soluble. They are designated by letters and, when their chemical constitution is known, by appropriate chemical names.

The classification, sources and effects of the most important vitamins will be found in text books.

The vitamin content of A and D in foods is measured in what are known as *international units.* This is a different quantity for each vitamin and depends on minimum requirements. Other vitamins are expressed in milligrams. The daily requirement of Vitamin A is about 4,000 I.U. while that of Vitamin D is about 400 I.U. Halibut oil, one of the richest sources of Vitamin A contains 5,000,000 I.U. per gram, while cod-liver oil contains only $1/50$ of that quantity.

There is little elementary practical work which can be conveniently done with vitamins and the following experiments will suffice.

EXPERIMENT 50. **To demonstrate the presence of Vitamin A.**—*Dissolve some cod-liver oil or halibut oil in about five times its volume of chloroform, and to a drop of this add a drop or two of a saturated solution of antimony trichloride in chloroform.* A bright blue colour develops.

EXPERIMENT 51. **To demonstrate the presence of Vitamin C.**—*Add one drop of dichlorophenol indophenol to 1 ml. of lemon or other fruit juice.* The pale blue solution turns pink and then fades owing to the reducing atcion of vitamin C.

EXPERIMENT 52. **To demonstrate the destructive action of copper and heat on Vitamin C.**—*Add one drop of 1 per cent. copper sulphate solution to 2 ml of lemon or other fruit juice and boil for ten minutes. Test for vitamin C as in experiment 51.* The pink colour persists showing that the vitamin C has been destroyed by heating in the presence of a copper salt.

PART V

PHYSIOLOGY

INTRODUCTORY NOTES

Experiments should be performed to illustrate the physiology of the systems. The following experiments are not elaborate and can be carried out at any suitable time. It is suggested, however, that the most appropriate time is during the study of the mammal.

The experiments are intended only to illustrate the working of the chief systems of the body and it is not suggested that they explain fully the details of the physiological processes. They are adequate, however, for the purpose for which they are intended and will occupy as much time as can be devoted to practical work of this kind. Those in *small type* may be omitted if time does not allow them to be performed.

Write the **object** of the experiment on the top of the page, then write a concise **method** or account of how the experiment was set up. Keep a record of any necessary readings and draw a *sectional* **diagram** of the apparatus, if any (*e.g.*, Fig. 261, p. 352). Lastly, when the experiment is finished, enter up the **result** together with any **observations** you have made and write a **conclusion.**

If an experiment has to be left for some time, carefully label it if necessary with the object (or number) of the experiment and keep a record of any other relevant information such as the name of the animal, date, time, temperature, etc.

Always set up a "control" experiment under opposite (or normal) conditions when practicable in order to show that the results you obtain are due to the conditions you have set up.

Reference should be made to the General Directions for Practical Work in the Introduction.

DIGESTION

THE ACTION OF ENZYMES

Enzymes are complex protein substances produced by living organisms which promote certain chemical reactions essential to the life of the organism. They survive after the reaction and may therefore be called *biological catalysts*.* Furthermore, they can function outside the organism of their origin. The chief characteristics of enzymes may be summarised as follows:—

(1) A small quantity of the enzyme will effect change in a large amount of the substrate.

(2) They act only within a certain range of temperature and are destroyed above 60°C.

(3) They act only in certain H-ion concentrations and each has its optimum *p*H.

(4) They are specific in their action.

(5) Some are inactive until combined with a co-enzyme.

(6) Their action varies with the concentration of the substrate.

The nomenclature of enzymes is rather confused by lack of co-ordination in practice, but the principle intended to be followed by biochemists is to add the suffix **-ase** to the name of the substrate: thus:—

Proteinases act upon proteins. ⎫ These are found in plants as
Amylases act upon starch.   ⎬ well as in animals.
Lipases act on fats.   ⎭

But many names which do not follow this rule are retained owing to long-established usage, for example:—

Ptyalin in saliva is an amylase, **pepsin** and **rennin** in gastric juice, **trypsin** from pancreatic juice and **erepsin** in intestinal juice are proteinases. **Invertase** hydrolyses cane sugar to invert sugar. **Diastase** is an amylase found in plants, while **zymase,** found in yeast, is an enzyme complex responsible for the fermentation of sugar.

Perform the following experiments in clean test-tubes, label them, plug them with cotton wool and place them in a thermostat (see Appendix II) kept at body temperature (38° to 40°C.). From the results you will discover which substances are digested by the different juices and the conditions under which the enzymes act.

---

* It should be noted that enzymes are not restricted to digestion. They play important roles in other metabolic processes *e.g.*, respiration.

DIGESTION IN THE MOUTH

Saliva

Chew a piece of rubber tubing for a few minutes: this will produce a copious supply of saliva. Transfer some to a test tube and dilute with an equal volume of distilled water. (It will act more quickly.) Filter a portion.

EXPERIMENT 1. *Test the reaction of saliva with a piece of litmus paper.*

EXPERIMENT 2. (*a*) *To some filtered saliva add a drop of dilute acetic acid.* Mucin, a viscous muco-protein, is precipitated. (*b*) *Perform Millon's test to show the presence of the protein, mucin.*

EXPERIMENT 3. *Add a fairly large quantity of saliva to a small quantity of* (*a*) *finely chopped lean meat and* (*b*) *some fat or olive oil. Label and place in the thermostat. Examine from time to time and observe whether the protein and fat are digested.*

The Action of Ptyalin

EXPERIMENT 4. *Add some diluted saliva to some starch sol, Label the tube and place it in the thermostat, kept at* 38°–°40C. *Shake periodically. Pour iodine into the cavities of a cavity tile. Later examine and test samples every five minutes by adding to the iodine in the cavity tile until no colour change is produced. Then test the residue with Fehling's solution.*

DIGESTION IN THE STOMACH

Gastric Juice

In the following experiments use Artificial Gastric Juice (see Appendix I).

EXPERIMENT 5. *Find the reaction of gastric juice with litmus.*

The Action of Pepsin and Rennin

EXPERIMENT 6. *Add a large quantity of gastric juice to a small quantity of* (*a*) *finely chopped lean meat or albumin,* (*b*) *starch sol,* (*c*) *a few drops of oil. Label the tubes and place them in the thermostat. Shake periodically. Examine a few hours later and note which have been digested. Then carry out the biuret test with the residue in* (*a*).

EXPERIMENT 7. *Repeat experiment 6* (*a*) *with gastric juice which has been made alkaline with a little caustic soda.*

EXPERIMENT 8. (*a*) *Add a little rennin* (*chymase*) *to some milk, diluted with an equal volume of water. Place in the thermostat and examine a few minutes later. (b) Relace the tube in the thermostat, leave it for an hour and then examine again.*

EXPERIMENT 9. *Repeat experiment 6 (a) or 8 but first heat the enzymes to about 70°C. for a few minutes. Note the effect of this temperature.*

DIGESTION IN THE DUODENUM

Pancreatic Juice

In the following experiments use Artificial Pancreatic Juice (see Appendix I).

EXPERIMENT 10. *Test the reaction of pancreatic juice with litmus.*

The Action of Trypsin, Amylase and Lipase

EXPERIMENT 11. *Repeat experiments 6 (a), (b) and (c), using pancreatic juice.*

Apply the biuret test to (a). Test (b) with iodine at intervals of five minutes on a white cavity tile as in experiment 4. When no colour change is observed, test the residual solution with Fehling's solution. Test the reaction of (c) with litmus when digestion appears to be complete.

EXPERIMENT 12. *Acidify some pancreatic juice with dilute hydrochloric acid and repeat Experiment 11.*

Bile Pigments

The two chief pigments of bile are breakdown products of hæmoglobin. They are **bilirubin** ($C_{33}H_{36}O_6N_4$) which is reddish-brown and **biliverdin** ($C_{33}H_{31}O_8N_4$), an oxidation product of bilirubin, which is green.

EXPERIMENT 13. **Gmelin's Test for Bile Pigments.** *To some ox bile, diluted with an equal volume of water, slowly add some fuming nitric acid, sloping the test tube so that the acid forms a lower layer.* A play of colours results at the junction of the two liquids. *Note these colours in the order of their appearance.*

RESPIRATION

TO DEMONSTRATE THE RESPIRATION OF ANIMALS

Expiration of Carbon Dioxide

EXPERIMENT 1. *Connect a Dreschel bottle containing caustic potash solution to a second Dreschel bottle containing lime water. Then connect the second bottle to the inlet-tube of a bell-jar or flask. This should be fitted with a short inlet-tube and a long outlet-tube reaching almost to the bottom. The latter should be connected to another Dreschel bottle of lime-water which, in turn should be connected to a water-pump by pressure tubing. If a bell-jar is used, the flange should be well greased with vaseline and it should stand on a ground-glass plate.*

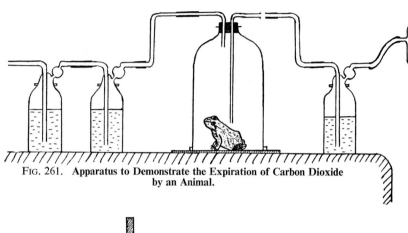

FIG. 261. **Apparatus to Demonstrate the Expiration of Carbon Dioxide by an Animal.**

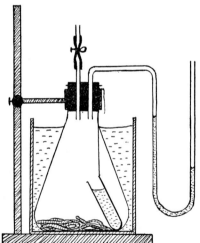

FIG. 262. **Apparatus to Demonstrate the Inspiration of Oxygen by an Animal.**

Now place a small animal (e.g., a frog) under the bell-jar or several large earthworms in the flask and set the pump working slowly.

The caustic potash in the first Dreschel bottle absorbs the carbon dioxide from the air which enters, as shown by the fact that the lime water in the next bottle remains clear. *After a short time examine the lime water in the last Dreschel bottle and draw your conclusion.*

Inspiration of Oxygen

EXPERIMENT 2. *Fit a conical flask with a cork through which passes a short piece of glass tubing fitted with a short rubber tube provided with a clip and a double right-angle tube bent further to form a U-tube manometer. Put some coloured water into the U-tube. Place a few large earthworms in the flask and lower a small tube of caustic potash solution into it. Replace the cork and stand the flask in a vessel of cold water to keep the temperature constant. Equalise the water levels in the manometer by opening the clip, closing it immediately.*

The carbon dioxide expired by the worms is absorbed by the caustic potash and the movement of the water in the manometer indicates a reduction in pressure due to the absorption of oxygen.

TO MEASURE THE RATE OF RESPIRATION AND TO FIND THE RESPIRATORY QUOTIENT

EXPERIMENT 3. *Connect a U-tube (A) containing soda-lime to a second U-tube (B) containing pumice soaked in concentrated sulphuric acid. This, in turn, should be connected to a light beaker fitted with a cork (C) into which the animal will be placed and should be large enough to take the animal at rest. This is the animal chamber. It should be connected to a pair of light U-tubes (D, E) containing pumice soaked in concentrated sulphuric acid. The second U-tube should be connected to another pair of light U-tubes, (F) containing soda-lime and (G) pumice soaked in concentrated sulphuric acid. The last U-tube must then be fitted with pressure tubing connected to a water pump. The animal chamber (C) and the U-tubes D, E, F and G should be provided with light wire handles so that they need not be touched by hand.*

The soda-lime in A will absorb the carbon dioxide from the air which enters and the sulphuric acid in B will absorb water. The sulphuric acid in D and E will absorb the water given off by the animal and the soda-line in F the carbon dioxide expired, while the sulphuric acid in G will absorb the water given off by the soda-lime in F.

Weigh the beaker C empty; then place a small animal in the beaker and weigh again in order to find the weight of the animal.

Weigh the U-tubes D and E together and F and G together. In all cases take care not to touch the glass by hand. Then connect up the apparatus and immerse the animal chamber in a vessel of cold water to keep the temperature as near constant as possible. Place a thermometer in the water.

Set the pump working slowly for about half an hour. Then disconnect as before, dry C and weigh it again. Weigh the U-tubes together as before. Record your weighings and calculate as follows:—

   (i) **Weight of water given off by the animal** = the increase in weight of D and E.

   (ii) **Weight of carbon dioxide expired by the animal** = the increase in weight of F and G.

   (iii) **Loss of weight of the animal** = the loss of weight of C.

(iv) **The weight of oxygen absorbed** = weight of water + weight of carbon dioxide expired — the loss of weight of the animal.

(v) **The respiratory quotient** = $\dfrac{\text{volume of } CO_2 \text{ expired}}{\text{volume of } O_2 \text{ inspired}}$

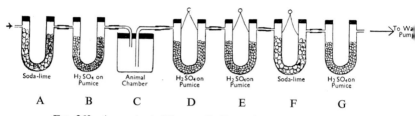

| Soda-lime | H_2SO_4 on Pumice | Animal Chamber | H_2SO_4 on Pumice | H_2SO_4 on Pumice | Soda-lime | H_2SO_4 on Pumice |
| A | B | C | D | E | F | G |

FIG. 263. **Apparatus to Measure the Rate of Respiration and the Respiratory Quotient.**

which obviously may be calculated as follows:—

$$\frac{\text{Weight of } CO_2 \text{ expired} \times 32}{\text{Weight of } O_2 \text{ inspired} \times 44}$$

Record the temperatures, the time during which the experiment was in progress, the weight of the animal at the outset and whether it was active or at rest.

TO DEMONSTRATE THE ACTION OF THE DIAPHRAGM

EXPERIMENT 4. *Securely tie a small rubber balloon on the end of a piece of glass tubing fixed through the cork of a bell-jar so that the*

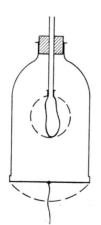

FIG. 264. **Apparatus to illustrate the action of the diaphragm.**
(*From Wallis—Human Biology*)

balloon is inside the jar. Now cover the base of the bell-jar with sheet rubber and tie it securely in position. A piece of string or tape should then be stuck to the centre of the rubber sheet on the outside.

The glass tube represents the trachea and bronchi, the rubber balloon the lungs, the bell-jar the wall of the thorax and the sheet rubber base the diaphragm.

Lower the diaphragm by gently pulling the string or tape and observe the effect on the "lungs". Then allow the diaphragm to return to its original position and again observe the effect. Explain the causes of the changes you see.

CIRCULATION AND PROPERTIES OF BLOOD

TO EXAMINE HEART BEAT

EXPERIMENT 1. *Mount a living specimen of* **Daphnia** *in a few drops of the water in which it is supplied and place on the microscope stage. The heart will be seen to be beating. Using a stop watch, count twenty heart beats and record the time. Repeat the experiment twice and calculate the average time for twenty beats.*

EXPERIMENT 2. *Pith a frog as directed in Appendix II and make a ventral dissection without ligaturing the anterior abdominal vein. Either cut through the centre of the pectoral girdle, exercising care so that the point of the scissors does not injure the heart, or better, cut through each side of the girdle afterwards removing the central portion. In either case be very careful not to cut any blood vessels. The heart will continue beating for at least an hour. Pull out the fore limbs so as to separate the two parts of the girdle and thus expose the heart as much as possible. Now remove the pericardium with great care in order to free the heart.*

Transfer the frog to a sufficiently large vessel and pour in enough Ringer's solution to cover the animal. Allow it to stand for five minutes or so to give the body time to adjust to the temperature of the solution. Then take the temperature of the solution and record it. Count twenty heart beats and, using a stop watch, observe the time taken. Record your result. Repeat twice and record the average time.

Then replace the Ringer's solution with another sample cooled to about 5°C. below that of the previous one and repeat the experiment, recording results as before. Again replace the Ringer's solution with other samples about 5°, 10°, and 15°C. above that of the original solution leaving the preparation for at least five minutes in each case before taking records as before.

From your recorded readings in the five experiments plot a graph of heart beat (vertical axis) against temperature (horizontal axis) and draw your conclusions from the results obtained.

TO DEMONSTRATE THE CIRCULATION OF THE BLOOD

EXPERIMENT 3. *Pith a frog, place it on a frog-plate (see Appendix II), stretch the web of the foot over the hole and fix the toes in position. Now put the frog-plate on the stage of the microscope and examine the web under the low power.*

Note the thin-walled **capillaries** and the **red blood corpuscles** flowing through them from the larger **arterioles** to the smaller capillaries and from these smaller capillaries to the larger **venules.**

TO EXAMINE THE PROPERTIES OF HÆMOGLOBIN

Hæmoglobin is a chromoprotein and is composed of globin combined with a base called **hæmatin** or **hæm** which contains ferrous iron ($C_{34}H_{32}N_4O_4FeOH$). It readily combines with oxygen to form the addition compound, **oxyhæmoglobin** (HbO_2).

EXPERIMENT 1. *Dilute about 5 ml. of fresh blood with about 100 ml. of water in a 150 or 200 ml. flask. Shake well with plenty of air and note the colour.* This is due to oxyhæmoglobin.

EXPERIMENT 2. *To the solution from Expt. 1 add a reducing agent such as ammonium sulphate or sodium hydrosulphite and observe the change of colour* as the oxyhæmoglobin is reduced to hæmoglobin.

TO INVESTIGATE THE CLOTTING OF BLOOD

EXPERIMENT 1. (*a*) *Clean the tip of a finger with alcohol, prick it with a sterilised needle and place a drop of blood on a microscopical slide. Leave it uncovered and examine later under the microscope.*

Note that the blood has clotted.

(*b*) *Put a few drops of 10 per cent. sodium citrate solution or 1 per cent. potassium oxalate on a microscopical slide. Introduce a drop of fresh blood from the finger. Put aside and examine under the microscope later.*

The sodium citrate or potassium oxalate, by removal of the calcium salts, will have prevented clotting from taking place.

(*c*) *To the "salted" blood from (b) add a few drops of calcium chloride solution.*

Observe that the blood clots, showing that clotting depends on the presence of calcium salts.

EXPERIMENT 2. (*a*) *If some freshly drawn blood is available, place it in a test tube and examine about twenty minutes later.*
Note that the blood has clotted, a reddish jelly-like mass of red corpuscles, the **clot,** being suspended in a yellowish fluid called **serum.**
(*b*) *Whip up some freshly drawn blood with twigs for a few minutes.*
Note that a stringy substance collects on the twigs. This is **fibrin.**
Leave the blood for about twenty minutes and examine again.
It will be seen that this defibrinated blood has not clotted.
(*c*) *Remove some of the fibrin and test it for protein.*

TO TEST FOR BLOOD

Dilute a few drops of blood with water, boil to destroy any oxidising enzymes and add about a couple of drops of an alcoholic solution of guaiacum. A precipitate is formed. *Add sufficient alcohol to dissolve this and add a little hydrogen peroxide.* A blue colour is produced.

TO IDENTIFY BLOOD GROUPS

If experiment (vii) on p. 309 on Blood Groups has not yet been performed it can conveniently be done now.

EXCRETION

Excretion means the elimination of the waste-products of metabolism from the body. The most important excretory products are water, carbon dioxide, urea, creatinine, uric acid and hippuric acid.

UREA

Urea $CO(NH_2)_2$ is an amide and is derived by deamination in the liver from the protein in the diet. It is the chief nitrogenous constituent in mammalian urine and is soluble in water.

EXPERIMENT 1. *Heat some crystals of urea in a test-tube.* The substance melts and then decomposes with the evolution of ammonia, *which can be detected by its odour and its action on red litmus paper. Carry out the biuret test (as used for proteins) with the residue.* Biuret is formed, giving a violet colour.

$$2CO(NH_2)_2 = NH_2.CO.NH.CO.NH_2 + NH_3.$$

EXPERIMENT 2. *Boil a solution of urea.* It is hydrolysed to carbon dioxide and ammonia. *Test for these gases.*

$$CO(NH_2)_2 + H_2O = CO_2 + 2NH_3.$$

Done ✓
With much
reactionary
Heat .

EXPERIMENT 3. *Add some alkaline sodium hypobromite (prepared by adding 2 c.c. of bromine to 23 c.c. of N. sodium hydroxide) to some urea.* A violent effervescence takes place with the liberation of nitrogen. ↲

$$CO(NH_2)_2 + 3NaOBr = 3NaBr + CO_2 + N_2 + 2H_2O.$$

EXPERIMENT 4. *To a few millilitres of urea solution in a watch glass add a few drops of concentrated nitric acid.* Urea nitrate $(CO(NH_2)_2.HNO_3)$ crystallises out. *Examine the crystals under the microscope.*

CREATININE

$$HN = C \underset{\overset{|}{CH_3}}{\overset{NH-CO}{<}}_{N-CH_2}$$

This orignates from phosphagen in muscle which is converted into creatine in muscular activity. Creatine is the precursor and the anhydride of the creatinine in urine.

EXPERIMENT 5. *Add a few drops of a freshly prepared dilute solution of sodium nitroprusside to an aqueous solution of creatinine or a few ml. of urine, then add a few ml. of sodium hydroxide.* Note that a deep red colour is formed and that this quickly changes to yellow. *Add dilute acetic acid and boil: a greenish-blue colour is obtained.*

URIC ACID

Uric acid, $C_5H_4N_4O_3$ is the chief nitrogenous excretory product of reptiles and birds. There is little in mammalian urine.

EXPERIMENT 6. **Murexide Test.** This test cannot be performed with urine: It is necessary to extract the uric acid from it. *Add a few drops of concentrated nitric acid to a few crystals of uric acid. Evaporate slowly in a fume chamber until no further fumes of nitric acid are evolved. When the yellowish-red residue is cool add a drop of very dilute ammonia.* Note a bright red colour is produced. *Now add some caustic soda and note that the colour changes to purple.*

HIPPURIC ACID

Hippuric acid or benzoyl glycine, $C_6H_5.CO.NH.CH_2.COOH$, is a normal constituent of the urine of herbivorous animals. Aromatic substances in the plant material are oxidised to benzoic acid in the animal and, in the liver, this undergoes condensation with glycine to hippuric acid.

EXPERIMENT 7. *Heat a few crystals of hippuric acid. Note that they melt and turn red.* The white sublimate is benzoic acid.

URINE

EXPERIMENT 8. (*a*) *Note the colour and transparency of normal urine.* (*b*) *Find the specific gravity of urine by means of a hydrometer or urinometer.*

EXPERIMENT 9. *Find the reaction of fresh urine by using litmus paper.*

EXPERIMENT 10. *Test for the presence of* **urea** *in urine by the sodium hypobromite test* (Experiment 3).

EXPERIMENT 11. *Test for* **creatinine** *in urine by the sodium nitro-prusside test* (Experiment 5). (N.B.—Acetone in urine will also give this reaction but it does not turn yellow and on the addition of acetic acid a purple colour is obtained. *Perform this test with acetone and verify this result.*)

EXPERIMENT 12. *Test for inorganic salts in urine as follows:*—

(i) **Chlorides.** *Add a few drops of concentrated nitric acid and then some silver nitrate solution. Note that a white precipitate is obtained. Add some ammonium hydroxide and the precipitate dissolves.*

(ii) **Phosphates.** *Acidify some urine with concentrated nitric acid and boil. Add this to some ammonium molybdate acidified with nitric acid and boil. A yellow precipitate is formed.*

(ii) **Sulphates.** *Add about two millilitres of concentrated hydrochloric acid to a little urine and then add excess of barium chloride solution. A white precipitate is obtained.*

NERVOUS RESPONSE

TO DEMONSTRATE REFLEX ACTION

Reflex actions are of two kinds:—

(i) **Unconditioned reflexes** which are natural, inherited reflex actions.

(ii) **Conditioned reflexes,** which are acquired during the life as result of frequently repeated stimuli.

The former only need be considered here.

EXPERIMENT 1. **Knee jerk.** *Sit on a chair and cross one leg over the other. Strike the free leg sharply with the edge of the hand on the tendon just below the knee-cap.*

Note the response, over which you have no control.

EXPERIMENT 2. **Iris Reflex.** *Get someone to close his eyes for a minute or two and on opening them, shine a torch on them.*

Observe the effect on the pupils.

EXPERIMENT 3. *Pith a frog* (see Appendix II) *and suspend it by its fore-limbs. Now apply as a stimulus to one of the legs or to the skin of the trunk a hot wire or a rod which has been dipped in acid. Note the response.* It should be noted that the brain, having been destroyed, the animal is not conscious and that the response is entirely reflex.

TO SHOW THAT THE NERVES CONVEY IMPULSES

EXPERIMENT 4. *Take the frog used in the last experiment and dissect one leg so as to expose a short length of the sciatic nerve. Sever the nerve. Now suspend the frog and stimulate as in* Experiment 3.

Observe that the leg in which the nerve has been severed remains motionless. The other will still respond.

HORMONES

(AUTACOIDS)

Autacoids are chemical substances which are the secretions of the ductless glands or **endocrine organs.** As most of these secretions act as stimulants they are known as **hormones,** but a few are inhibitory in their actions (*e.g.* those from the gonads which prevent the development of the secondary sexual characters of the opposite sex) and these are called **chalones.** So that the term **"autocoid"** is more applicable in general than the commonly used "hormone." As the endocrine organs are also known as internally secreting glands, hormones are also known as **internal secretions.** They are carried by the blood stream to other parts of the body and are responsible for chemical co-ordination and the control of metabolic activities generally. The chemical composition of some of the hormones is known as they have been isolated from the body (some have, in fact, been synthesised). They are of comparatively low molecular weight compared with vitamins. Like vitamins, minute quantities only are necessary for the organism, but unlike most vitamins they are produced within the animal. Like enzymes, many of them are specific in their action, but unlike them they are not destroyed by heat and are simpler in structure. Furthermore, most of them are destroyed by the digestive enzymes and they act in a different part of the body from that in which they are secreted and which they reach by means of the blood stream. It should be understood that the endocrine organs through the blood stream work together as a co-ordinated whole, thus constituting the **endocrine system.** Deficiency or superfluity of hormones may set up pathological conditions. (Growth-promoting hormones have been demonstrated in plants and are called **auxins.**)

It will be obvious from the nature of their actions that simple practical work on autacoids is not possible, but experiments 1 and 2 following will illustrate the actions of two of them.

EXPERIMENT 1. **To find the effect of pituitrin on the Skin of the Frog.**

Examine a demonstration specimen, in which a light-coloured frog has been injected with pituitary extract. (See Appendix II.)

Note the dark colour of the skin, caused by the expansion of the melanophores (black pigment cells). This is due to the action of **pituitrin,** the hormone of the secretion of the posterior lobe of the pituitary gland. *Compare this specimen with the uninjected frog similar in colour to the other frog before injection.*

EXPERIMENT 2. **To find the effect of Thyroid Extract on the Development of the Tadpole.**

Examine a demonstration in which tadpoles are kept in two separate aquaria, the one group being kept on their normal diet, while the other group is fed on thyroid extract. (See Appendix II.) *Examine after twenty-four hours or so.*

The tadpoles fed on thyroid metamorphose into minute frogs very quickly. This is due to the **thyroxin.**

EXPERIMENT 3. **To demonstrate the presence of iodine in thyroxin.**

To an alcoholic solution of thyroxin, add a few millilitres of chloroform and a few drops of chlorine water. Shake and allow to stand. The chloroform layer is coloured violet.

PART VI

GENETICS

INTRODUCTORY NOTES

The structure of *Drosophila* and its life-history have already been studied (pp. 125–126). Read through the external structure again. The differences between the sexes are repeated here.

MALE: **1st pair of legs:** The metatarsal, the first of the four podomeres of the tarsus, bears a **sex comb.**

> **Abdomen:** Darker and narrower than in female. Posterior end somewhat rounded in side view and last segment (6th) so modified, due to external genitalia (penis and two claspers), that it appears that there are only five.

FEMALE: Generally larger than male. **1st pair of legs:** There are no sex-combs.

> **Abdomen:** Paler in colour and broader than in male. Posterior end more pointed in side view and on the ventral side is the **vaginal plate.**

(Cultures of both sexes of the Wild Type and of the various mutants can be purchased from firms supplying biological material.—See Appendix V.)

Reference should be made to the General Directions for Practical Work in the Introduction.

EXPLANATION OF TERMS USED IN GENETICS

Chromosomes are paired bodies in the dividing nucleus and are responsible for the inheritance of characters. Those responsible for characters other than sex are known as **autosomes.** They occur in identical pairs in somatic cells, the nuclei of which are therefore said to be **diploid,** and these are known as **homologous chromosomes.** Those responsible for the inheritance of sex are known as **sex chromosomes** and form an identical pair in one sex (the female in all animals except Lepidoptera and Aves). This is known as the **homogenetic sex** and the chromosomes are referred to as **XX** while those in the other sex, the **heterogenetic sex,** are dissimilar and are referred to as **XY.** In the gametes, of course, since meiosis has taken place in their formation, only one member of each pair of chromosomes is present and the nuclei of the gametes are therefore said to be **monoploid** (or **haploid).**

Chromosomes are composed of units called **genes** which are the carriers of hereditary characters and contain D.N.A. There are two genes for every character, each occuring in identical places called **loci** on the separate members of a pair of chromosomes one paternal in origin the other maternal. When the two genes are identical for a particular character, the individual is a **homozygote** but if they are different, one for one character and the other for its alternative it is a **heterozygote**. The alternative characters are called **allelomorphs**

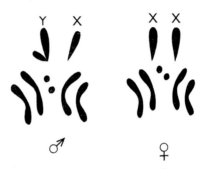

Fig. 265. **Drosophila melanogaster.** **Chromosomes.**

or **alleles.** If, when animals are mated, one character appears to the exclusion of its alternative allelomorph that character is said to be **dominant** and the other **recessive.** The dominant gene may be present on both chromosomes and these are **genotypes.** But individuals may appear identical for a particular character, regardless of their genetic constitution, and such are said to be **phenotypes.** These may thus be genotypically identical or different.

During gametogenesis interchange of corresponding genes may occur between the members of a paid of homologous chromosomes. This is known as **crossing over.** In some cases certain genes carried on the same chromosome remain together during inheritance. This is called **linkage** and when it is related to the sex of the individual it is known as **sex linkage.**

INHERITANCE IN DROSOPHILA MELANOGASTER

TECHNIQUE

I. PREPARATION OF APPARATUS AND MATERIALS

Culture Bottles

Bottles similar to half-pint milk bottles are best and most commonly used for keeping the cultures. Any wide-mouthed bottles of similar size will do equally well. In fact large specimen tubes can be used but they are rather unstable. The bottles must be plugged with cotton wool, preferably wrapped in muslin secured with thread.

Sterilisation

It is essential that the containers, the cotton-wool and muslin used as plugs and the paper used to absorb moisture and on which the flies can walk and pupate should be absolutely sterile. The absorbent type of toilet-paper or kitchen paper is ideally suitable.

(i) **Containers.**
These should be throughly washed, rinsed and then sterilised in a hot-air oven at about **65°–70°C.**

(ii) **Cotton-wool and Muslin.**
These should be heated in a hot-air oven **not above 180°C.** *to avoid charring.*

Food for the Cultures

Although *Drosophila* will thrive on rotting banana skins, it is better to provide the flies with a pabulum on which yeast will readily grow since the insect feeds on this. There are several formulæ for making such a pabulum but the most suitable is probably the *Maize/Molasses/Agar Medium* (see........). *The medium is melted by gentle head and poured into the culture bottles to a depth of a couple of inches or so. Some live yeast is then mixed with a little water and a few drops added to the medium which is then left for twenty-four hours before use.* Warning—Any mould growing on the medium will use the food and starve the flies. Fresh yeast should be added if this appears and if this does not successfully dispose of the infection, the medium should be replaced.

Ether Bottles

It will be necessary to *anaesthetise* the flies for examination and an ether bottle should be prepared for this purpose. The bottle should be identical to those used for the cultures but must be fitted with a cork on the inside of which a small cotton-wool pad wrapped in muslin must be stuck. When required for use a few drops of ether are put on to the pad.

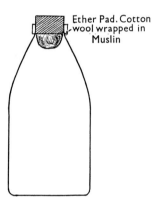

Ether Pad. Cotton wool wrapped in Muslin

FIG. 266. **Drosphilia. Ether Bottle.**
(From Wallis—Practical Biology)

Keeping the Cultures

On receipt of the cultures it is necessary to put them into the prepared culture bottles, keeping each type in a different bottle, suitably labelled. The label should state the Type, sex (if relevant) and the date of insertion into the bottle.

Having sterilised and prepared the culture bottles, insert a length of the sterilised paper, fixing the lower end into the medium its upper end resting against the side of the bottle. The cultures will contain both male and female flies and mating will, of course, take place. It is essential that only virgin females be used in the experiments. The following procedure should therefore be adopted.

Transfer the flies to a culture bottle, insert the plug and keep it at a constant temperature of about 24°C. When the metamorphosis has reached the pupal stage (about 4 days) transfer some of the pupæ to a new culture bottle and when the imagines emerge (about 4 days later) they must be transferred to the ether bottle. This is best done as follows:—Remove the cotton wool plug from the culture bottle and quickly invert the ether bottle on top. If necessary the two bottles must then be turned upside down together and the culture bottle tapped to ensure complete transference of the flies to the ether bottle.

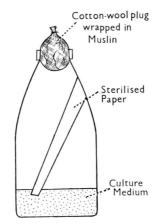

Cotton-wool plug wrapped in Muslin

Sterilised Paper

FIG. 267. **Drosophila. Culture Bottle.**
(*From Wallis—Practical Biology*)

Culture Medium

Anæsthetising the Flies

Add a few drops of ether to the cotton-wool of the ether bottle and insert the cork into it. Immediately the flies are motionless, remove the cork and transfer the flies to a white tile or into a Petri dish standing on a white tile. Anaesthesia is adequate when the flies are motionless. If they are left too long in the ether they will be killed. Should the wings be folded back and the legs folded or bunched together, anaesthesia has gone too far and the flies will probably be dead.

Separating the Sexes

The differences between males and females have already been given above and in the description of the imago in the account of the life-history.

Examine the flies with a lense and separate the males from the females using a camel-hair brush.

II. THE EXPERIMENTS

Mating

Suitable crossings are given below. The method is similar in all cases.

Transfer the females to a new culture bottle and introduce a few males of another type. When the flies have recovered from their anæsthesia mating will take place. Label the bottle stating the types mated and the date. It is important that the labels should not interfere with the observance of the flies. They should not, therefore, be stuck on the bottles. Loose labels tied with string can be attached to the necks of the bottles.

Examination of the Results of Mating

When the new imagines have emerged from the pupæ, anæsthetise as before. Then remove the flies and examine under a lens. Identify and arrange in batches on separate tiles or in separate Petri dishes according to type. Count the number in each batch. The larger the number of flies used the more accurate will be the result, of course. *Then tabulate your results. In studying sex-linked characters the imagines should be arranged in batches according to sex as well as to the other character.*

Suitable Crossings

There are several mutant types of Drosophila of which four are listed below providing suitable alternative allelomorphs.

The normal form of Drosophila with yellow striped body, long wings and red eyes is known as the **Wild Type.** Four of the many mutants are:—

1. **Vestigial Winged** (poorly developed wings, darker body and red eyes)
2. **Ebony body** (black body)
3. **White eyes**
4. **Brown Eyes**

Experiments can be performed crossing Wild Types with any of these mutants and carrying on further experiments by crossing the offspring. Back-crossings can also be performed. One example is given below and a similar procedure can be followed with the other types.

(1) Monohybrid Ratio

EXPERIMENT 1. (i) *Cross* **Wild Types (long-winged) (L)** *with* **Vestigial-Winged Types (1).**

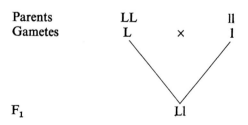

All the F1 generation will be long-winged, showing that long-wing is dominant to vestigial-wing.
(ii) *Interbreed the* F1 *generation.*
The F2 generation will consist of long winged and vestigial-winged types in the ratio of 3 : 1.

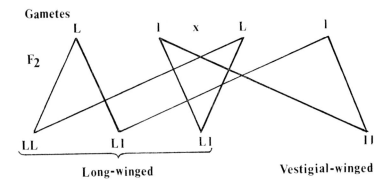

Long-winged   Vestigial-winged

Clearly the F1 generation was *heterozygous.* Two-thirds of the long-winged types in the F2 generation are also *heterozygous,* as will be seen in the above diagram. the other third being *homozygous* as are the vestigial-winged type. The homozygous types if interbred would breed true while the heterozygous long-winged types would give a similar result to that obtained in the F2 generation.
EXPERIMENT 2. *Back-cross the* **F1 (heterozygous) generation from Experiment 1 (i)** *with the* **homozygous dominant parental type.**

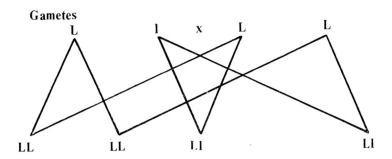

All the offspring will be long-winged but 50 per cent. of these are *homozygotes* and 50 per cent. are *heterozygotes.*

EXPERIMENT 3. *Back-cross the* (**heterozygous**) **F1 generation from Experiment 1 (i)** *with the* **homozygous recessive parental vestigial-winged type.**

Fifty per cent. of the offspring will be long-winged and 50 per cent. will be vestigial-winged.

Gametes

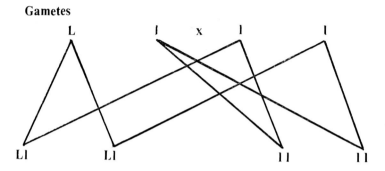

(2) Di-hybrid Ratio

EXPERIMENT 4. (i) *Cross the* **Wild (yellow-bodied) type (Y)** *with the* **mutant ebony-bodied type.** (y).

All the F1 generation are yellow-bodied, showing that yellow body (Y) is dominant to ebony body (y). It has already been seen that long-wing is dominant to vestigial-wing.

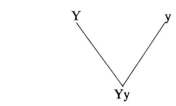

(ii) *Now cross* **yellow body vestigial-winged type (Yl)** *with* **ebony body long-winged type (yL).**

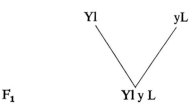

The F1 generation are all yellow-bodied, long-winged types but are, of course, *heterozygotes* LlYy).

(iii) *Finally interbreed this* **F1 generation.**

If sufficiently large numbers are taken four types will appear and in the following ratio:—

Yellow body long wings (YL) 9
Yellow body vestigial wings (Yl)......3
Ebony body long wings (yL) 3
Ebony body vestigial wings (yl) 1

This ratio can be worked out by the "chess-board" method.

	YL	Yl	yL	yl
YL	YL YL = YL	Yl YL = YL	yL YL = YL	yl YL = YL
Yl	YL Yl = YL	Yl Yl = Yl	yL Yl = YL	yl Yl = Yl
yL	YL yL = YL	Yl yL = YL	yL yL = yL	yl yL = yL
yl	YL yl = YL	Yl yl = Yl	yL yl = yL	yl yl = yl

Linkage

Non-Sex Linked Characters

EXPERIMENT 5. (i) *Cross* **long-winged yellow-bodied (LY) flies** *with* **vestigial-winged ebony-bodied flies (ly).**

All the F1 generation are long-winged yellow-bodied.

(ii) *Interbreed the* **F1 generation.**

The result will be as follows:—

Longwinged yellow bodies 3
Vestigial-winged ebony body 1

Obviously the two alleles for wing formation and body colour are inherited as if they were a single character. This is linkage.

Sex Linkage

EXPERIMENT 6. (i) *Cross pure bred* **red-eyed females** *with* **white-eyed males.**

All the Fl generation are heterozygous red-eyed males and females in equal numbers.

(ii) *Interbreed the F1 generation.*

The F2 generation should be in the following ratio:—

2 red-eyed females: 1 red-eyed male: 1 white-eyed male.

(iii) *Cross* **white-eyed females** *with* **pure red-eyed males.**

The result will be

Fifty per cent. white-eyed males.

Fifty per cent. red-eyed females.

(iv) *Interbreed these* **white-eyed males** *with* **red-eyed females.**

The offspring will be in the following ratio:—

1 red-eyed female	1 white-eyed female
1 red-eyed male	1 white-eyed male

In these experiments eye colour and sex are linked together and borne on the same chromosomes.

PART VII

VERTEBRATE EMBRYOLOGY

INTRODUCTORY NOTES

It should be remembered that during the development of the embryo undifferentiated cells during multiplication undergo specialisation and differentiation into structures which will ultimately be found in the adult. Consequently the appearance of the cells and developing organs in the embryo are quite different from those in the adult. It will be observed that the nuclei of embryonic cells are larger than in the adult and stages in mitosis may be visible. Furthermore it must be understood that, except in the earliest stages, various processes are going on concurrently.

In the study of embryology it is advisable to do the practical work with the aid of a text-book and plenty of diagrams and it must be borne in mind that owing to the very large number of sections cut through any particular region of an embryo, the specimens examined may not be identical with the diagrams studied.

Examination of embryological models, if available, is of great assistance but this must *not* replace the study of slides and specimens. Prepared slides will have to be used, of course, in all cases.

In drawing sections it is not necessary to fill in the individual cellular structure except where it is advisable in order to make a particular structure clearer. Outlines of the structures, adequately labelled, will suffice.

SEQUENCE OF DEVELOPMENT

The following sequence of stages occurs in the development of the embryo from the fertilised ovum:—

(1) **Segmentation** or **cleavage** of the zygote.

*(2) Formation of the **blastula,** a spherical structure composed of a single wall of cells enclosing a cavity, the **blastocœle.**

*(3) Formation of the **gastrula,** an elongated structure containing a fresh cavity, the **archenteron** or primitive gut, the blastocœle being obliterated. The wall is composed of two layers of cells, the **ectoderm** (which, in some cases, takes part in the formation of extra-embryonal structures) externally and what is often

---

*These stages are considerably altered in the chick and mammal.

called the **"endoderm"** internally. It is really better to refer to the latter as the **"inner layer"** since the endoderm and also the mesoderm and the notochord are derived from it.

(4) Formation of the **mesoderm,** the third of the germ-layers.

(5) Formation of the **neural tube** giving rise to the nervous system. This stage is referred to as the **neurala.**

(6) Formation of the **notochord** which, except in primitive Chordata in which it persists, is replaced by the vertebral column which develops around it from mesodermal cells.

(7) Formation of the **gut** from the archenteron.

(8) Development of **organs** and **systems.**

But it must be understood that, except in the earliest stages, many structures are developing at the same time.

The presence of yolk, however, considerably modifies these stages. Ova, such as that of Amphioxus, which contain a small amount of yolk evenly distributed, are said to be **microlecithal** (or, better, **isolecithal).** But in the majority of the Chordata the yolk is concentrated towards one pole of the ovum known as the **vegetative pole** to distinguish it from the **animal pole,** where the nucleus of the ovum is situated more or less free from yolk. Ova of this type are said to be **telolecithal.** In some ova, however, such as those of the invertebrate Astacus, the yolk is situated in the centre and these are known as **centrolecithal** ova.

There is no such thing as a completely yolkless ovum but the study of the cleavage and subsequent development of the zygote of Amphioxus (the ovum of which is microlecithal) serves as a basis from which the influence of yolk in higher types can be seen.

From the three primitive **germ layers** all the tissues and organs of the body are derived as shown on the next page *but it must be understood that many organs arise from more than one germ layer.*

Ectoderm	Mesoderm	Endoderm
1. Epidermis and structures derived from it.	1. Dermis	1. Pharynx.
2. Nervous system.	2. Muscles.	2. Respiratory tract.
3. Sensory epithelia—retina of the eye, membranous labyrinths of the ear, olfactory cavities and the lens of the eye.	3. Skeleton.	3. Alimentary canal (except the two extremities) and its associated organs.
4. Epithelium of the anterior and posterior ends of the alimentary canal.	4. Connective tissues.	4. Bladder.
	5. Blood.	
	6. Blood vessels.	
	7. Peritoneum. Pleura and pericardium.	
	8. Urino-genital organs.	
	9. Sclerotic of eye.	

In a late blastula and therefore before the formation of the gastrula (see below), certain **presumptive areas** from which parts of the adult animal arise, become evident in the frog and chick.

Furthermore, certain embryonic structures, such as the dorsal lip of the blastopore, appear to be responsible for the development of the cells adjacent to them. These structures are known as **primary organisers.** Later in development **secondary organisers** appear, such as the optic cup which is responsible for the development of the lens of the eye. This will be better understood as study progresses.

AMPHIOXUS (BRANCHIOSTOMA)

SEGMENTATION OR CLEAVAGE

Examine prepared slides of the **Fertilised Ova of Amphioxus in various stages of Cleavage.**

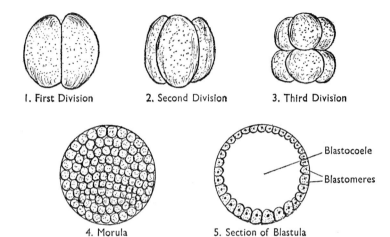

| 1. First Division | 2. Second Division | 3. Third Division |

4. Morula   5. Section of Blastula

Fig. 268. **Amphioxus. Segmentation of Zygote.**

(i) **Ova, Early Cleavage Stages.** Note the **holoblastic division** of the fertilised **microlecithal** (or **isolecithal**) **ovum** into cells more or less equal in size. It divides vertically into two, then into two again, also vertically but at right angles to the first division. The next division is horizontal. Thus, at this stage, there are eight cells. As this division takes place just above the equator, there are four smaller **micromeres** above and four slightly larger **megameres** below.

(ii) **Morula.** Note the mulberry-like mass of cells, resulting from repeated cell division.

BLASTULA

Examine a prepared slide of a section through the **Blastula of Amphioxus.**

This is a hollow single-layered sphere of cells called **blastomeres** with an enclosed **blastocœle** (or **segmentation cavity**). When the blastula is fully formed it is less spherical than it is in its earlier stages.

GASTRULATION

Examine prepared slides showing the **Formation of the Gastrula.**
Looking for stages in the formation of the gastrula—
(*a*) the invagination of the lower cells into the blastocœle (**embolic gastrulation**), gradually obliterating it, (*b*) the elongation of the embryo and the formation of the **two-layered gastrula** with the **ectoderm** externally and the **endoderm** internally, enclosing a new cavity, the **archenteron** (or **primitive gut**). The external opening of the archenteron is called the **blastopore**, having **dorsal, lateral** and **ventral lips.** The dorsal lip contains **chorda cells** which give rise later to the notochord.

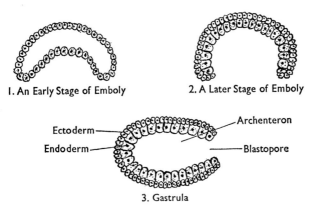

1. An Early Stage of Emboly   2. A Later Stage of Emboly

Ectoderm———
Endoderm———
———Archenteron
———Blastopore

3. Gastrula

Fig. 269. Amphioxus. Stages in Gastrulation.

FORMATION OF MESODERM, NEURAL TUBE, NOTOCHORD AND CŒLOM

Examine prepared slides of **T.S.** *and* **L.S. of the Larvæ of Amphioxus in various stages.**
Note the **ectoderm** (this becomes ciliated externally later), the **endoderm** and the metamerically arranged pairs of **mesodermal pouches.** These develop from **mesodermal grooves** which later become separated from the archenteron and the pouches enclose the **cœlom.** A cœlom formed in this way is said to be **enterocœlic.**
Note also the flattened ectodermal cells on the upper surface forming the **neural plate** on either side of which, owing to its sinking, ridges develop, the **neural folds,** which ultimately join and form the hollow **neural tube.** This is the **neurala.** The neural canal is connected to the blastopore by a short **neurenteric canal** at the posterior end (seen in L.S.). At the anterior end at first is an opening, the **neuropore,**

which later becomes closed. Find the **notochord** (in various stages of formation from the chorda cells) in the mid-dorsal region of the wall of the archenteron and composed of a rod of cells ventral to the neural tube and dorsal to the archenteron.

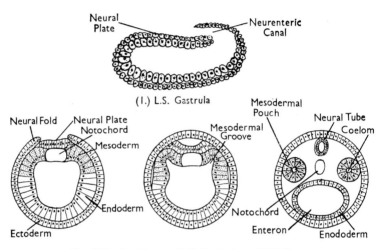

FIG. 270. **Amphioxus. L.S. Gastrula and T.S. Larva.**

RANA

FERTILISED OVUM

Examine **Ova of Rana** *with a hand lens.*

Note the spherical **telolecithal ovum** enclosed in a thin **vitelline membrane,** with **yolk** at the lower **vegetative pole** which is light in colour, the **animal pole** above being dark owing to the presence of pigment. The whole egg is enclosed in an **albuminous envelope.** This is thin immediately after laying but it swells up considerably during the first few hours after exposure to water and binds the eggs together to form the familiar frog spawn.

SEGMENTATION OR CLEAVAGE

In the following study note how the development of Rana is influenced by the presence of yolk, comparing it with that of Amphioxus.

Examine prepared slides of **Fertilised Ova of Rana in various stages of Segmentation.**

Segmentation is **holoblastic.** Note the early **cleavage stages,** the first two being vertical as in Amphioxus, the third horizontal but nearer the animal pole, forming four **micromeres** above and four larger **megameres** below. The micromeres divide more rapidly than the megameres owing to the quantity of yolk present.

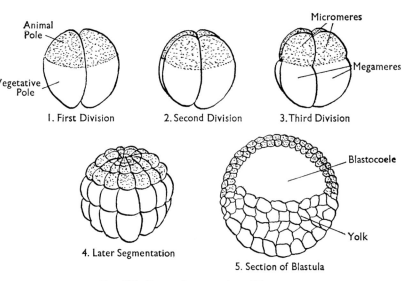

FIG. 271. **Rana. Segmentation of Zygote.**

BLASTULA

Examine a prepared slide of a **L.S. of the Blastula.**

Note the larger number and the smaller size of the upper (animal) cells and the mass of yolk cells occupying the lower part. The **blastocœle** is therefore in the upper hemisphere. Unlike the blastula of Amphiozus, the wall is two or three cells thick.

GASTRULATION

Examine prepared slides **showing the Formation of the Gastrula.**

The pigmented animal cells grow over the vegetative cells (**epibolic gastrulation**) and the blastopore is closed by the cellular **yolk-plug.** Note that the **ectoderm** and **endoderm** are many-layered at an early stage. A slit develops from the upper lip of the blastopore and passes inwards into the blastocœle which ultimately becomes obliterated and thus a new cavity, the **archenteron,** is formed. Look for the beginning of the development of the **mesoderm** which arises by the formation of cells from the lips of the blastopore which grow inwards between the ectoderm and endoderm. This begins before the completion of gastrulation. The dorsal lip of the blastopore is a primary organiser causing the cells of the gastrula to assume definite roles at an early stage.

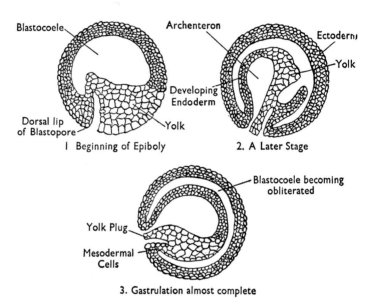

1 Beginning of Epiboly

2. A Later Stage

3. Gastrulation almost complete

FIG. 272. **Rana. Stages in Gastrulation.**

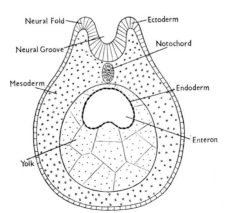

FIG. 273. **Rana. Formation of neural tube.**

FORMATION OF MESODERM, NEURAL TUBE AND NOTOCHORD

Examine prepared slides of T.S. *and* L.S. **of Larvæ before Hatching.**
Note the **ectoderm, endoderm** and **mesoderm** and the formation of
the **neural plate, neural folds, neural grooves** and finally the hollow
neural tube as in Amphioxus. (See Fig. 273, p. 378.)

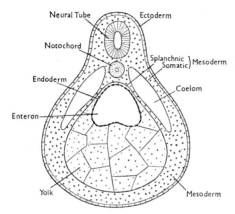

FIG. 274. **Rana. Neural tube complete. Formation of coelom.**

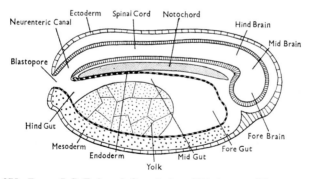

FIG. 275. **Rana. L.S. Embryo before closing of Blastopore. (Diagrammatic.)**

In longitudinal section also note the early differentiation of the
neural tube into the **spinal cord** and the three **cerebral vesicles** at its
anterior end which become the **fore, mid** and **hind brain**. A **neuropore**
and **neurenteric canal** will be found in the earlier stages. The solid
notochord develops from chorda cells originating from the dorsal
lip of the blastopore and finally forming a narrow strip in the roof

of the archenteron in a mid-dorsal position. The **fore, mid and hind gut** (enteron) will also be seen.

Continue your examination of **T.S. and L.S. Larvæ.**

The mesoderm separates into two layers, the outer of which, the **somatic mesoderm,** joins with the ectoderm to form the **somatopleure** while the inner, the **splanchnic mesoderm** joins with the endoderm to form the **splanchnopleure.** The cavity between is the **splanchnocœl** or **cœlom** which, owing to its method of formation, is said to be **schizocœlic.**

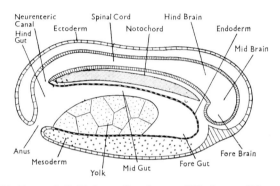

Fig. 276. **Rana. L.S. Embryo after closure of Blastpore.** (Diagrammatic.)

The mesoderm on each side of the notochord develops into paired metamerically segmented **mesodermal somites** which give rise to the muscles (myotome), the skeleton (sclerotome) and the dermis (dermatome) of the skin. The rest of the mesoderm remains unsegmented and is known as the **lateral plate mesoderm.** It extends ventrally between the ectoderm and the yolk cells.

In later longitudinal sections, note an invagination of the ectoderm, the **stomodæum** (the future mouth and buccal cavity), at the anterior end and another, the **proctodæum** (the future anus), at the posterior end.

LARVAL DEVELOPMENT

Examine freshly laid **Frog's Spawn** *and study the* **Development of the Embryo** (*kept in a suitable aquarium*) *from day to day with a hand lens.*

Note the **elongation of the embryo** (this occurs after gastrulation is completed) and the **hatching of the larva** (tadpole) about two weeks after fertilisation by a wriggling movement which frees it from the albuminous envelope. The larva attaches itself to a weed by a **cement gland** and continues to feed on yolk. Note the two pairs of branched **external gills,** followed soon by a third pair. By the end of about a week after hatching the stomodæum has opened into the pharynx and the **mouth** has developed. The tadpole, now shaped rather like a globe, swims about actively by means of a well-developed tail. Later (about two weeks after hatching), note the disappearance of the external gills and the development of the **operculum,** a fold of skin over the gill-slits which completely fuses with the body wall except on the left side, where an aperture, the **branchial aperture** or **spout,** is left. The development of internal gills is now complete.

The **hind-limb buds** appear at the base of the tail some weeks later (at about six or seven weeks), and it is not until the **hind-limbs** are almost fully developed (about two weeks later) that the **fore-limbs** are visible, their development having been hidden by the operculum.

During this period of development, the larva which has fed actively on a vegetable diet gradually changes to a carnivorous diet and grows considerably in size.

METAMORPHOSIS

Continue your examination of the **Larval Development.**

When the limbs are fairly well developed (about ten weeks after hatching), note that the larvæ frequently rise to the surface of the water for air; they now begin to respire by lungs. Feeding ceases and the skin is shed, the mouth and eyes enlarge and the shape of the body changes and becomes more frog-like. After a about fortnight when these changes are complete, the miniature frog with a short tail leaves the water. Finally, the tail is absorbed.

Feeding having been resumed (and this on an animal diet, chiefly insects), growth continues until, after about three years, the frog reaches its maximum size and is fully mature.

GALLUS
(The Chick)

THE OVUM

The **ovum** is large and is **telolecithal.** The orange-coloured **yolk** is enclosed in a **vitelline membrane** and is suspended in **dense albumen** surrounded by **clear albumen** by means of the twisted **chalaza** which extends to the membrane at each end of the egg. The whole is enclosed in two **shell membranes,** separating at the rounded end of the outer **shell** and enclosing the **air-sac.** On the upper side of the yolk is a clear protoplasmic area, the **germinal disc** (or **blastodisc**) in the centre of which is the nucleus or **germinal vesicle.**

INCUBATION

After laying, development will continue only if the egg is incubated at a suitable temperature (about 40°C.), but the rate of development (and thus the number of somites) is effected by the temperature of incubation, and this may vary slightly.

The age of the embryo after laying may be *roughly* estimated by counting the number of mesodermal somites present (see (iii) below). *The following table is therefore approximate:—*

Number of Somites. (Pairs)	Hours of Incubation.
2	20
6	24
9	28
12	33
16	38
27	48
36	72

TWO TO SIX PAIRS OF SOMITES

(First Day of Incubation)

The influence of yolk will be still more apparent in the study of the development of Gallus.

(i) *Examine prepared slides of* **the Blastoderm of the Chick showing the Segmentation of the Zygote in various early stages,** if available.

Segmentation takes place only in the germinal disc, and is therefore said to be **meroblastic.**

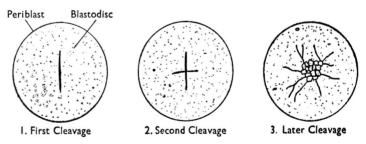

1. First Cleavage   2. Second Cleavage   3. Later Cleavage

FIG. 277. **Gallus. Cleavage of Zygote.**

Owing to the large amount of yolk, the division of the nucleus is not accompanied by complete division of the germinal disc. The first cleavage is shown by the appearance of a narrow groove or furrow and the second by a furrow at right angles to the first, neither of which extend right across the germinal disc. Cleavage continues in an irregular manner in the centre of the **blastodisc,** the outer part being referred to as the **periblast.**

(ii) *Examine a* **T.S. through the Blastoderm at a Late Stage of Segmentation.**

Soon a cavity appears below the central cells, separating the **blastoderm** (as the blastodisc is now called) from the yolk. This is the **subgerminal cavity** and it corresponds with the blastocœle of the simple types already studied, so we may consider this stage of development as the blastula stage.

A little later, the upper layer of cells divides into two layers, the **ectoderm** above and the **endoderm** below.

(iii) *Examine a slide of the* **Blastoderm with 2 pairs of somites (about 20 hours).**

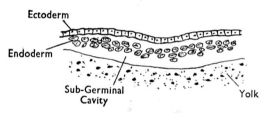

FIG. 278. **Gallus. T.S. Early Blastoderm.**

It will be observed that there is a comparatively clear region in the centre of the blastoderm; this is known as the **area pellucida** and it lies over the subgerminal cavity. Around it is the **area opaca,** darker in appearance and lying over the yolk. In the former will be seen a row of cells extending from about a quarter of the distance from the future anterior end almost to the posterior end in the mid-line at right angles to the long axis of the egg. This is the **primitive streak:** it swells at its anterior end to form the **primitive knot.** In the streak is a shallow groove, the **primitive groove,** which widens into the **primitive pit** in the knot. In front of the primitive streak the **notochord** will be seen.

A further development takes place from the anterior end of the primitive knot. This is the **notochordal process** from which the notochord develops. Observe also the block-like **mesodermal somites** on each side of the primitive streak. At this stage there are about two pairs.

It will be seen that the **mesoderm** extends along each side of the primitive streak and forwards on each side of the notochordal process and beyond it when the mesoderm of the two sides join. Anterior to this is a clearer region devoid of mesoderm and called the **pro-amnion.**

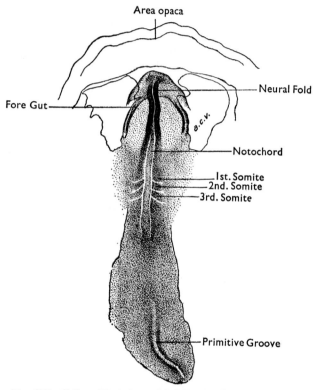

FIG. 279. Gallus. Blastoderm of 3-4 Somites (about 23 hours).
(*From Lillie's "Development of the Chick."*)

(iv) *Examine a slide showing* **Sections of the Blastoderm through the Primitive Streak with two pairs of Somites (about 20 hours).**

The cells have divided off from the primitive streak and they spread out between the ectoderm and endoderm and thus form the **mesoderm.**

Note the formation of the **neural** or **medullary plate** (in the region anterior to the notochordal process) and the **neural groove** formed by the **neural folds** which join to form the **neural tube.**

(v) *Examine a* **Blastoderm with six pairs of Somites (about 24 hours).**

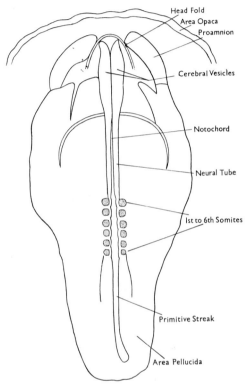

FIG. 280. **Gallus Embryo with 6 Somites (24 hours).**

Note that the primitive streak is shorter and identify the **head fold,** a curved structure between the pro-amniotic area and the **cerebral vesicles,** elongated swellings at the anterior end of the neural tube. Observe again and count the number of **mesodermal somites.**

Note also the plexus of small **blood vessels** in the central portion of the area opaca (formed by the union of groups of cells called **blood islands** which soon coalesce to form the **area vasculosa;** it is mesodermal in origin. Small vessels may be seen in the area pellucida running from the area vasculosa and as the blastoderm gets older, these spread very considerably.

(vi) *Examine a* **T.S. through an embryo with six pairs of Somites (about 24 hours).**

Note the **ectoderm, endoderm and mesoderm;** also the **neural plate, neural groove** and **notochord.**

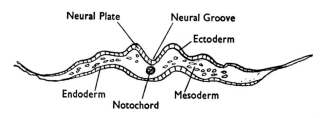

FIG. 281. **Gallus. T.S. Embryo with 6 Somites (24 hours).**

(vii) *Examine a* **Transverse Section of the Blastoderm of about seven Somites.**

The **mesoderm** has split into an upper **somatic layer** and a lower **splanchnic layer** forming a schizocœlic **cœlom** between them. The somatic layer with the ectoderm forms the **somatopleure** and the splanchnic layer with the endoderm forms the **splanchnopleure.** The **fore-gut** is formed as a pocket beneath the notochord. The **neural tube,** easily recognised by its dorsal position and its hollow form with thick walls, and the **notochord,** a small circular structure ventral to the neural tube, will also be seen. The loosely packed mesodermal cells are called **mesenchyme.**

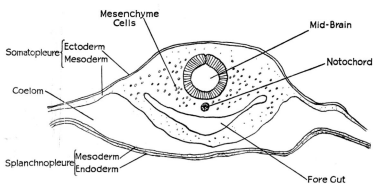

FIG. 282. **Gallus. T.S. through Mid-brain Region of Embryo with 7 Somites (about 25 hours).**

ABOUT SIX TO TWENTY-SEVEN PAIRS OF SOMITES

(Second Day of Incubation)

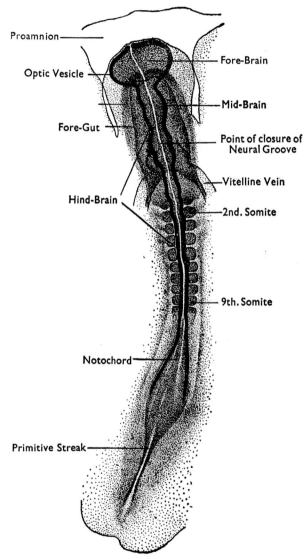

Fig. 283. **Gallus. Embryo with 10 Somites (Early Second Day).**
(*From Lillie's "Development of the Chick."*)

(i) *Examine* **Blastoderms of about ten to fifteen pairs of Somites (30 hours and 36 hours).**

The blastoderm is now much elongated.

The three **cerebral vesicles** are now clearly visible. Of these the **fore-brain vesicle (prosencephalon)** is the largest, swelling out on each side to form the **primary optic vesicle.** The other vesicles are

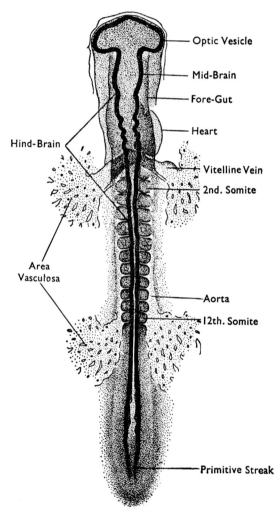

Fig. 284. **Gallus. Embryo with 12 Somites (about 33 hours).** (*From Lillie's "Development of the Chick."*)

the **mid-brain vesicle (mesencephalon)** and the **hind-brain vesicle (rhombencephalon)**. These are followed by the **spinal cord.** The **notochord** has matured from the notochordal process. The ventricle of the **heart** will now be visible bulging out to the right (it was formed from two endocardial tubes which develop at about twenty-five hours when there are seven somites and which fuse to form a single tube). From the posterior end of the heart two **vitelline veins** diverge while from the anterior end the two **aortic arches** arise, each at a later stage giving off a **vitelline artery** at a posterior trunk level. The blood vessels lead to and from the area vasculosa. *Count the* **mesodermal somites.**

In the later blastoderm, a small invagination, the **auditory** (or **otic) pit,** will be found level with the hindbrain. The head will be seen to have enlarged considerably and will be beginning to bend at the level of the mid-brain towards the right side.

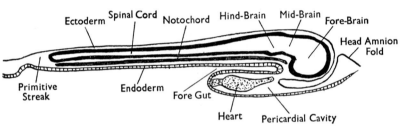

FIG. 285. **Gallus. L.S. Embryo of about 15 Somites (36 hours).**

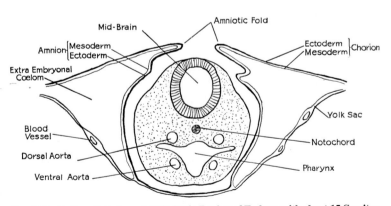

FIG. 286. **Gallus. T.S. through Mid-brain Region of Embryo with about 15 Somites (36 hours).**

(ii) *Examine a* **Longitudinal Section through the Anterior End of an Embryo of about fifteen somites.** See Fig. 285, p. 389.

Identify the **fore-brain** and, covering it, the **head amnion fold,** also the **mid-brain, hind-brain** and **spinal cord.** Find the **notochord** and below it, towards the anterior end, an invagination which is the **foregut,** while beneath this the **heart** will be seen in the **pericardial cavity.** Note the curved form of the head.

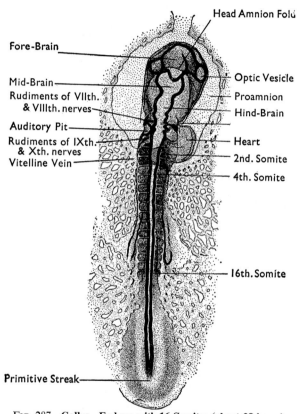

FIG. 287. Gallus. Embryo with 16 Somites (about 38 hours).
(*From Lillie's "Development of the Chick."*)

(iii) *Examine a* **Transverse Section through the mid-brain of an Embryo of about fifteen somites.** *See Fig. 286, p. 389.*

Note the two edges of the **amniotic fold** which have not yet met, and the **amnion** enclosing the **mid-brain** and, beneath it, the **notochord,** below which are the paired **dorsal aortæ** and **ventral aortæ**

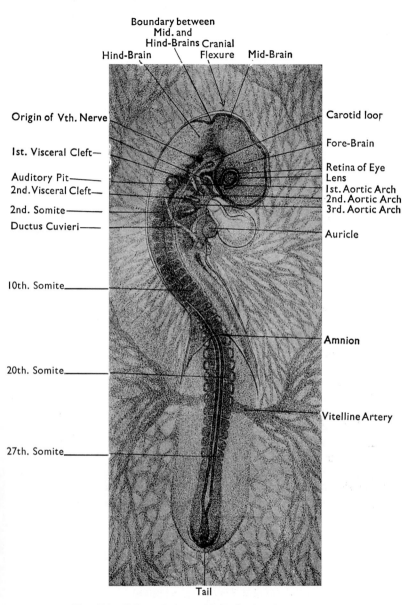

FIG. 288. **Gallus. Embryo with 27 Somites (about 48 hours).**
(*From Lillie's "Development of the Chick."*)

above and below the pharynx. The **chorion** composed of ectoderm externally and mesoderm internally is the outer layer of the head amnion fold while the **amnion,** already noted, composed of mesoderm externally and ectoderm internally, forms the inner layer. Between the two mesodermal layers is the **extra-embryonal cœlom.**

(iv) *Examine an* **Embryo with twenty-seven pairs of Somites (about 48 hours).** *See Fig.* 288, *p.* 391.

The structures already seen will be visible and it will be observed that the **amnion** stretches back to about the sixteenth somite. The **cranial flexure** is very marked and another is beginning to develop level with the posterior part of the hind-brain; this is the **cervical flexure.** At the posterior end of the embryo will be seen the **tail fold** behind which is the **tail amnion fold.** The primitive streak has almost disappeared. Examination of the fore-brain shows that it is now subdivided into an anterior **telencephalon** and a posterior **thalamencephalon** by a slight depression. Dorsal to the **heart,** three **visceral pouches** will be seen.

(v) *Examine* **Transverse Sections of the Embryo with about twenty-seven pairs of somites (about 48 hours).**

In **Sections through the Anterior End** one or more parts of the **brain** will be seen (remember that the head is curved downwards). The **optic cup** (or **secondary optic vesicle**), a secondary organism, formed by invagination of the outer part of the primary optic vesicle is now quite deeply invaginated and the **lens,** epidermal in origin, will be visible situated in the cavity. Identify the **notochord,** the **aortic arches,** the **amnion** and the **chorion,** and also what can be seen of the **yolk-sac,** a membrane which separates the embryo from the yolk and easily distinguished from the chorion by its possessing blood vessels.

In **Sections through the Trunk Region,** identify the **amnion,** the **chorion,** the **yolk-sac,** the **spinal cord,** the **notochord,** the **dorsal aorta** beneath the notochord, and the **gut** beneath the dorsal aorta. On either side of the spinal cord a **mesodermal somite** may be seen.

Anterior and **posterior liver diverticula** may be seen beneath the gut with part of the **vitelline veins** between them at the sides.

ABOUT TWENTY-SEVEN TO THIRTY-SIX PAIRS OF SOMITES

(Third Day of Incubation)

(i) *Examine an* **Embryo of about thirty pairs of Somites (about 60 hours).**

The **cranial flexure** will be seen as before and the **cervical flexure** is more marked. Note the general enlargement of the **embryo,** the prominent **ventricle of the heart** with the **bulbus cordis** running forward from it and the **sinus venosus** alongside it leading anteriorly into the **atrium.** The **dorsal aorta** curves round alongside the somites. The **head amnion fold** is now behind the level of the vitelline arteries and, continuous with it, are the **lateral amnion folds** and the **tail amnion fold.** In front of the latter is part of another membrane, the **allantois.** It arises from the gut and is the third of the embryonal membranes.

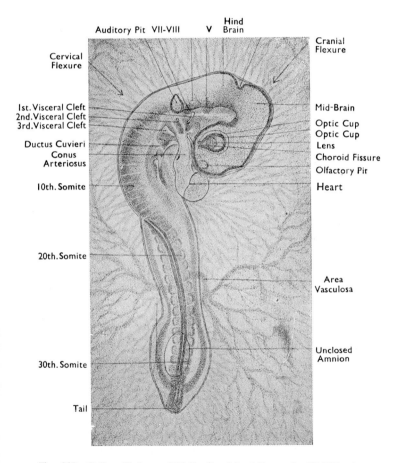

FIG. 289. **Gallus. Embryo of 31 Somites (about Two and a Half Days).**
(*From Lillie's "Development of the Chick."*)

(ii) *Examine a* **Transverse Section through the Hind Brain of an embryo with about thirty-four somites.**

At the anterior end of this section the **hind brain** will be seen and at the other end of the **spinal cord.** Between them are the **notochord** and the **dorsal aorta** and, on each side between the aorta and the hind brain, the **anterior cardinal veins.** The **auditory vesicles** (formed from the auditory pits seen earlier and which later in development become auditory sacs) are situated on either side of the hind brain at its posterior end. The enveloping membranes, **chorion, amnion** and **yolk-sac,** will also be seen.

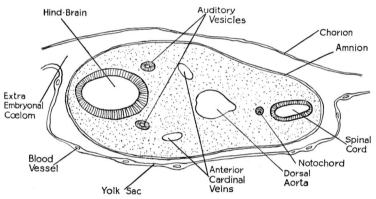

FIG. 290. **Gallus. T.S. through Hind-brain Region of Embryo with 34 Somites (about End of Third Day).**

(iii) *Examine a* **Transverse Section through the Pharyngeal Region of an embryo with about thirty-four somites.** *See Fig.* 291, *p.* 395.

Again the **hind brain** and **spinal cord** will be seen at the two extremites and the **notochord, dorsal aorta** and **anterior cardinal veins** as in the last section. In the centre of this section is a large cavity, the **pharynx,** with four **visceral pouches** at the edges. The elongated **laryngo-tracheal** groove (which gives rise to the larynx and trachea) extends from the pharynx posteriorly. The **amnion,** enclosing the **amniotic cavity,** will also be seen; likewise the **chorion** and **yolk-sac.**

(iv) *Examine a* **Transverse Section through the Heart Region of an embryo of about thirty-four somites.** *See Fig.* 292, *p.* 395.

The large **ventricle** occupies the centre of the section with the smaller **auricle** alongside it. Between the aorta and the ventricle is the **œsophagus** and, on either side of the aorta, the **posterior cardinal**

veins. Part of the **cœlom** will be visible. The **spinal cord** and **notochord** will also be seen at one end. At the opposite end lies the **forebrain**

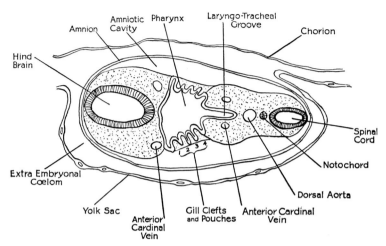

FIG. 291. Gallus T.S. through Pharyngeal Region of Embryo of about 34 Somites.

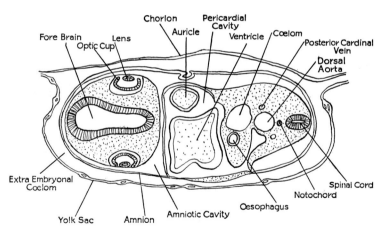

FIG. 292. Gallus T.S. through Heart Region of Embryo of about 34 Somites.

(owing to the cranial flexure) and on either side of it the **optic cup** (a secondary organiser developed from the optic vesicles seen previously) inside which the hollow **lens** of the eye will probably be seen. The **amnion** and **amniotic cavity** should also be noted.

(v) *Examine a* **Transverse Section through the Trunk Region of an Embryo with about thirty-four somites.**

Having identified the **amnion, chorion** and **yolk-sac,** note the large cavity in the centre of the embryo. This is the **hind-gut** and it bears an outgrowth on its ventral side, the **allantois.** The **spinal cord, notochord** and **aorta** will also be seen.

(vi) *Examine an* **Embryo of about thirty-six pairs of Somites (about 72 hours).**

Identify the structures and organs previously seen and note their greater development. Their are now four visceral pouches.

LATER DEVELOPMENT

Examine **Preserved Mounted Specimens of later Stages in Development.**

Fourth Day. The **urino-genital** system has developed and the other organs already seen have continued to develop. **Wing** and **leg buds** have appeared.

Fifth and Sixth Days. The cartilaginous skeleton develops.

Seventh to Ninth Days. The trunk has developed considerably, bringing it more into its final proportion with the head.

Tenth to Twentieth Days. Note the further development of the organs and systems and the diminution of the **yolk-sac** with the growth of the embryo in size. The **allantois** becomes a highly vascular sac enveloping the embryo, and both allantois and yolk sac develop stalks which, with associated blood vessels, form the **umbilical stalk.** This ultimately narrows and constricts at its base, thus sealing off the connection with the ventral wall of the embryo.

By the nineteenth or twentieth day, the head of the embryo is situated close to the air sac formed by the **shell membranes** and the yolk-sac has been absorbed.

Hatching occurs on the twenty-first day.

ORYCTOLAGUS

OVUM

Examine a **T.S. of Ovary of a Mammal containing Mature Ova.**

It will be recalled from histological studies that the **ovum** develops in a Graafian follicle in the stroma beneath the germinal epithelium of the ovary and that it is a spherical structure enclosed in the **zona pellucida** and is composed of **deutoplasm** or yolk (absent in the rabbit's ovum) in which is a large nucleus, the **germinal vesicle,** containing a nucleolus known as the **germinal spot.**

LATE FŒTUS

Examine a preserved mounted specimen of a **Late Fœtus attached to the Uterus.**

Note the vascular **allantoic placenta,** the organ of fœtal nutrition, respiration and excretion, to which the fœtus is attached by the **umbilical cord** containing two **umbilical arteries** and one **umbilical vein.**

THE PREPARATION
AND USES OF REAGENTS

MICROSCOPICAL REAGENTS

HARDENING AND FIXING AGENTS

Alcohol (Ethyl), Absolute or 70%
Use: *Stomach, pancreas, salivary glands.*

Aectic Acid (1 per cent.)

Acetic acid (glacial)	1 ml.
Distilled water	99 ml.

Acetic Alcohol (Carnoy's Fluid)

Acetic acid (glacial)	33 ml.
Alcohol (absolute)	99 ml.

Use: *Animal tissues generally. Good for mitosis.*

Bouin's Fluid (Picro-Formol)

Picric acid (sat. sol. aq.)	75 ml.
Formaldehyde	25 ml.
Acetic acid (glacial)	5 ml.

Use: *Animal tissues generally.*

Corrosive Acetic

Mercuric chloride (sat. sol. aq.) . . .	95 ml.
Acetic acid (glacial)	5 ml.

Use: *Animal tissues and small animals.*

Iodine Alcohol

Iodine	2·0 gm.
Alcohol 70%	98 ml.

Use: *Washing medium for tissues fixed in corrosive acetic.*

Mercuric Chloride (Corrosive)
See Corrosive Acetic.

Müller's Fluid

Potassium dichromate	2·5 gm.
Sodium sulphate	1·0 gm.
Distilled water	100 ml.

The yellow colour may be removed afterwards by soaking the tissue in 1% chloral hydrate.

Use: *Brain muscle, spleen, liver, kidney.*

Osmium Tetroxide ("Osmic Acid")

Osmium tetroxide	0·25 gm.
Distilled water	100 ml.

Use: *Protozoa.* Purchase ready prepared.

Picric Acid

Picric acid	sat. sol. aq.

Use: *Chitinous tissues.*

Picro-formol

See Bouin's Fluid.

Potassium Dichromate

Potassium dichromate	2·0 gm.
Distilled water	98 ml.

Use: *As for Müller's fluid. Also for brain and spinal cord.* It is generally quite as effective as Müller's fluid.

Ranvier's Alcohol

Alcohol (90%)	35 ml.
Distilled water	70 ml.

Use: *As for absolute alcohol, but it is a milder fixative.*

DECALCIFYING SOLUTIONS

Hydrochloric Acid

Hydrochloric acid (conc.).	10 ml.
Distilled water	90 ml.

Nitric Acid

For Large Bones:

Nitric acid (conc.)	10 ml.
Distilled water	90 ml.

For Young Bones:

Nitric acid (conc.)	1·0 ml.
Distilled water	99 ml.

MICROSCOPICAL STAINS

Many of these are best bought ready prepared, particularly in small laboratories. They can be purchased in solid form and in solution.

Acid Fuchsin

Acid fuchsin	0·5 gm.
Distilled water	100 ml.

Use: *Blood, connective tissue.*

Borax-Carmine (Grenacher's)

Borax	4 gm.
Distilled water	100 ml.
Add Carmine	3 gm.

Apply gentle heat until all dissolved.

Add Alcohol, 70%	100 ml.

Allow to stand for 2 or 3 days. Filter.
Use: *Animal tissues. A good general stain.*

Eosin (Aqueous)

Eosin Y	1·0 gm.
Distilled water	100 ml.

Use: *Animal tissues in general.*

Eosin (Alcoholic)

Eosin Y	1·0 gm.
Alcohol (70%)	100 ml.

Use: *Animal tissues in general.*

Hæmalum (Mayer)

Hæmatoxylin	0·25 gm.
Distilled water	250 ml.

When dissolved add

Sodium iodate	0·05 gm.
Alum	12·5 gm.

When dissolved, add

Chloral hydrate	12·5 gm.
Citric acid	0·25 gm.

Use: *A good general stain for animal tissues.*

Hæmatoxylin (Delafield)

Hæmatoxylin .	4	gm.
Absolute alcohol	25	ml.
Add Ammonium alum, sat. sol. aq. .	400	ml.

Leave exposed to light for three or four days. Filter.

Add Glycerine	100	ml.
Methyl alcohol	100	ml.

Allow to stand for 8 weeks. Filter.
Use: *A good general stain for animal tissues.*

Hæmatoxylin (Ehrlich)

Hæmatoxylin .	2	gm.
Absolute alcohol	100	ml.
Add Distilled water	100	ml.
Glycerine	100	ml.
Glacial acetic acid .	100	ml.
Alum .	in excess	

Leave exposed to light for six to eight weeks until it is dark red. Filter.
Use: *Animal tissues generally and Monocystis.*

Leishman's (Romanowsky) Stain

Best purchased ready prepared.

Methylene blue	1	gm.
Water	100	ml.
Add Sodium carbonate .	0·5	gm.
Water	100	ml.

Heat to 65°C. for 12 hours. Allow to stand for 10–12 days.

Add Eosin	0·2	gm.
Water	200	ml.

Allow to stand for 6–12 hours. Filter. Wash the ppt. until the washings are clear. Dry.

Leishman's stain (as above)	0·15	gm.
Methyl alcohol (acetone free) .	100	ml.

Use: *Blood.*

Methylene Blue

Methylene blue, sat. sol. in absolute alcohol	30	ml.
Add Potassium hydroxide	0·01	gm.
Distilled water	100	ml.

Use: *Blood.*

Methyl Orange.
0·04g → 100ml of 20% Alcohol.
pH = 2·8 – 4·6

Methyl violet

Methyl violet	1·0 gm.
Alcohol (70%)	100 ml.

Use: *Blood.*

Osmium Tetroxide ("Osmic Acid")

Purchase 2% solution.
Use: *Fats (Temporary stain).*

Picro-Carmine (Ranvier)

Pitric acid	sat. sol. aq.

Add Carmine (sat. sol. in ammonium hydroxide) *until saturated.*
Evaporate to $\frac{1}{5}$ of original volume on a water-bath. Filter. Evaporate to dryness.

Picrocarmine (as above)	1·0 gm.
Distilled water	100 ml.

Use: *Animal tissues generally.*

Sudan III

Sudan III	sat. sol in 70% alcohol

Use: *Fats (red).*

Sudan IV (Scharlach Red)

Sudan IV	5·0 gm.
Alcohol (70%)	100 ml.

Use: *Fats (red)*

Van Gieson

Acid fuchsin	0·25 gm.
Distilled water	25 ml.
Picric acid (sat. sol. aq.)	500 ml.

Use: *Connective tissue, muscle, epithelia.*

DIFFERENTIATING FLUID

Acid Alcohol

Alcohol, 70%	100 ml.
Hydrochloric acid conc.	2·5 ml.

DEHYDRATING AGENTS

(i) Ethyl alcohol

30% Alcohol.	*To prepare dilute solutions from* **Industrial**
50% Alcohol.	**Methylated Spirits** (=**95% ethyl alcohol),**
70% Alcohol.	*stand the spirit over anhydrous copper*
90% Alcohol.	*sulphate for three or four days, replacing*
Absolute Alcohol.	*the anhydrous copper sulphate as necessary*
	until it is no longer turned blue. Then
	prepare as follows:—

30% Alcohol

Alcohol, 95%	30 ml.
Distilled water	65 ml.

50% Alcohol

Alcohol, 95%	50 ml.
Distilled water	45 ml.

70% Alcohol

Alcohol, 95%	70 ml.
Distilled water	25 ml.

90% Alcohol

Alcohol, 95%	90 ml.
Distilled water	5 ml.

(ii) Cellosolve

= Ethylene glycol monoethyl ether. Purchase ready prepared.

CLEARING AGENTS

(i) Berlese's Medium

Chloral hydrate	16 gm.
Gum arabic	15 gm.
Water	20 ml.
Glucose syrup	10 gm.
Acetic acid (glacial)	10 gm.

Use: *Larvæ, chitinous structures and small insects.*

(ii) Cedar Wood Oil

Purchase *natural* oil of cedar wood.

Use: *Best clearing agent for animal tissues.*

(iii) **Clove Oil**
Purchase as such.
Use: *Animal tissues generally.*

(iv) **Xylene**
Use: *Animal tissues* but it is *not* recommended because it tends to cause shrinkage.

MOUNTING MEDIA

Permanent Mounts
(i) **Canada Balsam**
The best solvent is xylene.
Purchase ready prepared.

(ii) **Euparal** and **Euparal Vert**
Purchase ready prepared.

(iii) **Glycerine Jelly**
Best bought ready prepared .

Gelatin	25 gm.
Distilled water	150 ml.

Soak the gelatine in the water for a few hours. Then pour off the water, melt the gelatine and add:—

Glycerine	175 gm.

Add:

Phenol (5 % aq.)	a few drops
or, better, Thymol	a few crystals

Stir well while hot then heat gently and filter through glass wool in a hot-water funnel.

Temporary Mounts
(i) **Dilute Glycerine**

Glycerine	50 ml.
Distilled water	50 ml.

Add a few crystals of thymol.

(ii) **Physiological Saline***
(*a*) **For Invertebrate Tissues** (and **Vertebrate Blood**)

Sodium chloride	0·6 gm.
Distilled water	100 ml.

(*b*) **For Amphibian Tissues** (except blood)

Sodium chloride	0·75 gm.
Distilled water	100 ml.

* Also called Normal Salt Solution. This does *not* refer to its chemical normality.

(*c*) **For Mammalian Tissues (except Blood)**

Sodium chloride	0·9 gm.
Distilled water	100 ml.

Use: *Fresh animal tissues.*

(iii) **Ringer's Solution**

Sodium chloride	0·8 gm.
Calcium chloride	0·02 gm.
Potassium chloride	0·02 gm.
Sodium bicarbonate	0·02 gm.
Water	100 ml.

Use: *Amphibian tissues.*

(iv) **Locke's Solution**

Sodium chloride	0·9 gm.
Potassium chloride	0·042 gm.
Calcium chloride	0·048 gm.
Sodium bicarbonate	0·02 mg.
Glucose	0·2 gm.
Water	100 ml.

Use: *Mammalian tissues.*

EMBEDDING WAX

Paraffin Wax

Mix Paraffin wax, melting point 50°C. . .	2 parts
Paraffin wax, melting point 36°C. . .	1 part

Melt and stir.

M.P. of Mixture = 48°C.

RINGING CEMENT

Purchase Ringing Cement or Gold Size, Black Varnish or Black Enamel.

BIOCHEMICAL AND GENERAL REAGENTS

Acetic Acid, Dilute

Glacial acetic acid	23 ml.
Water	77 ml.

Adrenalin

Purchase 1 or 5 gm. ampoules.

Albumen
Purchase dried egg or blood albumen.

Ammonium Hydroxide

Ammonium hydroxide S.G. 0·88 . . .	25 ml.
Distilled water	75 ml.

Ammonium Molybdate

Dissolve Ammonium molybdate . . . . 15 gm.
  in ammonium hydroxide prepared as follows:—

Ammonium hydroxide, S.G. 0·88 . . .	10 ml.
Distilled water	5 ml.

Add Distilled water . . . . . . 120 ml.
Shake and add this to dilute nitric acid prepared as follows:—

Nitric acid (conc.)	18 ml.
Distilled water	95 ml.

Ammonium Sulphate

Ammonium sulphate . .	Sat. sol. in distilled water

Antimony Chloride (for Vitamin A Test)

Antimony trichloride . .	Sat. sol. in chloroform

Antimony = Sb
∴ 2Sb + 2HCl = 2 SbCl + H₂

Antiseptics
  Carbolic Acid

Phenol (cryst.)	5·0 gm.
Distilled water	95 ml.

  Corrosive Sublimate

Mercuric chloride	1·0 gm.
Distilled water	99 ml.

  Dettol

Dettol	15 ml.
Water	135 ml.

Barium Chloride

Barium chloride	3·0 gm.
Distilled water	97 ml.

Barfoed's Reagent

Copper acetate	4·5 gm.
Distilled water	100 ml.
Acetic acid (50%)	1·0 ml.

Calcium Chloride

Calcium chloride	2·0 gm.
Distilled water	98 ml.

Chloral Hydrate

Chloral hydrate	1·0 gm.
Distilled water	99 ml.

Chlorine Water

Prepare chlorine *by heating manganese dioxide with concentrated hydrochloric acid or by treating potassium permanganate with concentrated hydrochloric acid. Pass the chlorine into water until it is saturated. Store in an amber bottle.*

Chromic Acid

Chromium trioxide	10·0 gm.
Distilled water	90 ml.

or

Potassium dichromate	10·0 gm.
Distilled water	100 ml.
Sulphuric acid (conc.)	10 ml.

Use: *Cleaning glass apparatus.*

Copper Sulphate

Copper sulphate (cryst.)	10·0 gm.
Distilled water	90 ml.

Copper Sulphate 1 % (for Biuret Test)

Copper sulphate	1·0 gm.
Distilled water	99 ml.

2:6 Dichlorophenol-indophenol (for Vitamin C test)

Dissolve 0·2 gm. in 100 c.c. of distilled water, *allow to stand 24 hours. Filter.*
Or purchase tablets from B.D.H.

Dissolve one tablet in	
Distilled water	10 ml.

This can be used immediately.

Diphenylamine

Diphenylamine	20 gm.
Alcohol (95%)	80 ml.

Fehling's Solution
Solution A.
Copper sulphate	34·64 gm.
Water	500 ml.

Solution B.
Sodium potassium tartrate . . .	176·0 gm.
Potassium hydroxide	77·0 gm.
Water	500 ml.

Keep in separate bottles.
For use mix equal quantities of A and B.

Ferric Chloride
Ferric chloride	10·0 gm.
Distilled water	90 ml.

Formaldehyde 4%
Formalin (40% formaldehyde)	10 ml.
Water	90 ml.

(=10% Formalin).

Gastric Juice, Artificial
Pepsin	0·32 gm.
Distilled water	99·6 ml.
Hydrochloric acid	0·2 ml.

or use liquor pepticus, purchased ready prepared.

Guaiacum (for Blood Test)
Guaiacum	2·0 gm.
Alcohol (90%)	98 ml.

Hydrochloric Acid, Dilute
Hydrochloric acid (conc.).	26 ml.
Distilled water	73 ml.

Iodine
Potassium iodide	6·0 gm.
Water	100 ml.
Add Iodine	2·0 gm.

Lead Acetate
Lead acetate	10·0 gm.
Distilled water	90 ml.

Lime Water
Make a saturated solution in water. Shake well. Filter.

Lock's Solution
See Mounting Media, p. 405.

Millon's Reagent
Best purchased.
  Dissolve
  Mercury . . . . 50·0 gm.
  in Nitric acid (conc.) . . . . . . 35 ml.
  Add Distilled water . . . . . . 35 ml.

α-Naphthol
  α-Naphthol . . . . 2·0 gm.
  Alcohol (95%) . . . . . . 100 ml.

Osmium Tetroxide ("Osmic Acid")
  1% or 2% solution. Buy ready prepared.

Osmosis Solutions
  Sodium chloride
  Sodium chloride . . . . . . 5·0 gm.
  Distilled water . . . . . . 95 ml.
  Sucrose
  Sucrose . . . . 15·0 gm.
  Distilled water . . . . . . 100 ml.

Pancreatic Juice
  Pancreatin . . . . 1·8 gm.
  Distilled water . . . . . . 100 ml.
  or purchase liquor pancreatini.

Peptone
  Purchase commercial peptone.

Pituitrin
  Purchase ampoules.

Potassium Dichromate
  Potassium dichromate . . . . . 5·0 gm.
  Distilled water . . . . . . 95 ml.

Potassium Ferrocyanide

    Potassium ferrocyanide . . . . . 10·0 gm.
    Distilled water . . . . . . 90 ml.

Potassium Hydroxide

    Potassium hydroxide (pellets) . . . . 10·0 gm.
    Distilled water . . . . . . 90 ml.

Potassium Oxalate

    (To prevent blood from clotting.)
    Potassium oxalate . . . . . . 1·0 gm.
    Distilled water . . . . . . 99 ml.

Potassium Pyrogallate

    To Pyrogallol . . . . 5·0 gm.
    Distilled water . . . . . . 95 ml.
    Add Potassium hydroxide . . . . . 25·0 gm.
    Distilled water . . . . . . 15 ml.
    Use immediately.

Preserving Media. *See* Appendix II.

Rennin

Purchase commercial rennin ("rennet").

Ringer's Solution

See Mounting Media, p. 405.

Saline, Physiological

See Mounting Media, p. 404.

Silver Nitrate

    Silver nitrate . . . . 1·0 gm.
    Distilled water . . . . . . 99 ml.
    Keep in an amber bottle.

Sodium Citrate

    (To prevent blood from clotting.)
    Sodium citrate . . . . . . 10·0 gm.
    Distilled water . . . . . . 90 ml.

Sodium Hydroxide
  (i) **For General Use**
      Sodium hydroxide (pellets) . . . . 8·0 gm.
      Distilled water . . . . . . 92 ml.
  (ii) **For Insect Mouth Parts**
      Sodium hydroxide (pellets) . . . . 2·0 gm.
      Distilled water . . . . . . 98 ml.

Sodium Nitroprusside
      Sodium nitroprusside . . . . about 2·0 gm.
      Distilled water . . . . . . 100 ml.
  The solution must be freshly prepared for use.

Sulphuric Acid, Dilute
  Add Sulphuric acid (conc.) . . . . . 5 ml.
gradually to
      Distilled water . . . . . . 95 ml.
in a vessel standing in cold water.

Thymol
      Thymol . . . . 3·0 gm.
      Alcohol (90%) . . . . . . 97 ml.

Thyroid Extract
  Purchase 60 mg. tablets.

BIOLOGICAL METHODS

AMPHIOXUS—TO PREPARE FOR DISSECTION

Soak in the following solution:—

Nitric acid	20 ml.
Alcohol 70%	100 ml.

AQUARIA

(1) Use thoroughly washed flat-sided vessels, *e.g.*, good-class battery jars or museum jars and tanks.

(2) Place, if possible, in a north light.

(3) Cover the bottom with fine washed gravel and sand.

(4) Insert a few large stones to provide shaded shelter for the inmates. Some of them should come above the water level if Amphibia are included.

(5) Water-weed, *e.g.*, *Elodea*, *Anacharis*, *Vallisneria*—should be put into the water to provide oxygen, and water snails to keep the sides of the vessels clean and free from Algæ.

(6) Keep water-beetles and dragon-fly larvæ in vessels by themselves.

(7) Keep aquaria which contain Amphibia or water-beetles covered with perforated zinc.

(8) Do not overcrowd.

(9) Feed the inmates about three times a week with small earthworms or chopped meat. Ant "eggs" (pupæ) can be used occasionally. Do not overfeed and on no account allow food to remain in the water and putrefy.

(10) Immediately isolate any fish infected with *Saprolegnia* ("fish-fungus") which grows over the gills. Weak salt solution will sometimes cure this disease.

(11) Keep the water properly aerated.

(12) Do not change the water more often than is necessary.

For Micro-aquaria *see* below under M.

Aerating Apparatus

A simple and effective form of aerating apparatus for aquaria can be constructed as follows:—

Fit a 500 ml. 3-necked Woulffe's bottle with rubber stoppers. Through one of the outside necks fix a right-angle tube leading to the bottom of the bottle. This is the *water outlet tube* and should be fitted with rubber tubing leading to the sink. Through the other outside neck fit a short right-angle tube. This is the *air outlet* or *aerating tube* and should be connected by rubber tubing to a glass jet in the aquarium. The *water and air inlet-tube* is constructed as shown in the diagram, and is fitted in the centre neck. It is, of course, connected to the water-tap. By careful adjustment of the jet in the water inlet-tube sufficient pressure can be maintained in the Woulffe's bottle to aerate several aquaria and to siphon the waste water from the apparatus into the sink.

Several aquaria can be aerated if T-tubes, to each of which is attached a jet, are fitted where required in the aerating tube. Screw clips should be placed on the rubber connections to give finer adjustment. The advantages of this apparatus are (1) its simplicity and cheapness; (2) its constant efficiency; (3) the non-entry of water with the air into the aquaria.

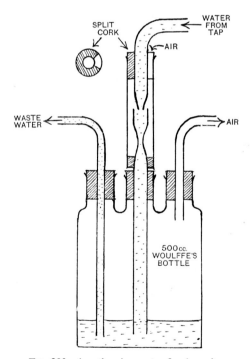

FIG. 293. **Aerating Apparatus for Aquaria.**

BLOOD—TO PREVENT CLOTTING

Add blood to $\frac{1}{10}$ of its volume of 10% sodium citrate or 1% potassium oxalate.

CLEANING OF GLASS APPARATUS

Wash with commercial hydrochloric acid or aqua regia (HCl, 3 parts: HNO_3, 1 part, by volume).

Very dirty apparatus may be cleansed with chromic acid which renders it chemically clean. This is prepared as follows:—

Dissolve

Potassium dichromate	10 gm.
in Water	100 ml.
Sulphuric acid.	10 ml.

Grease is best removed by caustic soda solution.

In all cases wash thoroughly with water after cleansing as above. For cleaning of microscopical slides see below under M.

CULTURE METHODS

Amœba*

Place one or two boiled (killed) wheat seeds in 100 ml. of water. Innoculate with Amœba from a culture by means of a pipette. Replace every 2 months. Sub-culture occasionally if successful.

Daphnia and Cyclops (For Feeding Hydra)

Mix Horlick's malted milk, 0·1 gm., into a paste with boiling water and dilute with 5·0 ml. of cold water. Add weekly to a vessel containing about 500 ml. of water containing Daphnia and Cyclops.

Hydra

Put a few decaying leaves or a little pond debris into water. Add Hydra by means of a pipette. Add Daphnia and Cyclops.

Lumbricus

Place earthworms in soil in a large box and add a quantity of leaf mould. Do not let the soil get too dry.

Paramœcium*

As for Amœba. Alternatively chop up some hay, add boiling water

* See also Micro-Aquaria.

and thus prepare hay infusion. Innoculate with Paramœcia when cold. Subculture every six weeks or so.

ENZYMES

Enzymes can be purchased in powder form.

The *Lenham Enzyme Kit* from Lenham Educational Ltd referred to in Part IV—Biochemistry—can be purchased from the suppliers of biological material.

FORMALDEHYDE 40 PER CENT.—TO PRESERVE

Keep some lumps of calcium carbonate in the bottle to neutralise the formic acid to which the formaldehyde is liable to be oxidised.

FROG PLATE

This can be conveniently made of a rectangular piece of ebonite. wood or brass about 6 in. long and 2 in. wide. A circular hole 1 in. in diameter should be cut with its centre in the mid-line about $1\frac{1}{2}$ in. from one end. Holes about $\frac{3}{16}$ in. in diameter may be drilled round the sides with their centres about $\frac{3}{4}$ in. apart.

Fig. 294. **Frog Plate.**

The frog plate should be placed on the stage of the microscope with the hole over the hole in the stage and clamped in a retort-stand. A pithed frog can then be placed on the plate, the web of the foot slightly stretched over the hole and kept in position by thread tied round the digits to the small holes at the edge of the plate or to small pins if the frog plate is made of wood.

GRAPHIC RECORDS—TO PRESERVE

Graphic records made on smoked glass or paper may be preserved for permanent storage by immersion in the Varnish given below, afterwards allowing them to dry in the air.

Methylated spirit	.	.	.	.	75%
White hard varnish .	.	.	.	.	25%

Stir well.

HORMONE EXPERIMENTS

The Effect of Pituitrin on the Melanophores of the Frog's Skin. Two similar light-coloured frogs should be anæsthetised and pithed. Into one inject a subcutaneous injection of 0·2 ml. of "Pituitrin" with a hypodermic syringe. When the desired effect is obtained, *i.e.*, when the melanophores have expanded and the skin consequently turned dark (this usually takes a little over an hour), the two frogs can be killed and preserved in 10% formalin, side by side, in a museum jar, the uninjected specimen being kept for comparison.

The Effect of Thyroid on the Metamorphosis of the Frog. A batch of tadpoles should be isolated in a separate aquarium when they are about 10 weeks old (i.e., when they are about to start metamorphosis), and fed on thyroid at the rate of 180 mg. for each dozen tadpoles. The tablets should be crushed in a mortar and the powder added to the water in the aquarium. A comparison with the unfed tadpoles within the next 24 hours to three days will serve as a control. Unsatisfactory results are obtained if the thyroid feeding is performed at an earlier stage of development.

INJECTION OF BLOOD-VESSELS

Successful injection of blood-vessels requires considerable care, practice and skill.

It can be done with an ordinary injection syringe, the nozzle of which should be filled with a piece of rubber tubing into which a glass cannula or jet can be fitted. Jets of various sizes should be made, very fine ones being required for small animals.

The operation must be performed on a *freshly killed* animal, otherwise the blood must first be washed out of the vessels by means of a solvent such as 10 per cent. sodium citrate. It is important that a jet of suitable size should be used. The injection masses must be freshly prepared and used immediately. A gelatine or a plaster of Paris medium (see below) may be used. If the gelatine method is adopted, the medium must be kept hot and the animal immersed in warm physiological saline.

Having inserted the nozzle of the syringe, first making an incision in the blood-vessel if necessary, tie a ligature (not too tightly) round the vessel to keep the nozzle in place and to prevent leakage at this point. Avoid too much pressure on the syringe or bursting of the blood vessels may occur.

Just before the completion of the operation, the vessels should be ligatured and the pressure maintained for a few minutes. When it is completed the animal should be washed in cold water and then allowed to remain in cold water for an hour or two before placing in the preserving fluid.

Places of injection are as follows:—

Crayfish.	Ostium of heart.	First wash out blood-vessels with physiological saline.
Frog.	Arteries—ventricle.	Use red mass.
	Veins—sinus venosus.	Use blue mass.
Dogfish.	Afferent arteries—conus arteriosus.	Use blue mass.
	Efferent and visceral arteries—caudal artery.	Use red mass. Cut off the end of the tail to expose the caudal artery.
	Visceral vein—caudal vein.	Use blue mass. Insert into the vein in the cut end of the tail.
Rabbit.	Arteries—root of the aortic arch.	Use red mass.
	Veins—sinus venosus.	Use blue mass. Drain the blood from the vessels before injection.

Injection Masses
Gelatin Medium

Soak 5·0 gm. of gelatine in hot water. When melted add sufficient of the colouring matter (see below) to give a bright red or blue colour. Stir well. Use hot.

Plaster of Paris Medium

Rub some plaster of Paris with a little water in a mortar to a thin cream. Add the colouring matter. Stir well. Use immediately.

Colouring Matter

Red.

Rub 2·5 gm. powdered carmine in water. Add ammonium hydroxide gradually until the carmine is dissolved and the solution transparent.

Blue.

Dissolve 1·0 gm. of Prussian blue in 50 ml. of water.

JOINTS—TO MAKE AIRTIGHT

Apply the wax mixture given below. This is more effective and cleaner to use than vaseline.

Beeswax	30 gm.
Vaseline	40 gm.

Melt and add—

Resin, powdered	15 gm.

Stir.

KILLING OF ANIMALS

Lethal Chambers

Small Animals

For small animals such as earthworms, cockroaches, locusts and frogs, a chemical desiccator is a suitable piece of apparatus. See that the flange and edge of the lid are greased with vaseline. Place cotton-wool in the bottom of the vessel and add chloroform. Replace the zinc gauze.

Place the animals on the gauze and replace the lid. This should be placed eccentrically at first so that a small space is left to admit air. The air-chloroform mixture will bring about anæsthesia. When active movement ceases, push the lid on completely so as to exclude the entry of air.

This apparatus is particularly advantageous in the case of earthworms as it prevents actual contact with the chloroform, which tends to make them brittle.

Large Animals

For larger animals like rabbits and rats, a large metal box with two holes in the lid provided with corks is most suitable.

Place the animal in the lethal chamber, remove both corks and insert a wad of cotton-wool soaked in chloroform through one hole. Alternatively pass coal-gas into the chamber—anæsthesia is rapid and death painless.

The animal will then breath a mixture of air and chloroform vapour (or coal gas) and will be anæsthetised. When active movement ceases, add more chloroform (or pass more gas) and close both holes with the corks to exclude air. In the case of caged rats, the cage containing the animals can be placed in the lethal chamber.

Rabbits recover rapidly when returned to the air if they are only anæthetised by coal-gas and it is therefore essential that they should be left in the chamber for a sufficient length of time (see below).

Lethal Times

It is obviously essential that the animals should be quite dead. To ensure that this is so, leave the animals for the length of time stated below *after anæsthesia has been obtained:*

Earthworm	5 minutes
Frog	20 ,,
Rabbit and rat	30 ,,
Cockroach and locust . . .	15 ,,

Methods for Specific Animals

Anodonta

Transfer the animal from cold into boiling water until killed (about 1 minute): then put into cold water. Do not use chloroform.

Cockroach

Either use the lethal chamber (chloroform) for small animals or place the insects into boiling water in a large beaker for a few seconds.

Crayfish

As for Anodonta.

Earthworm

Chloroform.

Frog

Chloroform.

Hydra

1 per cent. acetic acid.

Locust

Chloroform. Use the lethal chamber for small aminals.

Protozoa

1 per cent. acetic acid.

Rabbit and Rat

Chloroform or coal gas. See above.

Snail

Completely fill a jar with water previously boiled (and cooled) to expel the air. Place the snails in the water, replace the stopper and leave for a few days. This is the only satisfactory method to ensure that the animals are distended when dead. Do not place too many animals in the jar at the same time or they may be distorted.

LABELS

(1) All bottles should be clearly and neatly labelled in a uniform style. Books of printed labels can be purchased, but they are seldom sufficient or complete. If they are used, supplementary labels should be written in similar style. When all are done by hand the style known as "bold old face" is perhaps the most suitable, *e.g.:*

ALCOHOL
50%

All labels should be written in indian ink and should be varnished when on the bottle. A suitable varnish may be prepared as follows:—

Celluloid . . . . . 6·5 gm. (approx.)
Acetone . . . . . 100 ml.

The celluloid may be obtained from old photographic films which should first be soaked in hot water until the negative has been removed.

(2) Specimens being temporarily stored in bottles may be labelled internally by writing the name in pencil on a small strip of paper, which should be put into the preserving fluid in the bottle. This method is not really suitable for specimens in permanent storage as the labels are not readily visible.

(3) For microscopical slides gummed labels may be purchased but the self-adhesive "Microtabs" obtainable from Messrs. T. Gerrard & Co. are most convenient to use.

MICRO-AQUARIA

Glass vessels such as medium-sized crystallising dishes or beakers prove satisfactory. Petri dishes are suitable for such cultures as

Amœba but they are rather shallow and easily spilled. All vessels must be absolutely clean.

Tap water may be used, but sometimes rain or pond water gives better results.

The pabulum should be added to the water, which should be inoculated with a culture of the organisms by means of a pipette. Pure cultures can only be obtained by sub-culturing.

The vessels should be covered with a piece of glass and suitably labelled.

Some organisms are more difficult to culture successfully than others and in any case some of the cultures may prove to be failures.

For methods for specific organisms see Culture Methods above.

MICROSCOPICAL SLIDES—TO CLEAN

It is best to keep a vessel of methylated spirit into which permanent preparations no longer required can be placed. Leave the slides to soak and keep the jar covered.

Periodically remove, boil in water, with detergent, rinse well and dry. A number of coverslips will also be retrievable.

MUSEUM SPECIMENS

(1) Preserved Specimens

Flat-sided museum jars are better than the cylindrical ones for most purposes as the latter sometimes cause distortion. Either glass or plastic jars can be used.

Specimens can be fixed to a glass or plastic background by thread or fine catgut through holes drilled in it. This makes it possible for both sides of the specimen to be examined. Ebonite, giving a black background, is sometimes an advantage and is unaffected by formaldehyde. Specimens such as brains can be fixed to the background by Durofix, but the specimen must be dry at the time of fixation. It can afterwards be preserved in formaldehyde without ill effect on the Durofix.

For preserving fluids, see 'Preservation and Storage of Biological Material' pp. 422 *seq.*

The lids of *glass jars* should be securely sealed as follows:—

  ### (i) When the specimen is preserved in alcohol

Best glue   .   .   .   .   8 parts
         'Dissolve' in water.
Add Glycerine.   .   .   .   .   .   .   1 part
         Apply hot to the edge of the cover. Press.

(ii) **When the specimen is preserved in formalin**
Shellac varnish. Dissolve shellac in alcohol.
Apply to the edge of the cover and the jar. Press.
The lids of *plastic jars* are sealed with a special cement which can be purchased.
In all cases something heavy should be placed on top until the adhesive has set to ensure a close seal.
Specimens of similar organs of different animals (such as the brains of dogfish, frog and rabbit) may be mounted in the same jar. Dorsal and ventral views should be shown. Dissected systems and animals with injected blood vessels are most helpful to students.

(2) Bones

Disarticulated small bones, such as those of the frog or rabbit or the entire skeleton of the frog, may be stuck to the bottom of a wooden box made a suitable size, painted a dull black inside and fitted with a glass lid. Black passe-partout can be used to fix the glass in position.
Articulated skeletons (which are very instructive) and larger bones should be kept under glass or in museum cases to protect them from dirt and injury.

(3) Models

Embryological models, though expensive, often enable students to understand the more difficult parts of embryology. Enlarged models of the eye, ear, heart and other organs are also useful.

NERVES—TO SHOW UP DURING DISSECTION

Cover with picric acid (sat. sol. aq.) then wash well with water.

PITHING A FROG

Lightly anæsthetise (but do not kill) a frog with ether.
Cut through the skin and the occipito-atlantal membrane between the skull and spinal column on the dorsal side. Insert a seeker into the cranial cavity and move it about until the brain is completely destroyed. This can also be done without anæsthesia.

PRESERVATION AND STORAGE OF BIOLOGICAL MATERIAL

Material should be stored in wide-neck stoppered vessels for

preference and the bottles should clearly be labelled with the name
of the specimen and preserving fluid.

General

70% Alcohol

The material may be softened after preservation by immersing it
in the following solution for 24 hours:—

Glycerol—50%
Alcohol—50%

or 4% Formaldehyde (= 10% Formalin).

or Place the organs in the following solution until they are completely decolourised.

Potassium acetate	8·5 gm.
Potassium nitrate	4·5 gm.
Formalin	80 ml.
Water	400 ml.

Then, after allowing the fluid to drain off, place in 80% alcohol
until the natural colour returns.

Then preserve permanently in the following solution:—

Potassium acetate	200 ml.
Glycerine	300 ml.
Distilled water	900 ml.

Amphioxus

70% Alcohol.

To prepare for dissection soak in—

Nitric acid	20 ml.
Alcohol, 70%	100 ml.

Brains

Müller's fluid—see Appendix I, p. 399.

To remove the yellow colour afterwards soak in 1% chloral hydrate
solution

or 70% Alcohol,
or 4% Formaldehyde.

Cartilaginous Skeletons

Glycerine	300 ml.	ᵢ 50
Alcohol, 70%	700 ml.	350

Cockroaches and Locusts

70% alcohol. *First open up to allow preserving fluid to enter.*

Dogfish

If the animal is not already injected with formalin *open peritoneal cavity before placing the animal in* 4% Formaldehyde.

SKELETONS—TO PREPARE

(1) Cut away as much muscle as possible.
(2) Boil gently in water until the muscle and tendon is easily removed.
(3) Soak in chloroform to remove the fat from the bones.
(4) Place in dilute hydrogen peroxide to bleach the bones.
(5) Wash thoroughly in water.
(6) Allow to dry.

Bones may be joined by Durofix or one of the plastic cements. Cover the two surfaces to be joined with a thin layer of the cement and allow it to dry. Then apply another thin layer of cement and press the bones together. Any surplus cement can be removed from the surface when dry by means of a knife.

Bony skeletons may be articulated by Durofix or plastic cement and fine rustless wire, small holes being drilled in the bones where necessary.

Cartilaginous skeletons should be articulated with very fine wire.

STERILISATION

Glass: Dry heat: 150°C. for 1 hour. Use a hot air oven.
Pipettes and small glass apparatus: Rinse in mercuric chloride (0·1% aq.) *or* boil in water for $\frac{1}{4}$ hour.
　Used culture-tubes and petri dishes: Boil in water for 1 hour.
　Rubber: Steam for $\frac{1}{2}$ hour.
　Cotton-wool: Dry heat: not above 180°C.
　Instruments: Boil in water containing soda (to prevent rusting) for $\frac{1}{4}$ hour.
　Minor instruments (*e.g.*, platinum needles): Flame: heat to redness.
　Nutrient media: Sterilise on three consecutive days by steam for $\frac{1}{4}$ to $\frac{1}{2}$ hour.

THERMO-REGULATOR

To set Reichart's Thermo-Regulator.

Connect the top tube, A, to the gas-supply and the side-tube, B, to the burner as shown.

Turn the screw S full out. Light the gas at the burner. When the required temperature is reached, turn in the screw S sufficiently to close the jet J with mercury. Sufficient gas to maintain an even temperature will enter through the by-pass and the gas flame will be lowered. By contraction of the mercury when the temperature

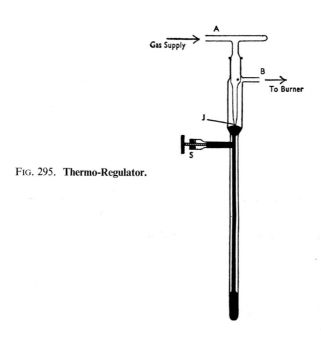

Fig. 295. **Thermo-Regulator.**

falls slightly (thus allowing more gas to enter) and by its expansion to the limit at which it was set (thus lowering the flame again) when the temperature rises, a constant temperature will be maintained for any length of time.

THERMOSTAT

A satisfactory thermostat for experiments on digestion can be made from an ordinary "steamer" as follows:—

Enlarge the existing holes in the bottom of the upper vessel sufficiently to take test-tubes. Fit a thermometer and a heat regulator through corks in two of the holes. A hole may be bored in the side of the lower vessel and a cork fitted into this, through which passes a piece of bent glass tubing to act as a water gauge. Stand the thermostat on a low tripod and place a small burner underneath.

Alternatively, a hot water oven fitted with a thermo-regulator and thermometer, may be used.

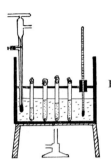

FIG. 296. **Thermostat.**

TRANSPARENCIES—TO SHOW THE ANIMAL SKELETON IN SITU

The skeletons of small vertebrate animals can be stained and the other tissues made transparent (thus showing the articulated skeleton *in situ*) by the following method:—

(1) Place in 90% alcohol for two days.

(2) Soak in 1% aqueous potassium hydroxide until the tissues are more or less transparent and the skeleton visible. *Do not leave the animal in this solution longer than is necessary to attain this condition.*

(3) Place in the following solution until the bones are stained deep red:—

Alizarin red	0·1 gm.
Potassium hydroxide (1% aqueous) . . .	1000 ml.

(4) Transfer to the following solution to clear:—

Potassium hydroxide	1·0 gm.
Distilled water	79 ml.
Glycerine	20 ml.

(5) When completely cleared, store in pure glycerine.

APPENDIX III

Equivalents

1 cm.	= 0·3937 in.	1 gm.	= 0·03527 oz. (Av.)
1 metre	= 39·37 in.		= 0·0022 lb.
1 litre	= 1·76 pints	1 kg.	= 2·205 lb. (Av.)
1 in.	= 2·54 cm.	1 oz. (Av.)	= 28·35 gm.
1 gallon	= 4·546 litres		

$1\mu = \dfrac{1}{25,400}$ in. or, more approximately, $\dfrac{1}{25,000}$

Conversion Table

	Multiply by
Cm. to in.	0·3937
Sq. cm. to sq. in.	0·155
Metres to feet	3·282
Ml. to cu. in.	0·06102
Litres to gallons	0·22
Grams to ounces	0·03527
μ (microns) to in.	0·00003937
Ins. to cm.	2·54
Sq. in. to sq. cm.	6·452
Feet to metres.	0·3048
Cu. in. to ml.	16·39
Gallons to litres	4·546
Ounces to grams.	28·35
Ins. to μ (microns)	25,400
°C to °F.	°C $\times \frac{9}{5} + 32$
°F to °C.	°F $- 32 \times \frac{5}{9}$

TREATMENT OF ACCIDENTS IN THE LABORATORY

Burns

(i) **By dry heat:** Treat with gentian violet jelly or tannic acid jelly or 1% picric acid. Bandage to exclude air.

(ii) **By acids:** Wash with plenty of water and then a saturated solution of sodium bicarbonate.

(iii) **By alkalis:** Wash with much water and then 1% acetic acid.

(iv) **Scalds:** Cover with gentian violet jelly; failing this, treat with sodium bicarbonate solution or boracic or zinc ointment. Bandage to exclude air.

Cuts

Wash thoroughly with T.C.P. or Dettol. Cover with elastoplast dressing. If large or deep cover with lint and bandage. Iodine should *not* be applied to an open wound.

Eye Accidents

(i) **Acid in the Eye.** Using an eye-bath, wash with weak (1%) sodium bicarbonate solution.

(ii) **Alkali in the Eye.** Wash with 1% boric acid.

Fainting (Syncope)

Lay the patient on his back; loosen the clothing. Administer sal volatile.

Poisons

(i) **Acids** or **alkalis.** Wash the mouth and drink much water. Drink a tumbler of lime-water if due to acid or 2% acetic acid if due to alkali.

(ii) **Mercuric chloride.** Take an emetic—a tablespoon of common salt in a glass of water.

First-Aid Cabinet

Every Laboratory should be fitted with a **First-aid Cabinet,** which should contain:—

Gentian violet jelly	
or Tannic acid jelly	
or Picric acid	1 % aqueous solution
Acetic acid	1 % aqueous solution
Acetic acid	2 % aqueous solution
Boric acid	1 % aqueous solution
Sodium bicarbonate . .	1 % aqueous solution
Sodium bicarbonate . .	Saturated aqueous solution

Sal volatile.	Sodium chloride.	Zinc or boracic ointment.
Lime-water.	Vaseline.	T.C.P. or Dettol.

Adhesive tape.	Lint.	Eye-bath.
Bandages, assorted.	Oiled silk.	Forceps.
Cotton-wool.	Elastoplast dressings	Safety pins.
Gauze.	(assorted sizes).	Scissors.

APPENDIX V

FIRMS SUPPLYING BIOLOGICAL APPARATUS AND MATERIAL*

BIOLOGICAL APPARATUS, MICROSCOPES, MICROSCOPICAL SLIDES, STAINS AND REAGENTS, DISSECTING INSTRUMENTS, CHARTS, FILM STRIPS, LANTERN SLIDES, BIOLOGICAL MODELS, CHEMICAL APPARATUS AND REAGENTS, ETC.

> **Baird & Tatlock (London) Ltd.,** Freshwater Road, Chadwell Heath, Essex.
> **A. Gallenkamp & Co. Ltd.,** 6 Christopher Street, E.C.2.
> **Gerrard & Haig Ltd.,** Gerrard House, Worthing Road, East Preston, Sussex.
> **Griffin & George Ltd.,** Ealing Road, Alperton, Wembley, Middx.

MICROSCOPICAL STAINS AND REAGENTS ONLY

> **Edward Gurr Ltd.,** 47 Upper Richmond Road, West, S.W.4.

MICROSCOPES AND ACCESSORIES ONLY

> **R. & J. Beck Ltd.,** Bushey Mill Lane, Watford, Herts.
> **E. Leitz (Instruments) Ltd.,** 30 Mortimer Street, W.1.
> **W. Watson & Sons Ltd.,** Barnet, Herts.

BIOLOGICAL MATERIAL (Living or Preserved)

> **Gerrard & Haig,** Address: see above.
> **Marine Biological Station,** Citadel Hill, Plymouth.

* This list is not exhaustive.

INDEX

HOW TO USE THIS INDEX

Animals are arranged under their *generic* names and also under their English names where applicable.

Animal organs and systems will be found under the generic names of the animals.

Animal tissues are placed in their alphabetical positions.

Biochemical substances and processes are in their alphabetical positions.

Biological methods (Appendix II) are arranged alphabetically.

Embryological structures will be found under Embryology.

Microscopical processes and other matter relating to this technique are in their alphabetical places.

Physiological processes and experiments in connection with them will be found under the names of those processes.

Reagents are not indexed except under general headings as they are already arranged in alphabetical order under those headings, in the Appendix.

Any subject not covered above should be sought in its alphabetical position.